Shadowing Function from Randomly Rough Surfaces

Related titles on radar:

Advances in Bistatic Radar Willis and Griffiths
Airborne Early Warning System Concepts, 3rd Edition Long
Bistatic Radar, 2nd Edition Willis
Design of Multi-Frequency CW Radars Jankiraman
Digital Techniques for Wideband Receivers, 2nd Edition Tsui
Electronic Warfare Pocket Guide Adamy
Foliage Penetration Radar: Detection and characterisation of objects under trees Davis
Fundamentals of Ground Radar for ATC Engineers and Technicians Bouwman
Fundamentals of Systems Engineering and Defense Systems Applications Jeffrey
Introduction to Electronic Warfare Modeling and Simulation Adamy
Introduction to Electronic Defense Systems Neri
Introduction to Sensors for Ranging and Imaging Brooker
Microwave Passive Direction Finding Lipsky
Microwave Receivers with Electronic Warfare Applications Tsui
Phased-Array Radar Design: Application of radar fundamentals Jeffrey
Pocket Radar Guide: Key facts, equations, and data Curry
Principles of Modern Radar, Volume 1: Basic principles Richards, Scheer and Holm
Principles of Modern Radar, Volume 2: Advanced techniques Melvin and Scheer
Principles of Modern Radar, Volume 3: Applications Scheer and Melvin
Principles of Waveform Diversity and Design Wicks *et al.*
Pulse Doppler Radar Alabaster
Radar Cross Section Measurements Knott
Radar Cross Section, 2nd Edition Knott *et al.*
Radar Design Principles: Signal processing and the environment, 2nd Edition Nathanson *et al.*
Radar Detection Di Franco and Ruby
Radar Essentials: A concise handbook for radar design and performance Curry
Radar Foundations for Imaging and Advanced Concepts Sullivan
Radar Principles for the Non-Specialist, 3rd Edition Toomay and Hannan
Test and Evaluation of Aircraft Avionics and Weapons Systems McShea
Understanding Radar Systems Kingsley and Quegan
Understanding Synthetic Aperture Radar Images Oliver and Quegan
Radar and Electronic Warfare Principles for the Non-specialist, 4th Edition Hannen
Inverse Synthetic Aperture Radar Imaging: Principles, algorithms and applications Chen and Marotella
Stimson's Introduction to Airborne Radar, 3rd Edition Baker, Griffiths and Adamy
Test and Evaluation of Avionics and Weapon Systems, 2nd Edition McShea
Angle-of-Arrival Estimation Using Radar Interferometry: Methods and applications Holder
Biologically-Inspired Radar and Sonar: Lessons from nature Balleri, Griffiths and Baker
The Impact of Cognition on Radar Technology Farina, De Maio and Haykin
Novel Radar Techniques and Applications, Volume 1: Real aperture array radar, imaging radar, and passive and multistatic radar Klemm, Nickel, Gierull, Lombardo, Griffiths and Koch
Novel Radar Techniques and Applications, Volume 2: Waveform diversity and cognitive radar, and target tracking and data fusion Klemm, Nickel, Gierull, Lombardo, Griffiths and Koch
Radar and Communication Spectrum Sharing Blunt and Perrins
Systems Engineering for Ethical Autonomous Systems Gillespie

Shadowing Function from Randomly Rough Surfaces

Derivation and applications

Christophe Bourlier and Hongkun Li

The Institution of Engineering and Technology

Published by SciTech Publishing, an imprint of The Institution of Engineering and Technology, London, United Kingdom

The Institution of Engineering and Technology is registered as a Charity in England & Wales (no. 211014) and Scotland (no. SC038698).

First published 2019

The Institution of Engineering and Technology
Michael Faraday House
Six Hills Way, Stevenage
Herts, SG1 2AY, United Kingdom

www.theiet.org

British Library Cataloguing in Publication Data
A catalogue record for this product is available from the British Library

ISBN 978-1-78561-535-1 (hardback)
ISBN 978-1-78561-536-8 (PDF)

Typeset in India by MPS Limited
Printed in the UK by CPI Group (UK) Ltd, Croydon

Contents

Acknowledgements ix
Introduction xi

1 Shadowing function from one-dimensional surface 1
Christophe Bourlier and Hongkun Li

1.1 Brief history and review 1
1.2 Monostatic shadowing function 2
1.2.1 Description of the problem 2
1.2.2 Statistical shadowing function 4
1.2.3 Determination of the function A 5
1.2.4 The Wagner formulation 5
1.2.5 The Smith formulation 6
1.2.6 Rigorous formulation 7
1.3 Case of any uncorrelated process 8
1.3.1 Introduction 8
1.3.2 The Wagner formulation 8
1.3.3 The Smith formulation 9
1.3.4 Rigorous formulation 9
1.3.5 Discussion 10
1.3.6 Average shadowing functions 12
1.3.7 Numerical result for a Gaussian process 13
1.4 Case of a correlated Gaussian process 17
1.4.1 Conditional joint PDF 17
1.4.2 The Wagner formulation 18
1.4.3 The Smith formulation 22
1.4.4 Numerical results 25
1.5 Validation: comparison with a Monte Carlo process 28
1.5.1 Random rough surface generation 28
1.5.2 Ray-tracing algorithm 32
1.5.3 Comparison of the average shadowing function 34
1.6 Generalization to the bitstatic case in reflection 35
1.6.1 Definition 36
1.6.2 Case of any uncorrelated process 37
1.6.3 Case of a correlated Gaussian process 40
1.6.4 Ray-tracing algorithm 41
1.6.5 Numerical results 42

1.7 Generalization to the bistatic case in transmission 45
1.7.1 Definition 46
1.7.2 Case of any uncorrelated process 46
1.7.3 Case of a Gaussian-correlated process 49
1.7.4 Ray-tracing algorithm 52
1.7.5 Numerical results for a Gaussian process 53
1.8 Non-Gaussian distribution 56
1.8.1 Gram–Charlier distribution 57
1.8.2 Monostatic shadowing function 58
1.8.3 Bistatic shadowing functions 61

2 Shadowing function from a two-dimensional surface **63**
Christophe Bourlier and Hongkun Li

2.1 Monostatic shadowing function 63
2.1.1 Correlation functions 64
2.1.2 Case of any uncorrelated process 66
2.1.3 Case of a correlated Gaussian process 73
2.2 Bistatic shadowing function 79
2.2.1 Introduction 79
2.2.2 Derivation 79
2.2.3 Average shadowing functions 82
2.2.4 Numerical results 84

3 Shadowing function with multiple reflections **97**
Hongkun Li and Christophe Bourlier

3.1 Monostatic case from a 1D surface 97
3.1.1 Introduction of the problem 97
3.1.2 Numerical ray-tracing algorithm 99
3.1.3 Empirical models defined by a cut-off angle 105
3.1.4 Statistical model 110
3.1.5 Improved statistical model 113
3.1.6 Extension to include multiple reflections 122
3.2 Bistatic case from a 1D surface 126
3.2.1 Introduction of the problem 126
3.2.2 Numerical tray-tracing algorithm 128
3.2.3 Model of Lynch and Wagner 138
3.2.4 Model of Li *et al.* 146
3.2.5 Extension to include multiple reflections 152
3.3 Extension to a 2D surface 154
3.3.1 Monostatic shadowing function with one reflection 154
3.3.2 Bistatic shadowing function with two reflections 157
3.4 Conclusion 159

4 Some applications **161**
Christophe Bourlier and Hongkun Li

4.1 Introduction 161
4.2 Microwave radar scattering 162
4.2.1 Unshaded Ament reflection coefficient 162
4.2.2 Illuminated height PDF 163
4.2.3 Ament reflection coefficient with shadow 169
4.2.4 Propagation factor 170
4.3 Sea surface infrared radiation 176
4.3.1 Zero-order emissivity from 2D sea surfaces 176
4.3.2 One-order emissivity from 1D sea surfaces 189
4.3.3 One-order reflectivity from 2D surfaces 194
4.3.4 Second-order reflectivity from 1D surfaces 201
4.3.5 Polarization effect 206
4.3.6 Multi-resolution model 207
4.4 Computer graphics 211
4.4.1 Introduction 211
4.4.2 Smith shadowing function versus that of Ashikmin *et al.* 211
4.4.3 One-order reflectivity versus the BRDF 213
4.4.4 Energy conservation 214

References **215**

Index **221**

Acknowledgements

I am grateful to Professor Joseph Saillard (retired) and Gérard Berginc who have initialized this work during my PhD Thesis. I would also like to thank the National Center for Scientific Research by whom I am employed and ONERA with whom I worked to include the resolution of the infrared camera.

Introduction

This book addresses the general problem of the derivation of the shadowing function from randomly rough surfaces. This topic is important for many applications like the wave scattering from rough surfaces both in optics and radar domains and also in the computer graphics for the calculation of the bidirectional reflectance distribution function.

From the theoretical point of view, if the high-frequencies theory of the geometrical optics (GO) approximation is valid to calculate the interaction of the incident wave (electromagnetic or acoustic) with the rough surface, for grazing angles (or near the horizon), this theory must be corrected by accounting for the fact that a part of the surface is not illuminated from the transmitter or not visible from the receiver due to the surface roughness.

If the transmitter and the receiver are the same, the geometry is named (in radar community) monostatic, whereas if they are distinct in the space, the geometry is named bistatic. As GO is assumed to be valid, the surface can be decomposed as a series of continuous smooth facets with continuous first derivatives between adjacent facets. The incident wave is then modelled as a collection of rays, which reflects on the facets in the local specular directions. From the Snell–Descartes laws and Fresnel coefficients, the reflected (and/or transmitted) field is then easy to derive and the main issue to solve is the derivation of the monostatic or /and bistatic shadowing functions.

Works on the derivation of the shadowing function began in the 1950s of the last century and has continued to progress to generalize them by introducing several reflections on the surface and by considering a two-dimensional (2D) surface.

This book presents an overview and recent advances of this topic by detailing this progress. First, the simpler problem is investigated, that is, monostatic case – one-dimensional (1D, or 2D problem) surface – one reflection, ending by the more complicated problem, that is bistatic case – 2D surface (or 3D problem) – multiple reflections. In addition, the authors will focus on the introduction of the simplifying assumptions to derive closed-from expressions of the shadowing function and will quantify their impact on the accuracy of the resulting models. In addition, applications of the shadowing function in problems encountered in physics will be addressed.

The problem of the derivation of the shadowing function from a rough surface is at the boundary between several scientific communities, each with its terminology. This book will make the link between these different communities and will help the reader to understand the theoretical aspects of this problem while giving practical applications.

Chapter 1

Shadowing function from one-dimensional surface

Christophe Bourlier[1] and Hongkun Li[1]

In this chapter, shadowing function from a randomly rough one-dimensional surface is derived by ignoring multiple reflections (Chapter 3).

First, a brief history and review of the derivation of the shadowing function is addressed.

Next, the special case of a monostatic configuration, for which the positions of the transmitter and receiver are the same, is investigated. This allows us to detail the rigorous derivation of the shadowing function and to discuss on the introduction of the simplifying assumptions need to obtain a closed-form expression of the statistical shadowing function. It depends then on the slope γ_0 and on the height ζ_0 of any arbitrary point on the surface, which are random variables.

Section 1.3 gives closed-form expressions of the statistical averages obtained by calculating the expected values of the statistical shadowing function over γ_0 or/and ζ_0. This requires the knowledge of the joint PDF (probability density function) of (ζ_0, γ_0) assumed to be uncorrelated.

In order to quantify the effect of the correlation between γ_0 and ζ_0, Section 1.4 deals with the case of a correlated process assumed to be Gaussian. This Gaussian assumption leads to closed-form expressions of the average shadowing functions.

Section 1.5 validates the different formulations by comparing them with a benchmark ray-tracing method, and a Monte Carlo (MC) process is applied to access the mean values.

The last two sections generalize the formulations to the bistatic cases in reflection (receiver above the surface) and transmission (receiver below the surface), for which the transmitter and receiver positions are distinct in the space.

1.1 Brief history and review

The problem of wave scattering the from a rough surface in the presence of shadowing was first considered analytically by Bass and Fuks [1] by means of the theory of random function overshoots developed in [2]. The statistical (this means that the average

[1]IETR Laboratory, CNRS, Nantes, France

over the surface slopes and heights is not performed) illumination function was then expressed from an infinite Rice series (for more details see [3]). The shadowing effect was rediscovered later, seemingly independently, with the Wagner [4], Smith [5], and Beckman [6] formulations, who retained the first term of the series. Moreover, Smith used the Wagner approach by introducing a normalization function.

For monostatic and bistatic configurations, these authors assumed a one-dimensional surface with an *uncorrelated* Gaussian process of surface heights and slopes. This means that the statistical illumination function is independent of the surface height autocorrelation function. More recently, for one- and two-dimensional surfaces with Gaussian statistics, Bourlier *et al.* [3,7] extended the Wagner and the Smith formulations by taking into account the correlation. For moderate incidence angles, they showed that the correlation could be neglected. Moreover, they showed by comparing the Wagner and the Smith formulations with an MC method that the Smith approach is more accurate than the Wagner one. For a 2D surface (3D problem), the correlation with respect to the azimuthal direction is ignored in [7], which can imply a discontinuity of the shadowing function when the transmitter and receiver are in the same plane. This issue has been recently solved by Heitz *et al.* [8].

1.2 Monostatic shadowing function

1.2.1 Description of the problem

As shown in Figure 1.1, a receiver measures the radiance of a randomly rough sea surface from the direction $\hat{s}$.[1] The observation direction $\hat{s}$ forms an angle $\theta \in [0°; 90°]$ with the zenith $\hat{z}$, which is named the zenith angle. Due to the surface roughness, only a part of the whole surface can be 'seen' by the receiver. Some parts of the surface may lie in the shadow of the receiver shown in dashed lines in Figure 1.1. The rays coming from the surface and propagating with respect to the direction $\hat{s}$ are blocked by the surface itself. This phenomenon is called the shadowing effect.

The shadowing effect depends on the observation direction $\hat{s}$ and the shape of the surface related to the roughness (heights, slopes, etc.). For instance, a flat surface is always fully illuminated in all observation directions $0° \leq \theta < 90°$.

From a pure geometrical point of view, the following features can be drawn:

1. In the case where $\theta = 0°$, which means that the emission ray is propagating vertically upward, no emission ray is blocked (see Figure 1.2(a)).
2. In the case where $\theta = 90°$, which means that the emission ray is propagating horizontally towards the receiver located at the level of the horizon, nearly all emission rays are blocked, except for the few ones at the edge of the surface (see Figure 1.2(b)).
3. The higher the point is, the more likely it is illuminated. The highest point of the surface is illuminated, as no other point can shadow it (see Figure 1.2(c), point H).

[1] In this book, the symbol $\hat{\bullet}$ stands for an unitary vector.

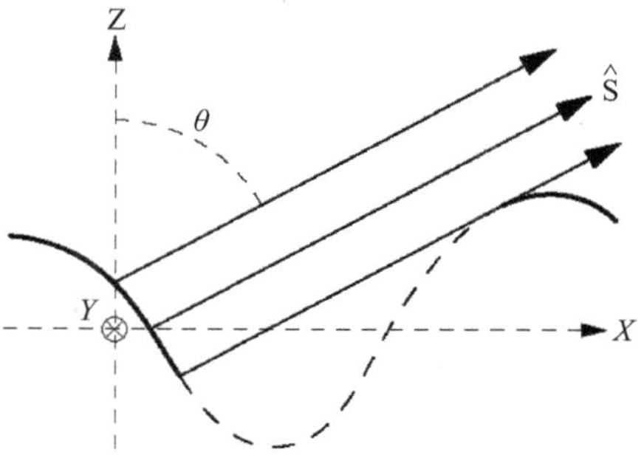

Figure 1.1 *Shadowing effect of rough surfaces. The observation direction forms an angle θ with the zenith. The $\hat{x}$ direction corresponds to the horizontal direction of the receiver. The dashed part of the surface lies in the shadow of the receiver, whereas the solid part of the surface is seen by the receiver*

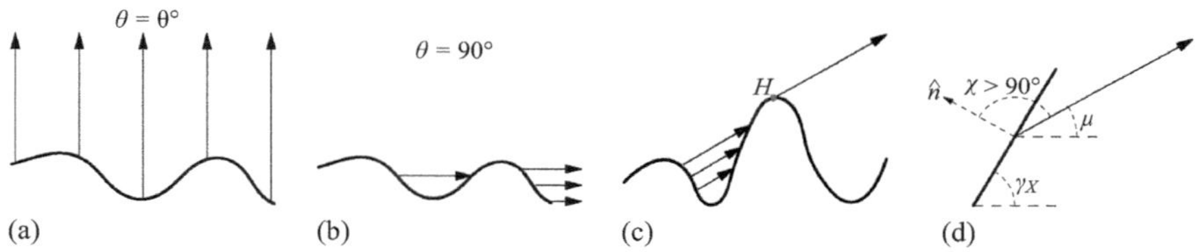

Figure 1.2 *Four physical features of the shadowing effect. (a) Case $\theta = 0$. (b) Case $\theta = 90°$. (c) Shadowing versus the height of the surface point. (d) Shadowing versus the slope of the surface point*

Indeed, if a surface point is in the shadow of the receiver, its emission ray along $\hat{s}$ reaches the surface at some other point.

4. The surface points with slopes γ in the $\hat{x}$ direction (corresponds to the horizontal direction of the receiver) being larger than the slope $\mu = \cot\theta$ of the emission ray lie in the shadow of the receiver, as the local angle of incidence[2] (the one between the normal to the facet $\hat{n}$ and the emission ray, see Figure 1.2(d)) $|\chi| > 90°$, which is not physical.

As predicted by items 1 and 2 in the above list, shadowing becomes more and more significant as the zenith observation angle θ increases. For large θ, shadowing is too significant to be neglected.

For instance, in measurements of sea surface radiation, the receiver located near the sea surface (e.g., on a ship or an airplane) satisfies this condition. In addition, if

[2]This book uses the term 'angle of incidence' and later 'plane of incidence' even though there is not a real incident ray. The emission ray is treated as if it was generated by a specular reflection of an incident ray, where the angle of incidence and the plane of incidence are defined.

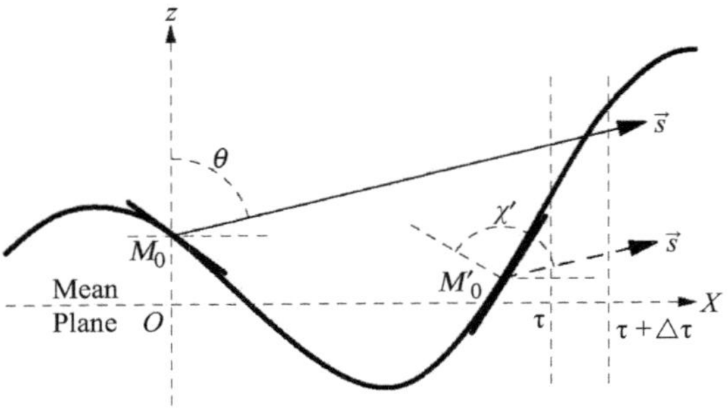

Figure 1.3 Emission rays from the surface propagate to the receiver along the observation direction $\hat{s}$. The emission ray from M_0 is blocked by the surface in the region $x \in [\tau; \tau + \Delta\tau]$. The emission ray from M'_0 has a local angle of incidence $|\chi'| > 90°$; thus, this emission ray does not exist

the wind speed above the sea surface is high, the surface slope standard deviation is also high (related to item 4 of the above list), which increases the shadowing effect.

An accurate estimation of the shadowing effect, named shadowing function or illumination function,[3] has to be developed to quantify this phenomenon.

1.2.2 Statistical shadowing function

This section reviews Wagner's [4] and Smith's models [5] for calculating the shadowing function without surface reflections S^0_{M} (superscript '0' for zero-order and subscript 'M' for monostatic), which gives the probability that an arbitrary point of the sea surface can be seen by the receiver along an observation direction $\hat{s}$.

Figure 1.3 shows a profile of the surface along the $\hat{x}$ direction. The shadowing function is derived by introducing a ray emitted from an arbitrary surface point M_0 along the observation direction $\hat{s}$. The probability that this ray leaves the surface without intersecting the surface equals the probability that M_0 is seen by the receiver.

The statistical shadowing function $S^0_{\mathrm{M}}(\mu, \zeta_0, \gamma_0, \tau)$ is the probability that the ray $M_0(\hat{s})$ emitted from M_0 does not intersect the surface in the interval $x \in [0, \tau]$. It is defined by the limit

$$S^0_{\mathrm{M}}(\mu, \zeta_0, \gamma_0, L_0) = \lim_{\tau \to L_0} S^0_{\mathrm{M}}(\mu, \zeta_0, \gamma_0, \tau), \tag{1.1}$$

where L_0 is the surface length.

The probability that the ray $M_0(\hat{s})$ does not cross the surface in the interval $x \in [0; \tau + \Delta\tau]$ is [4,5]

$$S^0_{\mathrm{M}}(\mu, \zeta_0, \gamma_0, \tau + \Delta\tau) = S^0_{\mathrm{M}}(\mu, \zeta_0, \gamma_0, \tau) Q(\Delta\tau|\mu, \zeta_0, \gamma_0, \tau), \tag{1.2}$$

[3]In the publications, the 'shadowing' term is employed. By definition, as shown in the next section, the term 'Illuminated' should be used. But, to avoid any confusion, the term 'shadowing' is employed.

where $Q(\Delta\tau|\mu, \zeta_0, \gamma_0, \tau)$ is the conditional probability that the ray does not intersect the surface in the range $x \in [\tau; \tau + \Delta\tau]$, given that it neither does in the interval $x \in [0; \tau]$. Expressing this term by its complementary probability [4,5]

$$Q(\Delta\tau|\mu, \zeta_0, \gamma_0, \tau) = 1 - g(\mu|\zeta_0, \gamma_0; \tau)\Delta\tau, \tag{1.3}$$

where $g(\mu|\zeta_0, \gamma_0; \tau)\Delta\tau$ is the conditional probability that the ray does intersect the surface in the range $x \in [\tau; \tau + \Delta\tau]$ (symbolically denoted by the event β) given that it does not in the interval $x \in [0; \tau]$ (symbolically denoted by the event α).

For $\Delta\tau$ sufficiently small, the Taylor series expansion of $S^0_{\mathrm{M}}(\mu, \zeta_0, \gamma_0, \tau + \Delta\tau)$ leads to the following ordinary differential equation [4,5]:

$$\frac{\mathrm{d}S^0_{\mathrm{M}}(\mu, \zeta_0, \gamma_0, \tau)}{\mathrm{d}\tau} = -g(\mu|\zeta_0, \gamma_0; \tau)S^0_{\mathrm{M}}(\mu, \zeta_0, \gamma_0, \tau). \tag{1.4}$$

The integration over $\tau \in [0; L_0]$ yields

$$S^0_{\mathrm{M}}(\mu, \zeta_0, \gamma_0, L_0) = A \exp\left[-\int_0^{L_0} g(\mu|\zeta_0, \gamma_0; \tau)\mathrm{d}\tau\right]. \tag{1.5}$$

To have a full derivation of (1.5), the functions A and $g(\mu|\zeta_0, \gamma_0; \tau)$ must be determined, which is the purpose of the following two subsections.

1.2.3 Determination of the function A

The function A is determined by imposing physical criteria. First of all, $S^0_{\mathrm{M}}(\mu, \zeta_0, \gamma_0, L_0)$ should be unity when $\theta = 0°$, since all the points on the surface are viewed by the receiver (feature 1; see Figure 1.2(a)) and the surface is assumed to be single-valued. Second, as M_0 is chosen randomly, the slope γ_0 of M_0 must be satisfied $|\chi_0| < 90°$ (feature 4; see Figure 1.2(d)), otherwise the emission ray would intersect the surface (ray $\hat{s}$ from M_0' in Figure 1.3). In that case, M_0 lies in the shadow of the receiver, and the statement above is neither valid nor necessary.

In the choice of A, Wagner and Smith agreed with each other, and A is chosen as the unit step function defined as

$$A = \Upsilon(\mu - \gamma_0) = \begin{cases} 0, & \gamma_0 > \mu \\ 1, & \gamma_0 < \mu \end{cases}. \tag{1.6}$$

However, for the derivation of the conditional probability $g(\mu|\zeta_0, \gamma_0; \tau)\Delta\tau$, the Wagner and Smith approaches differ and are both approximations.

1.2.4 The Wagner formulation

To derive the shadowing function, Wagner [4] assumed that the two events α and β (see (1.3) for their definition) are independent. In other words, the conditional probability $g(\mu|\zeta_0, \gamma_0; \tau)$ is approximated by the probability that the ray does intersect the surface in the range $x \in [\tau; \tau + \Delta\tau]$. Then, it is expressed as

$$p(\beta \mid \alpha) \approx p(\beta). \tag{1.7}$$

A ray intersecting the facet over the range $x \in [\tau; \tau + \Delta\tau]$ satisfies the following conditions:

$$\begin{cases} \zeta(\tau) < \zeta_0 + \mu\tau \\ \zeta(\tau + \Delta\tau) > \zeta_0 + \mu(\tau + \Delta\tau) \\ \gamma_0 > \mu \end{cases}, \tag{1.8}$$

where $\zeta(\tau)$ and $\zeta(\tau + \Delta\tau)$ are the heights of the surface at the abscissa $x = \tau$ and $x = \tau + \Delta\tau$, respectively. Assuming that the slope γ_0 of the facet in the region $x \in [\tau; \tau + \Delta\tau]$ is constant, the above conditions can be rewritten as

$$\begin{cases} \zeta_0 + \mu\tau > \zeta(\tau) > \zeta_0 + \mu\tau - (\gamma - \mu)\Delta\tau \\ \gamma_0 > \mu \end{cases}. \tag{1.9}$$

Then, $g(\mu|\zeta_0, \gamma_0; \tau)$ is expressed as (subscript 'W' for Wagner)

$$\begin{aligned} g_W(\mu|\zeta_0, \gamma_0; \tau)\Delta\tau &= \int_{\mu}^{+\infty} \int_{\zeta_0+\mu\tau-(\gamma-\mu)\Delta\tau}^{\zeta_0+\mu\tau} p(\zeta, \gamma|\zeta_0, \gamma_0; \tau) \mathrm{d}\zeta\, \mathrm{d}\gamma, \\ &\approx \Delta\tau \int_{\mu}^{+\infty} (\gamma - \mu) p(\zeta = \zeta_0 + \mu\tau, \gamma|\zeta_0, \gamma_0; \tau) \mathrm{d}\gamma, \end{aligned} \tag{1.10}$$

where $p(\zeta, \gamma|\zeta_0, \gamma_0; \tau)$ is the joint conditional PDF of the two points M and M_0 on the surface of heights and slopes (ζ, γ) and (ζ_0, γ_0), respectively, and separating by the horizontal distance τ. In addition

$$g_W(\mu|\zeta_0, \gamma_0; \tau) = \int_{\mu}^{+\infty} (\gamma - \mu) p(\zeta = \zeta_0 + \mu\tau, \gamma|\zeta_0, \gamma_0; \tau) \mathrm{d}\gamma. \tag{1.11}$$

To obtain a closed-form expression of g, the points M and M_0 can be assumed to be uncorrelated. Another way is to consider that the joint conditional PDF is Gaussian and next Gaussian and uncorrelated. These points will be investigated latter.

1.2.5 The Smith formulation

To derive the conditional probability $g(\mu|\zeta_0, \gamma_0; \tau)\Delta\tau$, Smith assumed that the condition 'the ray $M_0(\hat{s})$ does not intersect the surface in $x \in [0; \tau]$' can be changed by the condition 'the ray $M_0(\hat{s})$ is not shadowed by the point $x = \tau$.' Then

$$\zeta(\tau) < \zeta_0 + \mu\tau, \tag{1.12}$$

and symbolically denoted by the event α'. In other words, the conditional probability $g\Delta\tau$ is approximated as

$$p(\beta \mid \alpha) \approx p(\beta \mid \alpha'). \tag{1.13}$$

The conditional probability is then expressed as (subscript 'S' for Smith)

$$\begin{aligned}
g_S(\mu|\zeta_0,\gamma_0;\tau)\Delta\tau &= p(\beta|\alpha') = \frac{p(\alpha',\beta)}{p(\alpha')},\\
&= \frac{\displaystyle\int_{\mu}^{+\infty}\int_{\zeta_0+\mu\tau-(\gamma-\mu)\Delta\tau}^{\zeta_0+\mu\tau} p(\zeta,\gamma|\zeta_0,\gamma_0;\tau)\mathrm{d}\gamma\,\mathrm{d}\zeta}{\displaystyle\int_{-\infty}^{+\infty}\int_{-\infty}^{\zeta_0+\mu\tau} p(\zeta,\gamma|\zeta_0,\gamma_0;\tau)\mathrm{d}\gamma\,\mathrm{d}\zeta}\\
&\approx \frac{\displaystyle(\gamma-\mu)\Delta\tau\int_{\mu}^{+\infty} p(\zeta=\zeta_0+\mu\tau,\gamma|\zeta_0,\gamma_0;\tau)\mathrm{d}\gamma}{\displaystyle\int_{-\infty}^{+\infty}\int_{-\infty}^{\zeta_0+\mu\tau} p(\zeta,\gamma|\zeta_0,\gamma_0;\tau)\mathrm{d}\gamma\,\mathrm{d}\zeta},
\end{aligned}\tag{1.14}$$

and

$$\begin{aligned}
g_S(\mu|\zeta_0,\gamma_0;\tau) &= \frac{\displaystyle(\gamma-\mu)\int_{\mu}^{+\infty} p(\zeta=\zeta_0+\mu\tau,\gamma|\zeta_0,\gamma_0;\tau)\mathrm{d}\gamma}{\displaystyle\int_{-\infty}^{+\infty}\int_{-\infty}^{\zeta_0+\mu\tau} p(\zeta,\gamma|\zeta_0,\gamma_0;\tau)\mathrm{d}\gamma\,\mathrm{d}\zeta}\\
&= \frac{g_W(\mu|\zeta_0,\gamma_0;\tau)}{\displaystyle\int_{-\infty}^{+\infty}\int_{-\infty}^{\zeta_0+\mu\tau} p(\zeta,\gamma|\zeta_0,\gamma_0;\tau)\mathrm{d}\gamma\,\mathrm{d}\zeta}.
\end{aligned}\tag{1.15}$$

To obtain a closed-form expression of g, the points M and M_0 can be assumed to be uncorrelated. Another way is to consider that the joint conditional PDF is Gaussian and next Gaussian and uncorrelated. These points will be investigated latter.

1.2.6 Rigorous formulation

As shown in the previous two subsections, the Wagner and Smith formulations introduced assumptions to obtain a closed-form expression of g. In fact, the function g is expressed as a series of integrals [1,9–11], named the Rice series. Then (subscript 'R' for Rice)

$$g_R(\mu|\zeta_0,\gamma_0;\tau) = \sum_{n=0}^{n=\infty}(-1)^n I_n(\mu|\zeta_0,\gamma_0;\tau), \tag{1.16}$$

where for $n = 0$

$$I_0 = g_W, \tag{1.17}$$

and for $n \neq 0$

$$I_n(\mu|\zeta_0,\gamma_0;\tau) = \int_0^{\tau}\mathrm{d}\tau_1\int_{\tau_1}^{\tau}\mathrm{d}\tau_2\cdots\int_{\tau_{n-1}}^{\tau} W_n(\mu|\zeta_0,\gamma_0;\tau,\tau_1,\ldots,\tau_n)\mathrm{d}\tau_n, \tag{1.18}$$

in which

$$W_n(\mu|\zeta_0,\gamma_0;\tau,\tau_1,\ldots,\tau_n) = \int_\mu^\infty \left\{(\gamma-\mu)\left[\prod_{m=1}^{m=n}\int_\mu^\infty(\gamma_m-\mu)\right]\right.$$

$$\left.\times\, p_{2m+2}(\vec{Z},\vec{\gamma}|\zeta_0,\gamma_0;\tau,\tau_1,\ldots,\tau_m)\right\}\mathrm{d}\vec{\gamma}. \tag{1.19}$$

where $W_n d\tau d\tau_1 d\tau_2\cdots d\tau_n$ is the conditional probability that the incident ray of $Z_n = \zeta_0+\mu\tau_n$ ($n\in\mathbb{N}^*$ and $\zeta=\zeta_0+\mu\tau$) crosses the surface at $\zeta(\tau_n)=\zeta_n$ with $\mu\le\gamma_n=\gamma(\tau_n)$ in the ranges $[\tau;\tau+d\tau],[\tau_1;\tau_1+d\tau_1],[\tau_2;\tau_2+d\tau_2],\ldots,[\tau_n;\tau_n+d\tau_n]$ knowing $M_0(\zeta_0,\gamma_0)$. In addition, p_{2m+2} is the joint conditional PDF of random vectors $\vec{Z}=[\zeta_0\ \zeta\ Z_1\ Z_2\ \cdots\ Z_m]^T$ and $\vec{\gamma}=[\gamma_0\ \gamma\ \gamma_1\ \gamma_2\ \cdots\ \gamma_m]^T$ knowing $M_0(\zeta_0,\gamma_0)$. The symbol 'T' stands for the transpose.

As a conclusion, Wagner keeps only the first term of the series whereas Smith used the Wagner model divided by a normalization function.

1.3 Case of any uncorrelated process

1.3.1 Introduction

To compare the three (Wagner, Smith and Rice) formulations, first we assume that the joint conditional PDF is uncorrelated. For any PDF, this leads to

$$\begin{cases} p(\zeta,\gamma|\zeta_0,\gamma_0;\tau) & = p_\zeta(\zeta)p_\gamma(\gamma) \\ p_{2m+2}(\vec{Z},\vec{\gamma}|\zeta_0,\gamma_0;\tau,\tau_1,\ldots,\tau_m) & = p_\zeta(\zeta)p_\gamma(\gamma)\prod_{i=1}^{i=m}p_\zeta(Z_i)p_\gamma(\gamma_i) \end{cases}, \tag{1.20}$$

where p_ζ is the surface height PDF and p_γ that of the surface slopes. Since the correlation is omitted, the joint PDF is independent of the abscissa $\tau,\tau_1,\ldots,\tau_m$. Equation (1.19) becomes then

$$\begin{aligned} W_n(\mu|\zeta_0,\gamma_0;\tau,\tau_1,\ldots,\tau_n) &= g_W(\mu|\zeta_0,\gamma_0;\tau)\prod_{m=1}^{m=n}(\mu\Lambda)p_\zeta(Z_m) \\ &= \mu\Lambda p_\zeta(\zeta)\prod_{m=1}^{m=n}(\mu\Lambda)p_\zeta(Z_m), \end{aligned} \tag{1.21}$$

where

$$\Lambda = \frac{1}{\mu}\int_\mu^{+\infty}(\gamma-\mu)p_\gamma(\gamma)\mathrm{d}\gamma. \tag{1.22}$$

W_n depends on $\tau,\tau_1,\ldots,\tau_n$ because $Z_m=\zeta_0+\mu\tau_m$ ($m>0$ and $\zeta=\zeta_0+\mu\tau$).

1.3.2 The Wagner formulation

From (1.11) and (1.20), g_W is simplified as

$$g_W(\mu|\zeta_0,\gamma_0;\tau) = \mu\Lambda p_\zeta(\zeta). \tag{1.23}$$

The substitution of (1.23) and (1.6) into (1.5) leads to (the subscript 'W' is added)

$$S^0_{\mathrm{M,W}}(\mu, \zeta_0, \gamma_0, L_0) = \Upsilon(\mu - \gamma_0)\exp\left\{-\Lambda\left[P_\zeta(\zeta_0 + \mu L_0) - P_\zeta(\zeta_0)\right]\right\}, \quad (1.24)$$

where P_ζ is a primitive of p_ζ defined as

$$P_\zeta(\zeta) = \int p_\zeta(\zeta)\mathrm{d}\zeta. \quad (1.25)$$

1.3.3 The Smith formulation

From (1.15) and (1.20), g_S is simplified as

$$g_\mathrm{S}(\mu|\zeta_0, \gamma_0; \tau) = \frac{\mu\Lambda p_\zeta(\zeta_0 + \mu\tau)}{\displaystyle\int_{-\infty}^{\zeta_0+\mu\tau} p_\zeta(\zeta)d\zeta} = \frac{\mu\Lambda p_\zeta(\zeta_0 + \mu\tau)}{P_\zeta(\zeta_0 + \mu\tau) - P_\zeta(-\infty)}. \quad (1.26)$$

The substitution of (1.26) and (1.6) into (1.5) leads to (the subscript 'S' is added)

$$\begin{aligned}
S^0_{\mathrm{M,S}}(\mu, \zeta_0, \gamma_0, L_0) &= \Upsilon(\mu - \gamma_0)\exp\left[-\mu\Lambda\int_0^{L_0}\frac{p_\zeta(\zeta_0 + \mu\tau)\mathrm{d}\tau}{P_\zeta(\zeta_0 + \mu\tau) - P_\zeta(-\infty)}\right] \\
&= \Upsilon(\mu - \gamma_0)\exp\left[-\Lambda\int_{\zeta_0}^{\zeta_0+\mu L_0}\frac{\mathrm{d}P_\zeta(t)}{P_\zeta(t) - P_\zeta(-\infty)}\right] \\
&= \Upsilon(\mu - \gamma_0)\exp\left\{-\Lambda\ln\left[\frac{P_\zeta(\zeta_0 + \mu L_0) - P_\zeta(-\infty)}{P_\zeta(\zeta_0) - P_\zeta(-\infty)}\right]\right\} \\
&= \Upsilon(\mu - \gamma_0)\left[\frac{P_\zeta(\zeta_0) - P_\zeta(-\infty)}{P_\zeta(\zeta_0 + \mu L_0) - P_\zeta(-\infty)}\right]^\Lambda, \quad (1.27)
\end{aligned}$$

where the variable transformation $t = \zeta_0 + \mu\tau$ was applied.

1.3.4 Rigorous formulation

From (1.18), (1.20) and (1.21), I_1 is simplified as

$$\begin{aligned}
I_1(\mu|\zeta_0, \gamma_0; \tau) &= \int_0^\tau W_1(\mu|\zeta_0, \gamma_0; \tau, \tau_1)\mathrm{d}\tau_1 \\
&= (\mu\Lambda)^2 p_\zeta(\zeta_0 + \mu\tau)\int_0^\tau p_\zeta(\zeta_0 + \mu\tau_1)\mathrm{d}\tau_1 \\
&= \mu\Lambda^2 p_\zeta(\zeta_0 + \mu\tau)\left[P_\zeta(\zeta_0 + \mu\tau) - P_\zeta(\zeta_0)\right], \quad (1.28)
\end{aligned}$$

where Λ and P_ζ are defined from (1.22) and (1.25), respectively. Then

$$\begin{aligned}
\int_0^{L_0} I_1(\mu|\zeta_0, \gamma_0; \tau)\mathrm{d}\tau &= \mu\Lambda^2\int_0^{L_0} p_\zeta(\zeta_0 + \mu\tau)\left[P_\zeta(\zeta_0 + \mu\tau) - P_\zeta(\zeta_0)\right]\mathrm{d}\tau \\
&= \frac{\Lambda^2}{2}\left[P_\zeta(\zeta_0 + \mu L_0) - P_\zeta(\zeta_0)\right]^2, \quad (1.29)
\end{aligned}$$

where the variable transformation $t = \zeta_0 + \mu\tau$ and the identity $\int f'(t)f(t)dt = f^2(t)/2$ were applied.

From (1.18), (1.20) and (1.21) and using the same way as the derivation of I_1, I_2 is simplified as

$$\begin{aligned} I_2(\mu|\zeta_0,\gamma_0;\tau) &= \int_0^\tau \int_{\tau_1}^\tau W_2(\mu|\zeta_0,\gamma_0;\tau,\tau_1,\tau_2)\mathrm{d}\tau_1\mathrm{d}\tau_2 \\ &= (\mu\Lambda)^3 p_\zeta(\zeta_0+\mu\tau)\int_0^\tau \int_{\tau_1}^\tau p_\zeta(\zeta_0+\mu\tau_1)p_\zeta(\zeta_0+\mu\tau_2)\mathrm{d}\tau_1\mathrm{d}\tau_2 \\ &= \mu^2\Lambda^3 p_\zeta(\zeta_0+\mu\tau)\int_0^\tau p_\zeta(\zeta_0+\mu\tau_1)\left[P_\zeta(\zeta_0+\mu\tau)-P_\zeta(\zeta_0+\mu\tau_1)\right]\mathrm{d}\tau_1 \\ &= \mu\Lambda^3 p_\zeta(\zeta_0+\mu\tau)\left[P_\zeta(\zeta_0+\mu\tau)\,P_\zeta(\zeta_0+\mu\tau_1)\Big|_0^\tau - \frac{1}{2}\,P_\zeta^2(\zeta_0+\mu\tau_1)\Big|_0^\tau\right] \\ &= \mu\Lambda^3 p_\zeta(\zeta_0+\mu\tau)\frac{\left[P_\zeta(\zeta_0+\mu L_0)-P_\zeta(\zeta_0)\right]^2}{2}. \end{aligned} \tag{1.30}$$

Then

$$\begin{aligned} \int_0^{L_0} I_2(\mu|\zeta_0,\gamma_0;\tau)\mathrm{d}\tau &= \mu\Lambda^3\int_0^{L_0} p_\zeta(\zeta_0+\mu\tau)\frac{\left[P_\zeta(\zeta_0+\mu L_0)-P_\zeta(\zeta_0)\right]^2}{2}\mathrm{d}\tau \\ &= \frac{\Lambda^3}{6}\left[P_\zeta(\zeta_0+\mu L_0)-P_\zeta(\zeta_0)\right]^3. \end{aligned} \tag{1.31}$$

For $n \geq 0$, we can show by recurrence that

$$\int_0^{L_0} I_n(\mu|\zeta_0,\gamma_0;\tau)\mathrm{d}\tau = \frac{\Lambda^{n+1}}{(n+1)!}\left[P_\zeta(\zeta_0+\mu L_0)-P_\zeta(\zeta_0)\right]^{n+1}. \tag{1.32}$$

The substitution of (1.32) into (1.16) yields

$$\begin{aligned} -\int_0^{L_0} g_{\mathrm{R}}(\mu|\zeta_0,\gamma_0;\tau)\mathrm{d}\tau &= -X + \frac{X^2}{2} - \frac{X^3}{6} + \cdots + \frac{(-X)^n}{n!} \\ &= \exp(-X) - 1, \end{aligned} \tag{1.33}$$

where $X = \Lambda[P_\zeta(\zeta_0+\mu L_0) - P_\zeta(\zeta_0)]$ and n tends to infinity.

As a conclusion from (1.6) and (1.5), the statistical shadowing function is expressed as

$$S^0_{\mathrm{M,R}}(\mu,\zeta_0,\gamma_0,L_0) = \Upsilon(\mu-\gamma_0)\,e^{\exp\{-\Lambda[P_\zeta(\zeta_0+\mu L_0)-P_\zeta(\zeta_0)]\}-1}. \tag{1.34}$$

1.3.5 Discussion

From (1.24), (1.27) and (1.34), the statistical shadowing function is expressed as

$$S^0_{\mathrm{M}}(\mu,\zeta_0,\gamma_0,L_0) = \Upsilon(\mu-\gamma_0)\,\Psi_{\mathrm{W,S,R}}(\mu,\zeta_0,L_0), \tag{1.35}$$

where

$$\begin{cases} \Psi_{\mathrm{W}}(\mu, \zeta_0, L_0) = \exp\{-\Lambda\left[P_\zeta(\zeta_0 + \mu L_0) - P_\zeta(\zeta_0)\right]\} \\ \Psi_{\mathrm{S}}(\mu, \zeta_0, L_0) = \left[\dfrac{P_\zeta(\zeta_0) - P_\zeta(-\infty)}{P_\zeta(\zeta_0 + \mu L_0) - P_\zeta(-\infty)}\right]^{\Lambda} \\ \Psi_{\mathrm{R}}(\mu, \zeta_0, L_0) = \exp[\Psi_{\mathrm{W}}(\mu, \zeta_0, L_0) - 1] \end{cases} . \quad (1.36)$$

In (1.26), since the range of the denominator of g_{S} is [0; 1], this inverse is [1; ∞[, which means that $g_{\mathrm{S}} \geq g_{\mathrm{W}}$, and from (1.5), since $\exp(-g)$ is a decreasing function of g, we obtain $\Psi_{\mathrm{S}} \leq \Psi_{\mathrm{W}}$. Therefore, the statistical shadowing function of Smith is smaller than Wagner's for any assumed uncorrelated slope and height PDF. This arises from the fact that Smith introduced a normalization function in the denominator of (1.15). Since $\Lambda[P_\zeta(\zeta_0 + \mu L_0) - P_\zeta(\zeta_0)] \geq 0$, we have $\Psi_{\mathrm{R}} \geq \Psi_{\mathrm{W}}$. In conclusion, for any uncorrelated process, the statistical shadowing functions obey the following relationship:

$$0 \leq \Psi_{\mathrm{S}} \leq \Psi_{\mathrm{W}} \leq \Psi_{\mathrm{R}}. \quad (1.37)$$

Equation (1.35) also shows that the statistical shadowing effect carries a restriction over the surface slopes γ_0 within $\Upsilon(\mu - \gamma_0)$ and modifies the surface height distribution due to the Ψ function. We can note that the restriction over the slopes is independent of the formulation choice.

For sake of simplicity, L_0 is assumed to be infinite. In (1.36), the function $[P_\zeta(\zeta_0) - P_\zeta(-\infty)]^{\Lambda}$ makes a restriction on the surface heights ζ_0. The term $P_\zeta(\zeta_0) - P_\zeta(-\infty)$ tends to 1 when the point A is located at a high altitude ζ_0 ($\zeta_0 \to +\infty$), and the shadowing function is then maximum, that is to say, the shadowing effect is weak. Indeed, the higher the point A is, the less the probability that an incident or scattered wave in the upper medium crosses the surface before reaching it. Reversely, this term tends to 0 when the point A is located at a low altitude ζ_0 ($\zeta_0 \to -\infty$), and the shadowing function tends to 0 as well, that is to say, the shadowing effect is maximum. Indeed, the lower the point A is, the larger the probability that an incident or scattered wave in the upper medium crosses the surface before reaching it. This is illustrated in Figure 1.4(a), in which the point A' of lower altitude, than that of A, is in the shadow of the beam with slope μ_1.

Moreover, the height cumulative density function (CDF) $P_\zeta(\zeta_0) - P_\zeta(-\infty)$ is weighted by the term $\Lambda(\mu_1)$ which takes into account the surface slopes γ_0 that are greater than the absolute slope μ_1 of the wave of direction θ_1. When $\mu_1 \to 0$ (corresponding to a grazing angle), the function $\Lambda(\mu_1) \to +\infty$, then $\Psi \to 0$ (as $0 \leq P_\zeta(\zeta_0) - P_\zeta(-\infty) \leq 1$): The shadowing effect is maximum. Reversely, when $\mu_1 \to +\infty$ (corresponding to a zero angle), the function $\Lambda(\mu_1) \to 0$, then $\Psi \to 1$: the shadowing effect is minimum. Thus, this function holds for the fact that for a given surface point A, the lower the absolute slope of the beam of considered wave is, the higher the shadowing effect is, statistically. This is illustrated in Figure 1.4(b), in which the beam with slope μ_1' which is lower than the beam with slope μ_1 induces a more significant shadowing.

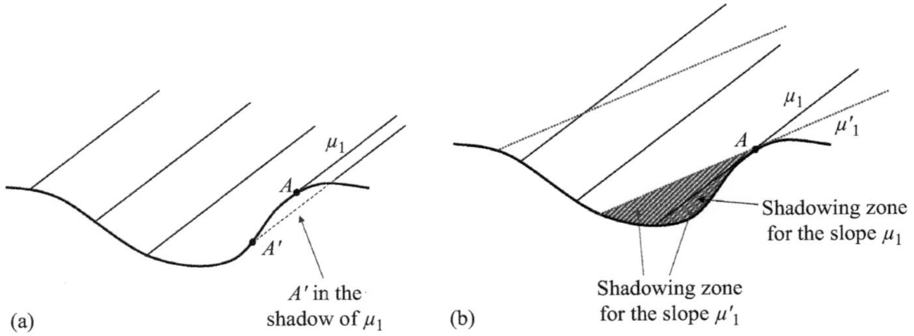

Figure 1.4 Influence of the height of the point A (a) and of the slope of the beam μ_1 (b) on the phenomenon of shadowing (propagation shadowing) of a random rough surface

1.3.6 Average shadowing functions

Since S_M^0 depends on the statistical variables ζ_0 and γ_0, expected values over γ_0, ζ_0 and both (γ_0, ζ_0) can be derived.

The expected value over γ_0 is defined as

$$\begin{aligned}\bar{\Gamma}(\mu, \zeta_0, L_0) &= \int_{-\infty}^{+\infty} S_M^0(\mu, \zeta_0, \gamma_0, L_0) p_\gamma(\gamma_0) d\gamma_0 \\ &= \Psi(\mu, \zeta_0, L_0) \int_{-\infty}^{+\infty} \Upsilon(\mu - \gamma_0) p_\gamma(\gamma_0) d\gamma_0 \\ &= \Psi(\mu, \zeta_0, L_0) \int_{-\infty}^{\mu} p_\gamma(\gamma_0) d\gamma_0,\end{aligned} \tag{1.38}$$

where p_γ is the surface slope PDF. The expected value over ζ_0 is defined as

$$\begin{aligned}\bar{Z}(\mu, \gamma_0, L_0) &= \int_{-\infty}^{+\infty} S_M^0(\mu, \zeta_0, \gamma_0, L_0) p_\zeta(\zeta_0) d\zeta_0 \\ &= \Upsilon(\mu - \gamma_0) \int_{-\infty}^{+\infty} \Psi(\mu, \zeta_0, L_0) p_\zeta(\zeta_0) d\zeta_0,\end{aligned} \tag{1.39}$$

where p_ζ is the surface height PDF. The expected value over both (γ_0, ζ_0) is defined as

$$\begin{aligned}\bar{\bar{S}}_M^0(\mu, L_0) &= \int_{-\infty}^{+\infty} \int_{-\infty}^{+\infty} S_M^0(\mu, \zeta_0, \gamma_0, L_0) p_{\gamma,\zeta}(\gamma_0, \zeta_0) d\gamma_0 d\zeta_0 \\ &= \int_{-\infty}^{+\infty} \Upsilon(\mu - \gamma_0) p_\gamma(\gamma_0) d\gamma_0 \int_{-\infty}^{+\infty} \Psi(\mu, \zeta_0, L_0) p_\zeta(\zeta_0) d\zeta_0 \\ &= \frac{\bar{Z}(\mu, \gamma_0, L_0)}{\Upsilon(\mu - \gamma_0)} \frac{\bar{\Gamma}(\mu, \zeta_0, L_0)}{\Psi(\mu, \zeta_0, L_0)},\end{aligned} \tag{1.40}$$

where the joint PDF $p_{\gamma,\zeta}(\gamma_0, \zeta_0) = p_\gamma(\gamma_0) p_\zeta(\zeta_0)$ for any correlated process.

For an infinite surface length L_0, from the Wagner formulation, $\bar{Z}$ is simplified as

$$\begin{aligned}\frac{\bar{Z}_{\mathrm{W}}(\mu,\gamma_0,\infty)}{\Upsilon(\mu-\gamma_0)} &= \int_{-\infty}^{+\infty} \exp\{-\Lambda\left[P_\zeta(\infty)-P_\zeta(\zeta_0)\right]\} p_\zeta(\zeta_0)\mathrm{d}\zeta_0 \\ &= \frac{1}{\Lambda}\exp\{-\Lambda\left[P_\zeta(\infty)-P_\zeta(\zeta_0)\right]\}\Big|_{-\infty}^{+\infty} \\ &= \frac{1-e^{-\Lambda}}{\Lambda},\end{aligned} \tag{1.41}$$

where $\int f'(t)e^{\Lambda f(t)}\mathrm{d}t = e^{\Lambda f(t)}/\Lambda$ and $P_\zeta(\infty)-P_\zeta(-\infty)=1$ by definition (from (1.25)).

For an infinite surface length L_0, from the Smith formulation, $\bar{Z}$ is simplified as

$$\begin{aligned}\frac{\bar{Z}_{\mathrm{S}}(\mu,\gamma_0,\infty)}{\Upsilon(\mu-\gamma_0)} &= \int_{-\infty}^{+\infty} \left[P_\zeta(\zeta_0)-P_\zeta(-\infty)\right]^{\Lambda} p_\zeta(\zeta_0)\mathrm{d}\zeta_0 \\ &= \frac{1}{1+\Lambda}\left[P_\zeta(\zeta_0)-P_\zeta(-\infty)\right]^{1+\Lambda}\Big|_{-\infty}^{+\infty} \\ &= \frac{1}{1+\Lambda}.\end{aligned} \tag{1.42}$$

For an infinite surface length L_0, from the Rice formulation, $\bar{Z}$ is simplified as

$$\begin{aligned}\frac{\bar{Z}_{\mathrm{R}}(\mu,\gamma_0,\infty)}{\Upsilon(\mu-\gamma_0)} &= e^{-1}\int_{-\infty}^{+\infty} e^{\exp\{-\Lambda[P_\zeta(\infty)-P_\zeta(\zeta_0)]\}} p_\zeta(\zeta_0)\mathrm{d}\zeta_0 \\ &= \frac{e^{-1}}{\Lambda}\int_{e^{-\Lambda}}^{1}\frac{e^t\mathrm{d}t}{t} = \frac{e^{-1}}{\Lambda}\left[\int_{e^{-\Lambda}}^{\infty}\frac{e^t\mathrm{d}t}{t}+\int_{\infty}^{1}\frac{e^t\mathrm{d}t}{t}\right] \\ &= \frac{e^{-1}}{\Lambda}\left[\int_1^{\infty}\frac{e^{e^{-\Lambda}t_1}\mathrm{d}t_1}{t_1}-\int_1^{\infty}\frac{e^t\mathrm{d}t}{t}\right] \\ &= \frac{e^{-1}}{\Lambda}\left[\mathrm{E}_1\left(-e^{-\Lambda}\right)-\mathrm{E}_1\left(-1\right)\right],\end{aligned} \tag{1.43}$$

where $t = e^{\Lambda[P_\zeta(\zeta_0)-P_\zeta(\infty)]}$, $\mathrm{d}t = \Lambda p_\zeta(\zeta_0)t\mathrm{d}\zeta_0$, $t_1 = t/e^{-\Lambda}$ and E_1 is the integral exponential (special) function defined as [12]

$$\mathrm{E}_1(x) = \int_1^{\infty}\frac{e^{-xt}\mathrm{d}t}{t}. \tag{1.44}$$

1.3.7 Numerical result for a Gaussian process

The height and slope PDF are assumed to be Gaussian and are defined as

$$p_\zeta(\zeta) = \frac{1}{\sigma_\zeta\sqrt{2\pi}}\exp\left(-\frac{\zeta^2}{2\sigma_\zeta^2}\right) \quad p_\gamma(\gamma) = \frac{1}{\sigma_\gamma\sqrt{2\pi}}\exp\left(-\frac{\gamma^2}{2\sigma_\gamma^2}\right), \tag{1.45}$$

where σ_ζ and σ_γ are the standard deviations of the surface heights and slopes, respectively. Then, from (1.38)

$$\frac{\bar{\Gamma}(\mu,\zeta_0,L_0)}{\Psi(\mu,\zeta_0,L_0)} = \int_{-\infty}^{\mu} p_\gamma(\gamma_0)\mathrm{d}\gamma_0 = \frac{1+\mathrm{erf}(v)}{2} \quad v = \frac{\mu}{\sigma_\gamma\sqrt{2}}, \tag{1.46}$$

and from (1.22)

$$\Lambda(v) = \frac{e^{-v^2} - v\sqrt{\pi}\,\mathrm{erfc}(v)}{2v\sqrt{\pi}}, \tag{1.47}$$

and erf, erfc are the error and error complementary functions, respectively, defined as [12]

$$\mathrm{erf}(v) = \frac{2}{\sqrt{\pi}}\int_0^v e^{-t^2}\mathrm{d}t \quad \mathrm{erfc}(v) = 1 - \mathrm{erf}(v). \tag{1.48}$$

Equations (1.46) and (1.47) show that the shadowing function depends only on v, which decreases the freedom degrees from two, (μ, σ_γ), to one, v. In other words, the shadowing function depends on the ratio of the incident slope $\cot\theta$ (angle θ defined from the vertical $\hat{z}$) over the surface slope standard deviation σ_γ.

In (1.36), from (1.25) and (1.45), the height CDF $P_\zeta(\zeta_0+\mu L_0) - P_\zeta(-\infty) = F_\zeta(\zeta_0+\mu L_0)$ is expressed as

$$\begin{aligned} F_\zeta(\zeta_0+\mu L_0) &= P_\zeta(\zeta_0+\mu L_0) - P_\zeta(-\infty) \\ &= \frac{1}{\sigma_\zeta\sqrt{2\pi}}\int_{-\infty}^{\zeta_0+\mu L_0} \exp\left(-\frac{\zeta^2}{2\sigma_\zeta^2}\right)\mathrm{d}\zeta \\ &= \frac{1}{2}\left[1+\mathrm{erf}\left(\frac{\zeta_0+\mu L_0}{\sigma_\zeta\sqrt{2}}\right)\right], \end{aligned} \tag{1.49}$$

and

$$\begin{aligned} P_\zeta(\zeta_0+\mu L_0) - P_\zeta(\zeta_0) &= P_\zeta(\zeta_0+\mu L_0) - P_\zeta(-\infty) - \left[P_\zeta(\zeta_0) - P_\zeta(-\infty)\right] \\ &= F_\zeta(\zeta_0+\mu L_0) - F_\zeta(\zeta_0) \\ &= \frac{1}{2}\left[\mathrm{erf}\left(\frac{\zeta_0+\mu L_0}{\sigma_\zeta\sqrt{2}}\right) - \mathrm{erf}\left(\frac{\zeta_0}{\sigma_\zeta\sqrt{2}}\right)\right]. \end{aligned} \tag{1.50}$$

Figure 1.5 plots the monostatic shadowed and unshadowed distributions of the surface slopes (versus the normalized surface slopes $s_0 = \gamma_0/(\sigma_\gamma\sqrt{2})$, (a)) and heights (versus the normalized surface heights $h_0 = \zeta_0/(\sigma_\zeta\sqrt{2})$, (b)) and for a given $v = \cot\theta/(\sigma_\gamma\sqrt{2})$ (given in the title of sub-figures). The surface length L_0 is infinite. From (1.35), (1.36), (1.49) and (1.50), they are defined, respectively, as

$$p_s(s_0) = \frac{1}{\sqrt{\pi}}\exp\left(-s_0^2\right)\Upsilon(v-s_0), \tag{1.51}$$

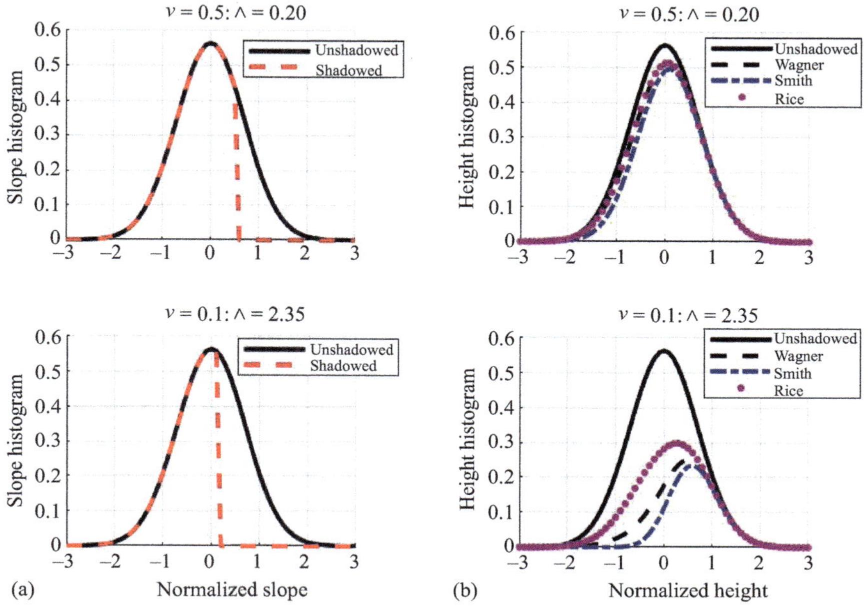

Figure 1.5 Monostatic shadowed and unshadowed distributions of the surface slopes (versus the normalized surface slopes $s_0 = \gamma_0/(\sigma_\gamma\sqrt{2})$, (a)) and heights (versus the normalized surface heights $h_0 = \zeta_0/(\sigma_\zeta\sqrt{2})$, (b)) and for a given $v = \cot\theta/(\sigma_\gamma\sqrt{2})$ (given in the title of sub-figures). The surface length L_0 is infinite

$$p_h(h_0) = \frac{1}{\sqrt{\pi}}\exp\left(-h_0^2\right) \times \begin{cases} \exp\left[-\dfrac{\Lambda}{2}\mathrm{erfc}(h_0)\right] & \text{Wager} \\ \left[\dfrac{1}{2} + \dfrac{1}{2}\mathrm{erf}(h_0)\right]^{\Lambda} & \text{Smith} \\ \exp\left\{\exp\left[-\dfrac{\Lambda}{2}\mathrm{erfc}(h_0)\right] - 1\right\} & \text{Rice} \end{cases}. \quad (1.52)$$

As seen in Figure 1.5(a), the area of p_s is inversely proportional to v due to the unit step function $\Upsilon(v - s_0)$ (same for the three formulations). For $v = 0$, only the range over $s_0 \in]-\infty; 0]$ contributes. As depicted in Figure 1.5(b), for any v, we observe that $p_{h,\mathrm{S}} \leq p_{h,\mathrm{W}} \leq p_{h,\mathrm{R}}$ (Wagner, dashed curve; Smith, broken curve; Rice, dotted curve), which is in agreement with (1.37). Moreover, the shadowing effect on the surface height distribution increases when v decreases due to the fact that Λ grows. For small values of $x = \Lambda\mathrm{erfc}(h_0)/2$, we have $\exp(e^{-x} - 1) \approx \exp(-x)$, which explains why the deviation between the Wagner and Rice results is small for $h_0 > 1$ ($\mathrm{erfc}(1)/2 \approx 0.0786$) and $\Lambda < 1$.

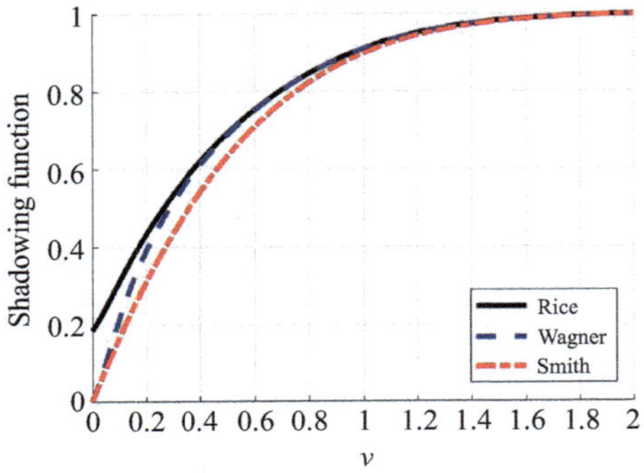

Figure 1.6 Average monostatic shadowing function over (γ_0, ζ_0) *versus the parameter* $v = \cot\theta/(\sigma_\gamma\sqrt{2})$

Figure 1.6 plots the average monostatic shadowing functions ((1.41), (1.42) and (1.43), in which Λ is computed from (1.47)) over (γ_0, ζ_0) versus the parameter $v = \cot\theta/(\sigma_\gamma\sqrt{2})$.

We observe that Smith's results (broken curve) are smaller than Wagner's (dashed curve), meaning that (1.37) is also satisfied for the average shadowing function. The Wagner and Rice (full curve) results are equal for $v > 0.6$, whereas they differ for smaller values, corresponding to incidence angles close to 90°. Indeed

$$\bar{\bar{S}}_{\mathrm{M}}^{0}(0, \infty) = \begin{cases} e^{-1}/2 \approx 0.184 & \text{Rice} \\ 0 & \text{Wagner} \end{cases}. \tag{1.53}$$

Physically, the shadowing function is equal to zero at a grazing angle of 90°. Thus, when the correlation is not included, the Rice results do not give satisfying results at grazing angles, whereas the Wagner ones are correct but overestimate the shadowing function (see Section 1.5).

Figure 1.6 also shows that for $v > 2 = v_0$, the average shadowing function is close to unity, which means that the whole surface is illuminated. Then, a limit angle, θ_{lim}, can be defined below which the shadow can be ignored. It is defined as

$$\theta_{\mathrm{lim}} = \operatorname{arccot}\left(v_0\sqrt{2}\sigma_\gamma\right) \approx \operatorname{arccot}\left(2\sqrt{2}\sigma_\gamma\right). \tag{1.54}$$

Since the correlation is not introduced, the statistical shadowing function does not depend on the surface height autocorrelation function. To study this dependence, the next section presents the Smith and Wagner formulations for a Gaussian surface height and slope joint process. The complexity of (1.16) does not allow us to analytically and/or numerically compute the function g_{R}.

1.4 Case of a correlated Gaussian process

In this section, the correlation between the surface heights and slopes is investigated for a Gaussian process and for any surface height autocorrelation function. Then, the Wager and Smith formulations are extended by accounting for the correlation.

1.4.1 Conditional joint PDF

Equations (1.11) and (1.15) required the knowledge of the conditional joint PDF $p(\zeta, \gamma|\zeta_0, \gamma_0; \tau)$ ($\zeta = \zeta_0 + \mu\tau$) of the surface heights (ζ, ζ_0) and slopes (γ, γ_0) of two points separated by the distance τ. For a Gaussian process and from the Bayes theorem, it is expressed as

$$\begin{aligned} p(\zeta, \gamma|z_0, \gamma_0; \tau) &= \frac{p_4(\zeta_0, \zeta, \gamma_0, \gamma; \tau)}{p_{\zeta,\gamma}(\zeta_0, \gamma_0)} \\ &= \frac{\sigma_\gamma \sigma_\zeta}{2\pi\sqrt{\det(C_4)}} \exp\left[-\frac{1}{2}\vec{V}_4^T \bar{C}_4^{-1} \vec{V}_4 + \frac{\zeta_0^2}{2\sigma_\zeta^2} + \frac{\gamma_0^2}{2\sigma_\gamma^2}\right], \end{aligned} \tag{1.55}$$

where $\vec{V}_4$ is the vector of components $[\zeta_0\ \zeta\ \gamma_0\ \gamma]$, $[C_4]$ is the covariance matrix of elements $C_{4,ij} = \langle V_{4,i} V_{4,j}\rangle$ ($i = \{1, 4\}$ and $j = \{1, 4\}$ and $V_{4,i}$ denotes the component i of the vector $\vec{V}_4$), where the symbol $\langle \ldots \rangle$ stands for the ensemble average operator. In addition, the symbol det gives the determinant of the matrix. We can show that the covariance matrix is expressed as

$$\bar{C}_4 = \begin{bmatrix} \sigma_\zeta^2 & C_{\zeta,0}(\tau) & 0 & C_{\zeta,1}(\tau) \\ C_{\zeta,0}(\tau) & \sigma_\zeta^2 & -C_{\zeta,1}(\tau) & 0 \\ 0 & -C_{\zeta,1}(\tau) & \sigma_\gamma^2 & -C_{\zeta,2}(\tau) \\ C_{\zeta,1}(\tau) & 0 & -C_{\zeta,2}(\tau) & \sigma_\gamma^2 \end{bmatrix}, \tag{1.56}$$

where

$$\begin{cases} C_{\zeta,0}(\tau) = \langle \zeta(\tau_0)\zeta(\tau_1)\rangle \\ C_{\zeta,1}(\tau) = \langle \zeta(\tau_0)\gamma(\tau_1)\rangle = \dfrac{\partial}{\partial \tau_1}\langle \zeta(\tau_0)\zeta(\tau_1)\rangle = \dfrac{\partial C_{\zeta,0}(\tau_1 - \tau_0)}{\partial \tau_1} \\ \qquad = \dfrac{dC_{\zeta,0}(\tau)}{d\tau} = C'_{\zeta,0}(\tau) \\ C_{\zeta,2}(\tau) = \langle \gamma(\tau_0)\gamma(\tau_1)\rangle = \dfrac{\partial^2}{\partial \tau_0 \partial \tau_1}\langle \zeta(\tau_0)\zeta(\tau_1)\rangle = \dfrac{\partial^2 C_{\zeta,0}(\tau_1 - \tau_0)}{\partial \tau_0 \partial \tau_1} \\ \qquad = -\dfrac{d^2 C_{\zeta,0}(\tau)}{d\tau^2} = -C''_{\zeta,0}(\tau) \end{cases}, \tag{1.57}$$

where $\tau = \tau_1 - \tau_0$. Since the Gaussian process is assumed to be stationary (or homogeneous in space), the surface height autocorrelation function $C_{\zeta,0}$ depends only on the difference $\tau = \tau_1 - \tau_0$. In addition, $C_{\zeta,1}$ is the intercorrelation function between the elevation $\zeta_0 = \zeta(\tau_0)$ and the slope $\gamma = \gamma(\tau_1)$ and $C_{\zeta,2}$ is the surface slope autocorrelation function. These two functions are expressed from the first and second derivatives of $C_{\zeta,0}$ over τ, respectively. By definition, $C_{\zeta,0}$ is an even function, which implies that $C_{\zeta,1}$ is an odd function, that $C_{\zeta,2}$ is an even function and that the covariance matrix is symmetric.

From (1.56), inverting the covariance matrix $\bar{C}_4$, the joint conditional PDF (1.55) is expressed as

$$p(\zeta,\gamma|z_0,\gamma_0;\tau) = \frac{\sigma_\gamma\sigma_\zeta}{2\pi\sqrt{\det(C_4)}}\exp\Big[-\frac{C_{11}(\zeta_0^2+\zeta^2)+C_{33}(\gamma_0^2+\gamma^2)}{2\det(C_4)}+\frac{\zeta_0^2}{2\sigma_\zeta^2}+\frac{\gamma_0^2}{2\sigma_\gamma^2} - \frac{2C_{12}\zeta_0\zeta+2C_{34}\gamma_0\gamma+2C_{13}(\zeta_0\gamma_0-\zeta\gamma)+2C_{14}(\zeta_0\gamma-\zeta\gamma_0)}{2\det(C_4)}\Big], \tag{1.58}$$

where the elements of $\bar{C}_4^{-1}$ (inverse of $\bar{C}_4$ denoted as $\{C_{ij}\}$) are

$$\begin{cases} C_{11} = \sigma_\zeta^2\left(\sigma_\gamma^4 - C_{\zeta,2}^2\right) - C_{\zeta,1}^2\sigma_\gamma^2 \\ C_{12} = C_{\zeta,0}\left(C_{\zeta,2}^2 - \sigma_\gamma^4\right) - C_{\zeta,1}^2 C_{\zeta,2} \\ C_{13} = -C_{\zeta,1}\left(C_{\zeta,0}\sigma_\gamma^2 + C_{\zeta,2}\sigma_\zeta^2\right) \\ C_{14} = C_{\zeta,1}\left(C_{\zeta,1}^2 - C_{\zeta,0}C_{\zeta,2} - \sigma_\zeta^2\sigma_\gamma^2\right) \\ C_{33} = \sigma_\gamma^2\left(\sigma_\zeta^4 - C_{\zeta,0}^2\right) - C_{\zeta,1}^2\sigma_\zeta^2 \\ C_{34} = C_{\zeta,2}\left(\sigma_\zeta^4 - C_{\zeta,0}^2\right) + C_{\zeta,1}^2 C_{\zeta,0} \\ \det(C_4) = \dfrac{C_{33}^2 - C_{34}^2}{\sigma_\zeta^4 - C_{\zeta,0}^2} \end{cases}. \tag{1.59}$$

If the correlation is neglected, then $C_{\zeta,0} = C_{\zeta,1} = C_{\zeta,2} = 0$, which implies that $C_{11} = \sigma_\zeta^2\sigma_\gamma^4$, $C_{12} = C_{13} = C_{14} = C_{34} = 0$, $C_{33} = \sigma_\zeta^4\sigma_\gamma^2$ and $\det(C_4) = \sigma_\zeta^4\sigma_\gamma^4$. Then, $C_{11}/(2\det(C_4)) = 1/(2\sigma_\zeta^2)$ and $C_{33}/(2\det(C_4)) = 1/(2\sigma_\gamma^2)$, which implies from (1.58) that $p(\zeta,\gamma|z_0,\gamma_0;\tau) = p_{\zeta,\gamma}(\zeta,\gamma) = p_\zeta(\zeta)p_\gamma(\gamma)$. The correlation between ζ and γ vanishes because $\langle\zeta\gamma\rangle = C_{\zeta,1}(0) = 0$ since $C_{\zeta,1}$ is an odd function.

1.4.2 The Wagner formulation

Using the following equation [12],

$$\int_\mu^\infty (\gamma-\mu)\,e^{-A\gamma^2-2B\gamma}\,d\gamma = \frac{e^{-\mu(\mu A+2B)}}{2A}\Big[1 - e^{\kappa^2}\kappa\sqrt{\pi}\,\text{erfc}(\kappa)\Big], \tag{1.60}$$

where $\kappa = (B+\mu A)/\sqrt{A}$ and collecting the term over γ in (1.58), the integration over γ in (1.11) leads to

$$g_W(\mu|\zeta_0,\gamma_0;\tau) = \frac{\sigma_\gamma\sigma_\zeta e^{-D-\mu(\mu A+2B)}}{4\pi A\sqrt{\det(C_4)}}\Big[1 - e^{\kappa^2}\kappa\sqrt{\pi}\,\text{erfc}(\kappa)\Big], \tag{1.61}$$

where

$$\begin{cases} A = \dfrac{C_{33}}{2\det(C_4)}, & B = \dfrac{\zeta_0 C_{14} - \zeta C_{13} + \gamma_0 C_{34}}{2\det(C_4)} \qquad \zeta = \zeta_0 + \mu\tau \\ D = \dfrac{(\zeta_0^2 + \zeta^2)C_{11} + 2\zeta_0\zeta C_{12} + 2\gamma_0(\zeta_0 C_{13} - \zeta C_{14}) + \gamma_0^2 C_{33}}{2\det(C_4)} \\ \quad -\dfrac{\zeta_0^2}{2\sigma_\zeta^2} - \dfrac{\gamma_0^2}{2\sigma_\gamma^2} \end{cases}. \quad (1.62)$$

To express $S_M^0(\mu, \zeta_0, \gamma_0, L_0)$ defined by (1.5) versus $v = \mu/(\sigma_\gamma\sqrt{2})$, the following variable transformations are introduced

$$C_{\zeta,0} = \sigma_\zeta^2 f_0 \qquad C_{\zeta,1} = -\sigma_\gamma\sigma_\zeta f_1 \qquad C_{\zeta,2} = -\sigma_\gamma^2 f_2. \quad (1.63)$$

Then

$$\begin{cases} \dfrac{C_{11}}{2\det(\bar{C}_4)} = \dfrac{1}{2\sigma_\zeta^2}\dfrac{f_{11}}{f_M} & \dfrac{C_{33}}{2\det(\bar{C}_4)} = \dfrac{1}{2\sigma_\gamma^2}\dfrac{f_{33}}{f_M} \\ \dfrac{C_{13}}{2\det(\bar{C}_4)} = \dfrac{1}{2\sigma_\zeta\sigma_\gamma}\dfrac{f_{13}}{f_M} & \dfrac{C_{12}}{2\det(\bar{C}_4)} = \dfrac{1}{2\sigma_\zeta^2}\dfrac{f_{12}}{f_M} \\ \dfrac{C_{34}}{2\det(\bar{C}_4)} = \dfrac{1}{2\sigma_\gamma^2}\dfrac{f_{34}}{f_M} & \dfrac{C_{14}}{2\det(\bar{C}_4)} = \dfrac{1}{2\sigma_\zeta\sigma_\gamma}\dfrac{f_{14}}{f_M} \\ \det(\bar{C}_4) = f_M(\sigma_\zeta\sigma_\gamma)^4 \end{cases}, \quad (1.64)$$

where

$$\begin{cases} f_{11} = 1 - f_2^2 + f_1^2 & f_{33} = 1 - f_0^2 - f_1^2 \\ f_{13} = f_1(f_0 - f_2) & f_{12} = f_0 f_2^2 + f_1^2 f_2 - f_0 \\ f_{34} = f_0^2 f_2 + f_1^2 f_0 - f_2 & f_{14} = f_1(1 - f_1^2 - f_0 f_2) \\ f_M = (f_{33}^2 - f_{34}^2)/(1 - f_0^2) \end{cases}. \quad (1.65)$$

For Gaussian and Lorentzian surface height autocorrelation functions expressed in Table 1.1, Figure 1.7 plots the functions f_{ij} versus y. f_{ij} are also plotted for the uncorrelated case. If the correlation is neglected, then $f_{ij} = \delta_{ij}$, where δ_{ij} is the Kronecker symbol defined as $\delta_{ij} = 1$ for $i = j$, otherwise 0.

Figure 1.7 shows that the functions $\{f_{12}, f_{34}, f_{13}, f_{14}\}$ $(i \neq j)$ equal zero when $y \geq y_G = 3$ and $y \geq y_L = 4$ for the Gaussian (broken curve) and Lorentzian (chain curve) cases, respectively, whereas $\{f_{11}, f_{33}\}$ $(i = j)$ becomes independent of y and tend towards unity. For the uncorrelated case (full curves), $\{f_{ij}\}$ is equal to either zero or one. Therefore, in the range $y \in]y_{G,L}; \infty[$, the correlation can be neglected and equations valid for any uncorrelated process can be applied. While for the range $y \in [0, y_{G,L}]$, the function g has to be computed from Table 1.2.

Table 1.1 Expressions of $C_{\zeta,0}$, $C_{\zeta,1}$, $C_{\zeta,2}$ and $\{f_{0,1,2}\}$ (defined from (1.63)) for a Gaussian and Lorentzian surface height autocorrelation functions

	Gaussian	**Lorentzian**
$C_{\zeta,0}$	$\sigma_\zeta^2 \exp(-\tau^2/L_c^2)$	$\sigma_\zeta^2/(1+\tau^2/L_c^2)$
$C_{\zeta,1}$	$-2\tau\sigma_\zeta^2 \exp(-\tau^2/L_c^2)/L_c^2$	$-2\tau\sigma_\zeta^2/(1+\tau^2/L_c^2)^2/L_c^2$
$C_{\zeta,2}$	$-(2\sigma_\zeta^2/L_c^2)(1/(1-2\tau^2/L_c^2))$ $\times \exp(-\tau^2/L_c^2)$	$-(2\sigma_\zeta^2/L_c^2)(1/1-3\tau^2/L_c^2)$ $\times(1/(1+\tau^2/L_c^2)^3)$
f_0	$\exp(-y^2)$	$1/(1+y^2)$
f_1	$y\sqrt{2}\exp(-y^2)$	$y\sqrt{2}/(1+y^2)^2$
f_2	$(1-2y^2)\sqrt{2}\exp(-y^2)$	$(1-3y^2)/(1+y^2)^3$
σ_γ	$\sigma_\zeta\sqrt{2}/L_c$	$\sigma_\zeta\sqrt{2}/L_c$
$\eta = \sigma_\gamma L_c/(\sigma_\zeta)$	$\sqrt{2}$	$\sqrt{2}$

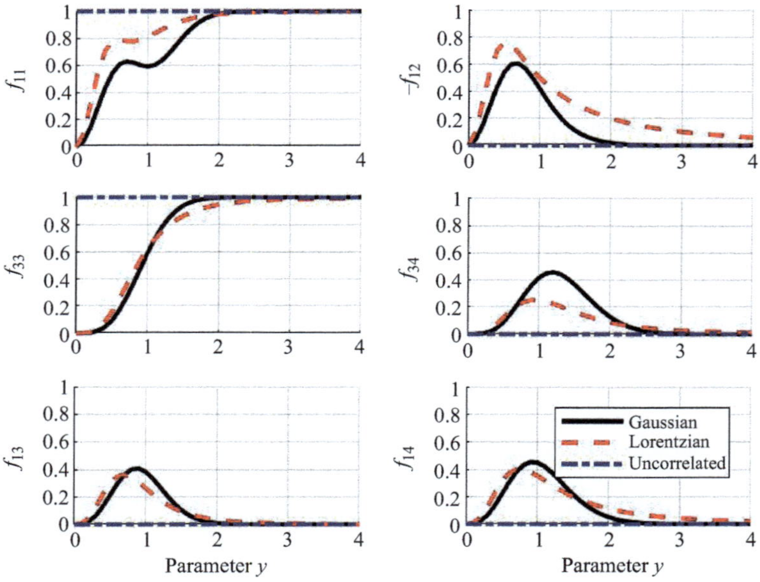

Figure 1.7 Functions f_{ij} versus y for a Gaussian and Lorentzian surface height autocorrelation functions defined in Table 1.1. f_{ij} are also plotted for the uncorrelated case

Table 1.2 Wagner and Smith monostatic statistical shadowing function for a correlated Gaussian process

Statistical shadowing function	$S_{\mathrm{M}}^{0}(v,h_0,s_0,y_0)=\Upsilon(v-s_0)\exp\left[-L_c\int_0^{y_t}g(v,h_0,s_0,y)\mathrm{d}y+G_{\mathrm{W,S}}(v,h_0,y_t,y_0)\right]$
Wagner: $L_c g_{\mathrm{W}}$	$(\eta\sqrt{f_M}/2\pi f_{33})e^{-D-v(vA+2B)}\left[1-\exp(\kappa^2)\kappa\sqrt{\pi}\mathrm{erfc}(\kappa)\right]$ $\kappa=(B+vA)/\sqrt{A}\qquad \eta=\sigma_\gamma L_c/\sigma_\zeta=\text{constant}$ $D=((h_0^2+h^2)f_{11}+2h_0hf_{12}+2s_0(h_0f_{13}-hf_{14})+s_0^2f_{33})/f_M-h_0^2-s_0^2$ $v(vA+2B)=(v^2f_{33}+2v(f_{34}s_0+h_0f_{14}-hf_{13}))/f_M\qquad h=h_0+yv\eta$ $\kappa=(h_0f_{14}-hf_{13}+s_0f_{34}+vf_{33})/\sqrt{f_{33}f_M}$
Wagner: G_{W}	$\Lambda(v)\left[\mathrm{erf}(h_0+y_tv\eta)-\mathrm{erf}(h_0+y_0v\eta)\right]/2, y_0>y_t$
Smith: $L_c g_{\mathrm{S}}$	$(\eta/\pi)(\sqrt{f_M}/f_{33})(\exp[-D-v(vA+2B)-h_0^2-s_0^2]\left[1-\exp(\kappa^2)\kappa\sqrt{\pi}\mathrm{erfc}(\kappa)\right])/$ $(\exp[B_1^2/A_1-D_1]\left[1+\mathrm{erf}\left(\sqrt{A_1}h+B_1/\sqrt{A_1}\right)\right])$ $D_1=(h_0^2(1-f_1^2)+s_0^2(1-f_0^2)+2h_0s_0f_0f_1)/f_{33}$ $\frac{B_1}{\sqrt{A_1}}=-(h_0f_0+s_0f_1)/\sqrt{f_{33}},\qquad \sqrt{A_1}=1/\sqrt{f_{33}}$
Smith G_{S}	$\ln\left[(1+\mathrm{erf}(h_0+y_tv\eta))/(1+\mathrm{erf}(h_0+y_0v\eta))\right]^{\Lambda(v)}, y_0>y_t$

Thanks to this property, we can write with $y_0=L_0/L_c>y_t=y_{G,L}$ (normalized observation length) that

$$-L_c\int_0^{y_0}g(v,h_0,s_0,y)\mathrm{d}y=-L_c\int_0^{y_t}g(v,h_0,s_0,y)\mathrm{d}y+G(v,h_0,y_t,y_0),\quad (1.66)$$

where

$$G(v,h_0,y_t,y_0)=\begin{cases}0 & y_0\le y_t\\ -L_c\displaystyle\int_{y_t}^{y_0}g(v,h_0,s_0,y)\mathrm{d}y & y_0>y_t\end{cases}.\quad (1.67)$$

From the following additional variable transformations,

$$h_0=\frac{\zeta_0}{\sqrt{2}\sigma_\zeta}\qquad s_0=\frac{\gamma}{\sqrt{2}\sigma_\gamma}\qquad y=\frac{\tau}{L_c}\qquad v=\frac{\mu}{\sqrt{2}\sigma_\gamma},\quad (1.68)$$

Bourlier *et al.* [3] showed that the Wagner statistical monostatic function (1.5) is expressed in Table 1.2.

If the correlation is neglected, then $f_{ij} = \delta_{ij}$, which implies in Table 1.2 that $\kappa = v$, $D = h^2$, $v(vA + 2B) = v^2$ and $L_c g_W = \eta \Lambda v e^{-h^2}/\sqrt{\pi}$, in which Λ is expressed from (1.47) and $h = h_0 + y v \eta$. The function G defined from (1.67) is then

$$\begin{aligned} G_W(v, h_0, y_t, y_0) &= -\frac{\eta \Lambda v}{\sqrt{\pi}} \int_{y_t}^{y_0} e^{-(h_0 + y v \eta)^2} \mathrm{d}y = -\frac{\Lambda}{\sqrt{\pi}} \int_{h_0 + y_t v \eta}^{h_0 + y_0 v \eta} e^{-h^2} \mathrm{d}h \\ &= \frac{\Lambda}{2} \left[\operatorname{erf}(h_0 + y_t v \eta) - \operatorname{erf}(h_0 + y_0 v \eta) \right]. \end{aligned} \tag{1.69}$$

For Gaussian and Lorentzian autocorrelation functions, Table 1.1 shows that the slope standard deviation σ_γ is equal (which implies that $\eta = \sqrt{2}$ is also equal). If the correlation is neglected, then the function G is also equal and the statistical shadowing function is also equal. This means that surfaces having different autocorrelation functions (the surface heights differ) can have the same shadowing function if the correlation is neglected.

1.4.3 The Smith formulation

For a Gaussian process and from the Bayes theorem, the joint conditional PDF $p(\zeta, \gamma | z_0, \gamma_0; \tau)$ is expressed as

$$p(\zeta, \gamma | z_0, \gamma_0; \tau) = \frac{p_4(\zeta_0, \zeta, \gamma_0, \gamma; \tau)}{p_{\zeta,\gamma}(\zeta_0, \gamma_0)}. \tag{1.70}$$

In (1.15), the integration over the slope γ gives

$$\begin{aligned} \int_{-\infty}^{+\infty} \int_{-\infty}^{\zeta_0 + \mu\tau} p(\zeta, \gamma | \zeta_0, \gamma_0; \tau) \mathrm{d}\gamma \mathrm{d}\zeta &= \int_{-\infty}^{\zeta_0 + \mu\tau} \left[\int_{-\infty}^{+\infty} \frac{p_4(\zeta_0, \zeta, \gamma_0, \gamma; \tau)}{p_{\zeta,\gamma}(\zeta_0, \gamma_0)} \mathrm{d}\gamma \right] \mathrm{d}\zeta \\ &= \int_{-\infty}^{\zeta_0 + \mu\tau} \frac{p_3(\zeta_0, \zeta, \gamma_0; \tau)}{p_{\zeta,\gamma}(\zeta_0, \gamma_0)} \mathrm{d}\zeta. \end{aligned} \tag{1.71}$$

The joint PDF $p_3(\zeta_0, \zeta, \gamma_0; \tau)$ is expressed as

$$p_3(\zeta_0, \zeta, \gamma_0; \tau) = \frac{1}{(2\pi)^{3/2} \sqrt{\det(C_3)}} \exp\left[-\frac{1}{3} \vec{V}_3^T \bar{C}_3^{-1} \vec{V}_3 \right], \tag{1.72}$$

where $\vec{V}_3$ is the vector of components $[\zeta_0 \ \zeta \ \gamma_0]$, $[C_3]$ is the covariance matrix of elements $C_{3,ij} = \langle V_{3,i} V_{3,j} \rangle$ ($i = \{1, 3\}$ and $j = \{1, 3\}$ and $V_{3,i}$ denotes the component i of the vector $\vec{V}_3$), where the symbol $\langle \ldots \rangle$ stands for the ensemble average operator. In addition, the symbol det gives the determinant of the matrix. We can show that the covariance matrix is expressed as

$$\bar{C}_3 = \begin{bmatrix} \sigma_\zeta^2 & C_{\zeta,0}(\tau) & 0 \\ C_{\zeta,0}(\tau) & \sigma_\zeta^2 & -C_{\zeta,1}(\tau) \\ 0 & -C_{\zeta,1}(\tau) & \sigma_\gamma^2 \end{bmatrix}. \tag{1.73}$$

From (1.73), inverting the covariance matrix $\bar{C}_3$, the joint conditional PDF (1.72) is expressed as

$$p_3(\zeta_0, \zeta, \gamma_0; \tau) = \frac{1}{(2\pi)^{3/2}\sqrt{\det(C_3)}} \exp\Big[-\frac{C_{11}\zeta_0^2 + C_{22}\zeta^2 + C_{33}\gamma_0^2}{2\det(C_3)} - \frac{C_{12}\zeta_0\zeta + C_{13}\zeta_0\gamma_0 + C_{23}\zeta\gamma_0}{\det(C_3)}\Big], \quad (1.74)$$

where the elements of $\bar{C}_3^{-1}$ (inverse of $\bar{C}_3$ denoted as $\{C_{ij}^s\}$) are

$$\begin{cases} C_{11}^s = \sigma_\zeta^2\sigma_\gamma^2 - C_{\zeta,1}^2 & C_{12}^s = -C_{\zeta,0}\sigma_\gamma^2 \\ C_{13}^s = -C_{\zeta,0}C_{\zeta,1} & C_{22}^s = \sigma_\zeta^2\sigma_\gamma^2 \\ C_{23}^s = \sigma_\zeta^2 C_{\zeta,1} & C_{33}^s = \sigma_\zeta^4 - C_{\zeta,0}^2 \\ \det(C_3) = \sigma_\gamma^2\left(\sigma_\zeta^4 - C_{\zeta,0}^2\right) - \sigma_\zeta^2 C_{\zeta,1}^2 \end{cases}. \quad (1.75)$$

Using the following equation [12],

$$\int_{-\infty}^{\zeta'} \exp(-A_1\xi^2 - 2B_1\xi)\,\mathrm{d}\zeta = \frac{1}{2}\sqrt{\frac{\pi}{A_1}}\left[1 + \operatorname{erf}\left(\frac{A_1\zeta' + B_1}{\sqrt{A_1}}\right)\right] e^{(B_1^2/A_1)}, \quad (1.76)$$

and in (1.74) collecting the terms over ζ, in (1.71) the integration over ζ leads to

$$\begin{aligned} \text{Equation (1.71)} &= \frac{1}{p_{\zeta,\gamma}(\zeta_0, \gamma_0)} \int_{-\infty}^{\zeta_0+\mu\tau} p_3(\zeta_0, \zeta, \gamma_0; \tau)\mathrm{d}\zeta \\ &= \frac{1}{4\pi\sqrt{2}p_{\zeta,\gamma}(\zeta_0, \gamma_0)\sqrt{A_1\det(C_3)}}\left[1 + \operatorname{erf}\left(\frac{A_1\zeta' + B_1}{\sqrt{A_1}}\right)\right] e^{(B_1^2/A_1) - D_1}, \end{aligned} \quad (1.77)$$

where

$$\begin{cases} A_1 = \dfrac{C_{22}^s}{2\det(C_3)} \\ B_1 = \dfrac{C_{12}^s\zeta_0 + C_{23}^s\gamma_0}{2\det(C_3)} \\ D_1 = \dfrac{C_{11}^s\zeta_0^2 + C_{33}^s\gamma_0^2 + 2C_{13}^s\zeta_0\zeta}{2\det(C_3)} \\ \zeta' = \zeta_0 + \mu\tau \end{cases}. \quad (1.78)$$

If the correlation is neglected, then $C_{\zeta,0} = C_{\zeta,1} = C_{\zeta,2} = 0$, $C_{12}^s = C_{13}^s = C_{23}^s = 0$, $C_{11}^s = C_{22}^s = \sigma_\zeta^2\sigma_\gamma^2$ and $C_{33}^s = \sigma_\gamma^4$. In addition, $A_1 = 1/(2\sigma_\zeta^2)$, $B_1 = 0$, $D_1 = \zeta_0^2/(2\sigma_\zeta^2) + \gamma_0^2/(2\sigma_\gamma^2)$ and $\det(C_3) = \sigma_\gamma^2\sigma_\zeta^4$. Equation (1.77) becomes then

$$\text{Equation (1.77)} = \frac{1}{2}\left[1 + \operatorname{erf}\left(\frac{\zeta'}{\sqrt{2}\sigma_\zeta}\right)\right]. \quad (1.79)$$

As expected, if $\zeta' \to \infty$, the above equation tends to unity.

The substitution of (1.77) and (1.61) into (1.15) leads to

$$g_S(\mu|\zeta_0, \gamma_0; \tau) = \frac{\sqrt{2A_1 \det(C_3)}}{2\pi A\sqrt{\det(C_4)}} \times \frac{e^{-D-\mu(\mu A+2B)-(\zeta_0^2/2\sigma_\zeta^2)-(\gamma_0^2/2\sigma_\gamma^2)}\left[1 - e^{\kappa^2}\kappa\sqrt{\pi}\,\mathrm{erfc}(\kappa)\right]}{\left[1 + \mathrm{erf}\left(\dfrac{A_1\zeta' + B_1}{\sqrt{A_1}}\right)\right] e^{(B_1^2/A_1)-D_1}}, \quad (1.80)$$

where

$$p_{\zeta,\gamma}(\zeta_0, \gamma_0) = \frac{1}{2\pi\sigma_\zeta\sigma_\gamma}\exp\left(-\frac{\zeta_0^2}{2\sigma_\zeta^2} - \frac{\gamma_0^2}{2\sigma_\gamma^2}\right) = p_\zeta(\zeta_0)p_\gamma(\gamma_0). \quad (1.81)$$

Moreover, $\kappa = (B + \mu A)/\sqrt{A}$ and $\{A, B, D, A_1, B_1, D_1, \zeta'\}$ are expressed from (1.62) and (1.78).

If the correlation is neglected, then from (1.59), (1.62), (1.75) and (1.78), $A = 1/(2\sigma_\gamma^2)$, $B = 0$, $D = \zeta^2/(2\sigma_\zeta^2)$, $\det(C_4) = \sigma_\gamma^4\sigma_\zeta^4$, $A_1 = 1/(2\sigma_\zeta^2)$, $B_1 = 0$, $D_1 = \zeta_0^2/(2\sigma_\zeta^2) + \gamma_0^2/(2\sigma_\gamma^2)$, $\det(C_3) = \sigma_\gamma^2\sigma_\zeta^4$. From (1.80), this leads to

$$\begin{aligned} g_S(\mu|\zeta_0, \gamma_0; \tau) &= \frac{1}{\pi}\frac{e^{-(\zeta^2/2\sigma_\zeta^2)-(\mu^2/2\sigma_\gamma^2)}\left[1 - e^{\kappa^2}\kappa\sqrt{\pi}\,\mathrm{erfc}(\kappa)\right]}{1 + \mathrm{erf}\left(\zeta'/\sqrt{2}\sigma_\zeta\right)} \\ &= \frac{2v\sqrt{\pi}\Lambda(v)}{\pi}\frac{e^{-(\zeta^2/2\sigma_\zeta^2)}}{1 + \mathrm{erf}\left((\zeta'/\sqrt{2}\sigma_\zeta)\right)} = \frac{\mu\Lambda(v)p_\zeta(\zeta)}{(1/2)\left[1 + \mathrm{erf}\left((\zeta/\sqrt{2}\sigma_\zeta)\right)\right]}, \end{aligned} \quad (1.82)$$

where $\zeta' = \zeta = \zeta_0 + \mu\tau$, $\kappa = \mu\sqrt{A} = \mu/(\sigma_\gamma\sqrt{2}) = v$, Λ is expressed from (1.47) and p_ζ is the surface height PDF assumed to be Gaussian and expressed from (1.45). For an uncorrelated process assumed to be Gaussian, (1.26) is retrieved, in which the cumulative height function $P_\zeta(\zeta_0 + \mu\tau) - P_\zeta(-\infty)$ is expressed from (1.49).

Using the variable transformations given from (1.63) and (1.68), the function g_S is expressed in Table 1.2. If the correlation is neglected, then $f_{ij} = \delta_{ij}$ and $f_0 = f_1 = 0$, which implies in Table 1.2 that $\kappa = v$, $D = h^2$, $v(vA + 2B) = v^2$, $D_1 = h_0^2 + s_0^2$, $B_1/\sqrt{A_1} = 0$, $\sqrt{A_1} = 1$ and $L_c g_S = 2\eta\Lambda v e^{-h^2}/\sqrt{\pi}/[1 + \mathrm{erf}(h)]$, in which Λ is expressed from (1.47) and $h = h_0 + y v \eta$. The function G defined from (1.67) is then

$$G_S(v, h_0, y_t, y_0) = -\frac{2\eta \Lambda v}{\sqrt{\pi}} \int_{y_t}^{y_0} \frac{e^{-(h_0+yv\eta)^2}}{1+\text{erf}(h_0+yv\eta)} \text{d}y$$

$$= -\frac{2\Lambda}{\sqrt{\pi}} \int_{h_0+y_t v\eta}^{h_0+y_0 v\eta} \frac{e^{-h^2}}{1+\text{erf}(h)} \text{d}h = -\Lambda \left[\ln\left\{1+\text{erf}(h)\right\}\right]_{h_0+y_t v\eta}^{h_0+y_0 v\eta}$$

$$= \ln\left\{\left[\frac{1+\text{erf}(h_0+y_t v\eta)}{1+\text{erf}(h_0+y_0 v\eta)}\right]^{\Lambda}\right\}. \tag{1.83}$$

1.4.4 Numerical results

In this section, the Wagner and Smith formulations are compared with and without correlation.

1.4.4.1 Numerical implementation

The computation of the average shadowing function over the heights and slopes requires 3-fold numerical integrations over the normalized surface heights h_0 and slopes s_0 and over the normalized distance $y = \tau/L_c$. It is defined as

$$\begin{aligned} \bar{\bar{S}}_{\text{M}}^0(v, y_0) &= \int_{-\infty}^{+\infty} \int_{-\infty}^{+\infty} S_{\text{M}}^0(v, h_0, s_0, y_0) p_{s,h}(s_0, h_0) \text{d}s_0 \text{d}h_0 \\ &= \frac{1}{\pi} \int_{-\infty}^{v} \exp(-s_0^2) \text{d}s_0 \left\{ \int_{-\infty}^{+\infty} \exp(-h_0^2) \right. \\ &\times \exp\left[-L_c \int_0^{y_t} g(v, h_0, s_0, y) \text{d}y + G(v, h_0, y_t, y_0)\right] \text{d}h_0 \Big\}, \end{aligned} \tag{1.84}$$

where $L_c g$ and G are expressed in Table 1.2.

In addition, s_0 ranges from $s_{0,\text{min}}$ to v with a sampling step Δs_0, h_0 ranges from $h_{0,\text{min}}$ to $h_{0,\text{max}}$ with a sampling step Δh_0 and y ranges from 0 to y_t with a sampling step Δy. Typically, $s_{0,\text{min}} = -3$ since $\exp(-s_{0,\text{min}}^2) \approx 0$, $h_{0,\text{min}} = -3$, $h_{0,\text{max}} = 3$ and $\Delta h_0 = \Delta s_0 = 0.01$ and $\Delta y = 0.1$. In addition, if $v > s_{0,\text{max}} = 3$, then the upper bound over s_0 is $s_{0,\text{max}}$. The accuracy of the three numerical integrations increases as the sampling steps decreases and are done from a conventional trapezoidal rule.

For large positive values of x, the function $1 - x\sqrt{\pi}\text{erfc}(x)\exp(x^2)$ is undefined as $\lim_{x\to\infty} \text{erfc}(x) = 0$ and $\lim_{x\to\infty} \exp(x^2) = \infty$, which implies numerical problems in the evaluation of the functions $L_c g_W$ and $L_c g_S$. Then, the following expansion is applied

$$1 - x\sqrt{\pi}\text{erfc}(x)\exp(x^2) = \begin{cases} \dfrac{1}{2x^2} & \text{for } x > 12.2 \\ -2x\sqrt{\pi}\exp(x^2) & \text{for } x < -2.2 \end{cases}. \tag{1.85}$$

For large negative values of x, the function $1+\text{erf}(x)=1+\text{erf}(-|x|)=1-\text{erf}(|x|)=\text{erfc}(|x|)$ tends toward zero, which implies numerical problems in the evaluation of the function $L_c g_S$. Then, the following expansion is applied

$$1+\text{erf}(x)\approx-\frac{\exp(-x^2)}{x\sqrt{\pi}}\quad\text{for}\quad x<-5.6. \tag{1.86}$$

1.4.4.2 Average over the surface heights

From (1.41) and (1.42), for an infinite surface length L_0 and an uncorrelated Gaussian process, the average shadowing function $\bar{Z}$ over the heights is expressed as

$$\bar{Z}(v,s_0,\infty)=\Upsilon(v-s_0)\begin{cases}\dfrac{1-\exp[-\Lambda(v)]}{\Lambda(v)} & \text{Wagner}\\ \dfrac{1}{1+\Lambda(v)} & \text{Smith}\end{cases}. \tag{1.87}$$

The above equation shows, for both Wagner's and Smith's formulations, all points with normalized slopes $s_0<v$ are shadowed equally for all s_0. However, this is true only when the correlation between the heights and the slopes of the surface is neglected. It is interesting to have a close look at the angular illumination behaviour. For the correlated case, it is computed from (1.84), in which the integration over s_0 is not performed and it is replaced by the unit step function $\Upsilon(v-s_0)$.

Figure 1.8 plots the average shadowing function over the heights ζ_0 versus s_0 computed from the Wagner (Figure 1.8(a)) and the Smith (Figure 1.8(b)) models, where $v=0.3$.

As we can see, both Wagner's and Smith's models with correlation bend down significantly as s_0 approaches v (for increasing s_0), or equally as γ_0 approaches μ, whereas the uncorrelated ones remains constant for $s_0<v$. Next, the strengths with and without correlation jump to 0 at $s_0>v$ due to the $\Upsilon(v-s_0)$ function. Then, the correlation implies that all points with slope $\gamma_0<\mu$ are not equally shadowed.

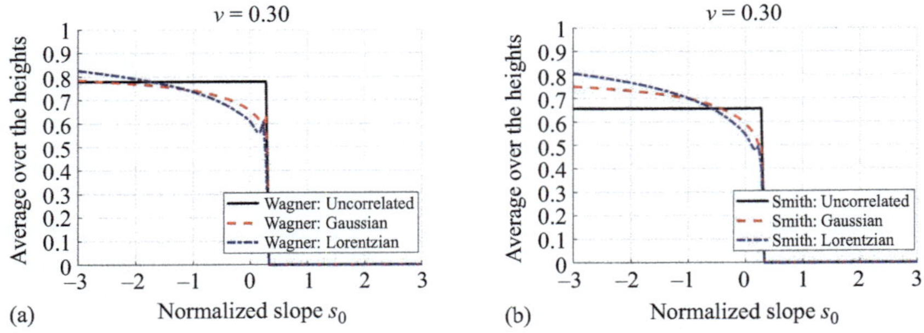

Figure 1.8 Average shadowing function over the heights ζ_0 versus s_0 computed from the Wagner (a) and the Smith (b) models, where $v=0.3$

Figure 1.8 also shows that the Smith results predict a larger difference with the uncorrelated results than the Wagner ones, and the strengths depend slightly on the surface height autocorrelation function (Gaussian or Lonrentzian; see Table 1.1 for their definition).

1.4.4.3 Average over the surface heights and slopes

Figure 1.9 plots the average shadowing function over the normalized slope s_0 and height h_0 versus v computed from the Wagner (a) and the Smith (b) models. At the bottom, the difference against the uncorrelated results is plotted.

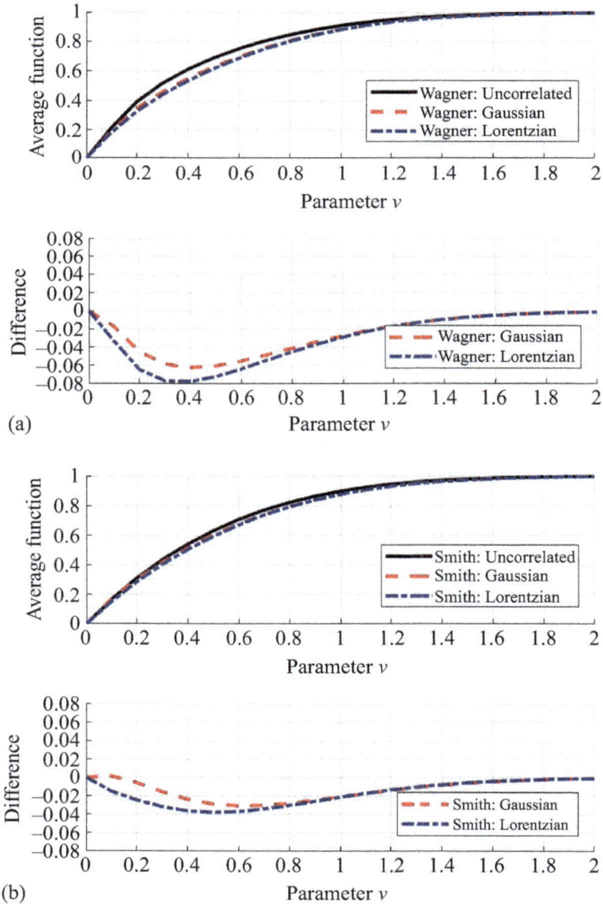

Figure 1.9 Top: Average shadowing function over the normalized slope s_0 and height h_0 versus v computed from the Wagner (a) and the Smith (b) models. Bottom: Difference against the uncorrelated results

As we can see, the strengths predicted with correlation are smaller than those without correlation and the difference is larger for the Wagner formulation. Moreover, the results are few sensitive to the surface height autocorrelation function.

1.5 Validation: comparison with a Monte Carlo process

To select the formulation as accurate as possible, the different approaches are compared with a benchmark method (any assumption is introduced). It is computed from the algorithm developed by Brokelman and Hagfors [13], in which a random rough surface profile must be generated.

1.5.1 Random rough surface generation

In this section, we describe how to generate realizations of a random rough surface. We assume that the surface is Gaussian, which means that the height PDF follows a Gaussian process or a normal law. We assume that the surface height profile $z(x)$ $(\tau = x)$ is univocal and follows a stationary Gaussian random process.

1.5.1.1 Statistical parameters

For a centred Gaussian process, the surface height PDF is given by

$$p_\zeta(\zeta) = \frac{1}{\sigma_\zeta\sqrt{2\pi}} \exp\left(-\frac{\zeta^2}{2\sigma_\zeta^2}\right), \tag{1.88}$$

and checks

$$\begin{cases} \langle 1 \rangle = \int_{-\infty}^{\infty} p_\zeta(\zeta)d\zeta = 1 \\ \langle \zeta \rangle = \int_{-\infty}^{\infty} \zeta p_\zeta(\zeta)d\zeta = 0 \\ \langle(\zeta - \langle\zeta\rangle)^2\rangle = \langle\zeta^2\rangle = \int_{-\infty}^{\infty} \zeta^2 p_\zeta(\zeta)d\zeta = \sigma_\zeta^2 \end{cases}, \tag{1.89}$$

where

$$\langle\bullet\rangle = \int_{-\infty}^{\infty} (\bullet) p_\zeta(\zeta)d\zeta. \tag{1.90}$$

The real number σ_ζ stands for the surface height standard deviation and the surface height mean value $\langle\zeta\rangle$ is zero. Since the height PDF is Gaussian, the derivative $d^n\zeta(x)/dx^n$ also follows a Gaussian process.

Full characterization of the random rough surface height ζ needs to know the correlation between two surface heights of abscissa x_1 and x_2. For a real ζ, the surface height autocorrelation function is then defined as

$$\langle\zeta(x_1)\zeta(x_1 + x)\rangle = C_{\zeta,0} = C_\zeta(x). \tag{1.91}$$

Since the process is stationary, C_ζ depends only on the abscissa difference $x = x_2 - x_1$ between two points of the surface. Then, a Gaussian process is fully characterized by its height PDF, p_ζ, and its surface height autocorrelation function, C_ζ.

The power spectral density (PSD) or the surface height spectrum is defined as

$$\hat{C}_\zeta(k) = \text{FT}[C_\zeta(x)] = \int_{-\infty}^{+\infty} C_\zeta(x)e^{-jkx}dx, \tag{1.92}$$

and

$$C_\zeta(x) = \frac{1}{2\pi}\text{FT}^{-1}[\hat{C}_\zeta(k)] = \frac{1}{2\pi}\int_{-\infty}^{+\infty} \hat{C}_\zeta(k)e^{jkx}dk, \tag{1.93}$$

where FT denotes the Fourier transform.

From (1.93), (1.92), (1.91) and (1.89), we have

$$\sigma_\zeta^2 = C_\zeta(0) = \frac{1}{2\pi}\int_{-\infty}^{+\infty} \hat{C}_\zeta(k)dk. \tag{1.94}$$

In addition, we have shown in a previous subsection that the surface slope autocorrelation function $C_\gamma = C_{\zeta,2}$ is defined from the surface height autocorrelation function C_ζ as

$$C_\gamma(x) = -\frac{d^2C_\zeta}{dx^2} = \frac{1}{2\pi}\int_{-\infty}^{+\infty} k^2\hat{C}_\zeta(k)e^{jkx}dx, \tag{1.95}$$

and then the slope variance is

$$\sigma_\gamma^2 = C_\gamma(0) = \frac{1}{2\pi}\int_{-\infty}^{+\infty} k^2\hat{C}_z(k)dk. \tag{1.96}$$

In addition, (1.96) shows that the surface slope spectrum is $\hat{C}_\gamma(k) = k^2\hat{C}_\zeta(k)$.

1.5.1.2 Generation of a random profile

At the input of a linear filter, if e is a stationary process (of second order) of PSD $\hat{C}_e$, then the output signal s of PSD $\hat{C}_s$ satisfies [14]

$$\hat{C}_s = \left|\hat{C}_g\right|^2 \hat{C}_e, \tag{1.97}$$

where $\hat{C}_g = \text{FT}(g)$ is the PSD of g, where g is the impulse response of the filter. In addition, if $\hat{C}_g \in \mathbb{R}^+$, then

$$\hat{C}_g = \sqrt{\frac{\hat{C}_s}{\hat{C}_e}}. \tag{1.98}$$

Since the system is assumed to be linear, we have [14]

$$s = g * e = \text{FT}^{-1}[\text{FT}(g)\text{FT}(e)] = \text{FT}^{-1}\left[\sqrt{\frac{\hat{C}_s}{\hat{C}_e}}\,\text{FT}(e)\right], \tag{1.99}$$

where the symbol * stands for the convolution product.

Since we want to generate a surface height Gaussian process $\zeta = s$, a Gaussian white noise of unitary variance is applied at the input of the filter, which implies that $\hat{C}_e = 1 \; \forall k$. Then

$$\zeta = \mathrm{FT}^{-1}\left[\sqrt{\hat{C}_\zeta}\;\mathrm{FT}(e)\right]. \tag{1.100}$$

Numerically, the convolution product is calculated in the Fourier domain because the complexity of a fast Fourier transform is $\mathcal{O}(N \log N)$ instead of $\mathcal{O}(N^2)$ if the convolution product is calculated from its definition as

$$\zeta(i) = g(i) * e(i) = \frac{1}{N}\sum_{n=1}^{N} g(n)e(n+i), \tag{1.101}$$

where N is the number of samples of both g and e. Since the surface height ζ is real, from (1.100), the function inside the operator FT^{-1} must satisfy $f^*(-k) = f(k) \; \forall k$, with * the complex conjugate operator. As shown further, $\hat{C}_\zeta$ is real and an even function of k. Thus, $\mathrm{FT}(e) = \hat{e}$ is a complex Gaussian white noise, which must satisfy $\hat{e}(-k)^* = \hat{e}(k)$. For more details, see [15] (Chapter 4).

For surface height Gaussian, Lorentzian and exponential autocorrelation functions is defined as

$$C_\zeta(x) = \sigma_\zeta^2 \begin{cases} \exp\left(-\dfrac{x^2}{L_c^2}\right) & \text{Gaussian} \\ \dfrac{1}{1+(x^2/L_c^2)} & \text{Lorentzian} \\ \exp(-|x|/L_c) & \text{Exponential} \end{cases}, \tag{1.102}$$

the surface height spectra (or PSD) are given from (1.92) by

$$\hat{C}_\zeta(k) = \sigma_\zeta^2 \begin{cases} L_c\sqrt{\pi}\exp\left(-\dfrac{k^2L_c^2}{4}\right) & \text{Gaussian} \\ L_c\pi\exp(-|k|L_c) & \text{Lorentzian} \\ \dfrac{2L_c}{1+k^2L_c^2} & \text{Exponential} \end{cases}. \tag{1.103}$$

The profile can be generated from the MATLAB® codes provided in [16].

1.5.1.3 Numerical results

Figure 1.10 plots the normalized surface heights $h_0 = \zeta_0/(\sigma_\zeta\sqrt{2})$ over $x/L_c \in [0; 20]$ versus the normalized abscissa x/L_c, in the middle, the normalized height histogram versus h_0 and at the bottom, the normalized surface height autocorrelation function C_ζ/σ_ζ^2 versus x/L_c. In Figure 1.10(a), C_ζ is Gaussian, whereas in Figure 1.10(b), it is exponential. The correlation length is $L_c = 200$, the variance is $\sigma_\zeta = 1$ and the number of samples is $N = 1{,}000{,}000$.

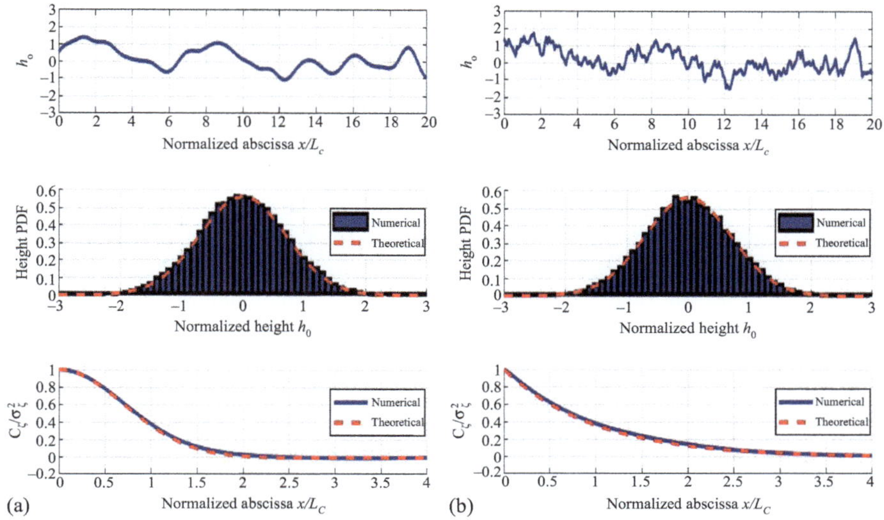

Figure 1.10 *Top: Normalized surface heights $h_0 = \zeta_0/(\sigma_\zeta\sqrt{2})$ versus the normalized abscissa $x/L_c \in [0;20]$. Middle: Normalized height histogram versus h_0. Bottom: Normalized surface height autocorrelation function C_ζ/σ_ζ^2 versus x/L_c. In (a), C_ζ is Gaussian, whereas in (b), it is exponential. The correlation length is $L_c = 200$, the variance is $\sigma_\zeta = 1$, and the number of samples is $N = 1{,}000{,}000$*

The top of Figure 1.10 shows that the normalized heights h_0 (more precisely, erf(2) = 99.5% of h_0) range from -2 to $+2$ and rapid variations occur for the exponential profile. Indeed, the surface height spectra expressed from (1.103) show that as the wavenumber k (dual of x in the Fourier domain) increases, the spectrum decreases more rapidly for a Gaussian profile as it depends of the exponential function. In other words, the contribution of the high frequencies is larger for an exponential profile than that of a Gaussian.

In the middle of Figure 1.10, the PDF computed from the generated surface follows a Gaussian process, and it matches well with the theoretical one.

The bottom of Figure 1.10 shows a very good agreement between the theoretical autocorrelation function and that computed from the generated surface with the help of the following equation:

$$C_\zeta(i) = \frac{1}{N}\sum_{j=1}^{j=N} \zeta(j)\zeta(j+i), \tag{1.104}$$

where $\zeta(j)$ is the surface height of the sample j.

The numerical values of $\tilde{\sigma}_\zeta$ and $\tilde{\sigma}_\gamma$ are $\{1.000, 0.992\}$ and $\{0.00700, 0.02180\}$ for a Gaussian and exponential profiles, respectively. Then, $\tilde{\sigma}_\zeta$ is in agreement with the theoretical value $\sigma_\xi = 1$ and $\tilde{\sigma}_\gamma$ is in agreement with the theoretical value $\sigma_\gamma = \sqrt{2}\sigma_\xi/L_c = 0.00707$ (from (1.96) and (1.102)) of a Gaussian profile.

For an exponential profile, from (1.102), the surface slope standard deviation σ_γ is not defined because the derivative of the function $\exp(-|x|/L_c)$ is not defined at $x = 0$. In the spectral domain, from (1.96) and (1.103), σ_γ tends to infinity because for high values of k, $k^2\hat{C}_\xi(k)$ behaves as $2\sigma_\xi^2/L_c$. A means to have a finite value of σ_γ is to introduce an upper bound wave number $k_{\max}$. Numerically, $k_{\max} = \pi/\Delta x$ (π instead of 2π because a bilateral spectrum is applied, that is $k \in [-k_{\max}; +k_{\max}]$) exists, and it is expressed from the Fourier transform rule. Then, σ_γ depends on $k_{\max}$.

Simulations, not displayed here, show that the surface heights computed from Gaussian and Lorentzian profiles are very similar.

As the correlation length decreases, simulations not displayed here show that the surface is more irregular (the horizontal distance between two consecutive extrema decreases) because the surface slope variance increases. For instance, for a Gaussian autocorrelation function, $\sigma_\gamma = \sqrt{2}\sigma_\xi/L_c$; if L_c is divided by 2, then the slope variance is multiplied by 2.

1.5.2 Ray-tracing algorithm

From a given realization of a rough surface and a collection of incident rays, which propagate from the receiver to the rough surface with respect to the observation direction $\hat{s}(\theta)$, the illuminated points can be performed by applying a ray-tracing algorithm.

The heart of the task lies in finding the points tangent to the incident ray, beginning from which the shadowing line is drawn. The points under these lines are in the shadow of the receiver and satisfy the following criterion:

$$(\gamma_i - \mu)(\gamma_{i+1} - \mu) < 0, \tag{1.105}$$

where μ is the slope of the incident ray and γ_i is the slope of the ith point on the surface.

The algorithm is shown in Figure 1.11 [3,13]. Note that the incident rays comes from the right.

Figure 1.12 plots the surface normalized heights h_0 of the whole (dashed curve) and illuminated points (full curve) versus the sample index. At the top, case of a Gaussian autocorrelation function, and at the bottom, case of a Lorentzian autocorrelation function. $L_c = 200$, $\sigma_\xi = 1$ and the parameter $v = \mu/(\sigma_\gamma\sqrt{2}) = 0.3$.

From the shadowed points, it is then possible to calculate the average shadowing function over the surface height and slope expressed as

$$\bar{\bar{S}}_{\mathrm{M}}^{0}(v) = \frac{1}{N}\sum_{i=1}^{i=N} b_i, \tag{1.106}$$

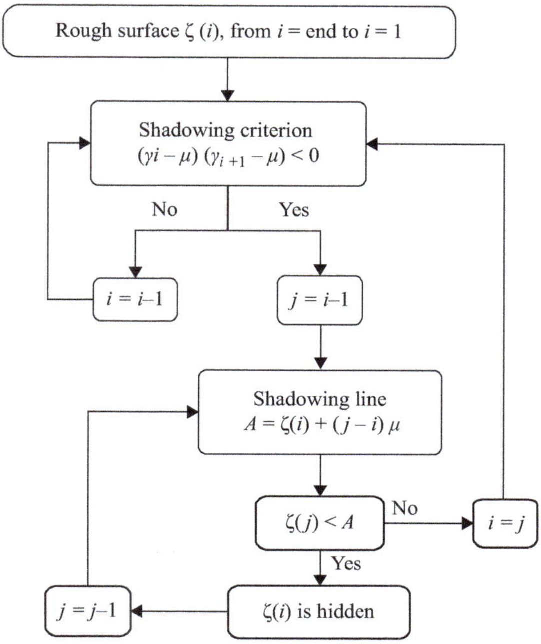

Figure 1.11 Ray-tracing algorithm for a monostatic configuration

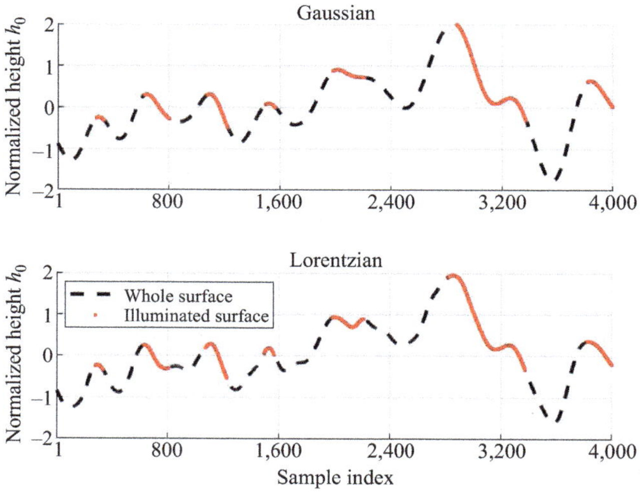

Figure 1.12 Surface normalized heights h_0 of the whole and illuminated points versus the sample index. Top: Gaussian autocorrelation function. Bottom: Lorentzian autocorrelation function. $L_c = 200$, $\sigma_\zeta = 1$ and the parameter $v = \mu/(\sigma_\gamma\sqrt{2}) = 0.3$

where the Boolean b_i is defined as

$$b_i = \begin{cases} 0 & \text{if the point is shadowed} \\ 1 & \text{otherwise} \end{cases}. \tag{1.107}$$

As shown in [3], any expected value (or ensemble average) can be performed and the marginal slope and height PDFs of the illuminated points can also be computed.

In general, to compute the ensemble average of a random variable, α, two ways can be applied:

- If α is known analytically (from a closed-form expression), its ensemble average can be computed analytically by assuming, for instance, that the surface heights obey a Gaussian process. For instance, this way is applied to calculate the average of the shadowing function over ζ_0 or/and γ_0 from the Wagner and Smith formulations.
- If α is not known analytically, but obtained numerically from, for instance, a ray-tracing method, an MC process is then applied to evaluate the ensemble average $\langle\alpha\rangle$. It consists in computing α from different realizations (same statistics but the seeds are different) of the surface, and then the operator $\langle\ldots\rangle$ is evaluated by taking the mean value. The surface length L must be enough large to have a good representation of the surface statistics. Typically, $L > 10L_c$.

For our problem, the surface length can be very large (a one-dimensional surface is considered) because no additional calculations are done on this surface. For the calculation of the scattered field from rough surfaces [16], it is not the case because from the method of moments, a square matrix of sizes $N \times N$ must be inverted, where N is the number of samples of the generated surface. Thus, to access to the ensemble average is very useful to apply an MC process, for which N does not exceed 5,000 (this value depends on the performances of the PC).

1.5.3 Comparison of the average shadowing function

Figure 1.13 compares the average shadowing function over the normalized slope s_0 and height h_0 versus v computed from the Wagner (a) and the Smith (b) models with that obtained from the ray-tracing algorithm (MC). At the bottom, the difference against the MC results is plotted.

Figure 1.13 shows that the correlation improves the results and the Smith formulation is better than the Wagner one. According to the MC results, the Smith and Wagner models overestimate the strengths and the Smith results with correlation match well with those of MC.

Figure 1.14 plots the average shadowing function over the normalized slope s_0 and height h_0 versus v computed from an MC process and by considering Gaussian and Lorentzian autocorrelation functions. At the bottom, the difference against the results computed from a Gaussian autocorrelation function is plotted. As we can see, the shadowing strengths are not very sensitive to the surface height

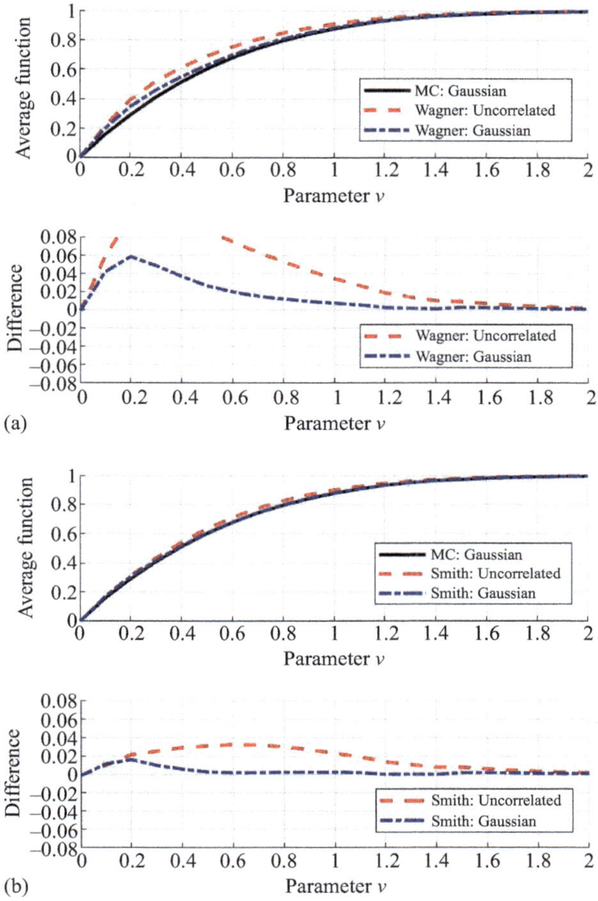

Figure 1.13 *Top: Average shadowing function over the normalized slope s_0 and height h_0 versus v computed from the Wagner (a) and Smith (b) models and an MC process. Bottom: Difference against the MC results*

autocorrelation function, which is consistent with the Wagner and Smith results (Figure 1.9).

1.6 Generalization to the bitstatic case in reflection

In this section, the concept of the shadowing function is extended to the case, for which the surface is illuminated by a transmitter and observed by a receiver. In other words, as shown in Figure 1.15, this function gives the probability that an arbitrary

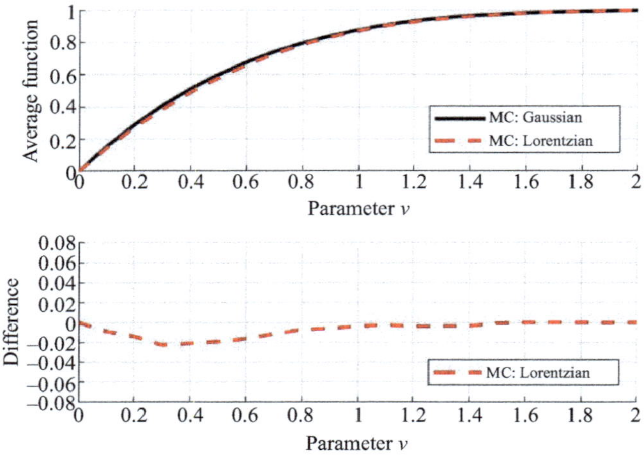

Figure 1.14 Top: Average shadowing function over the normalized slope s_0 and height h_0 versus v computed from an MC process. Bottom: Difference against the MC results by considering a Gaussian autocorrelation function

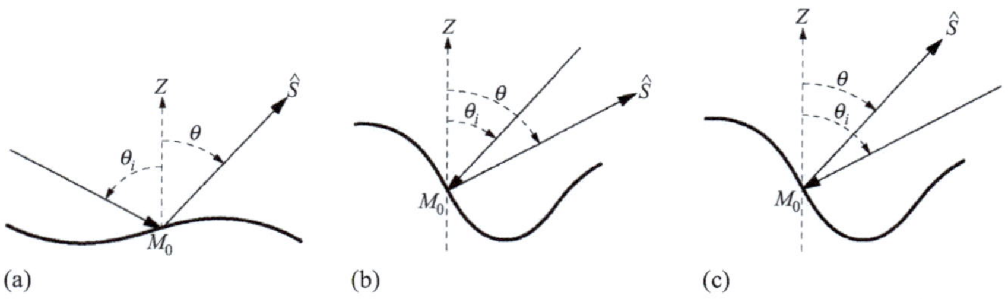

Figure 1.15 Three cases for the bistatic case. (a) Transmitter and the receiver are on different sides of M_0; (b) transmitter and the receiver are on the same side with θ_i smaller; (c) transmitter and the receiver are on the same side with θ smaller. Zenith angles are oriented, with the direction of θ being the positive direction

point of the surface M_0 is seen by both the transmitter and the receiver (above the surface).

1.6.1 Definition

Figure 1.15 shows three cases of single reflection. The event 'the ray $M_0(\hat{s})$ does not intersects the surface' is denoted as a and that 'the ray $M_0(\hat{s}_i^-)$ does not intersects the surface' is denoted as b. M_0 is an arbitrary point of the surface with height ζ_0 and slope γ_0. The bistatic illumination function with one surface reflection S^1_{B} is given by [4]

$$S^1_{\mathrm{B}} = p(ab) = p(a)p(b|a) = p(b)p(a|b). \tag{1.108}$$

The probability $p(a)$ equals the monostatic shadowing function without surface reflection S^0_{M}.

In Figure 1.15(a), the transmitter and the receiver are on different sides (with respect to the zenith direction) of the point of interest M_0 (note that θ_i is negative here as it is in the opposite direction of θ). The events a and b are assumed to be uncorrelated. Then, the bistatic illumination function is given for the first case by [4]

$$\begin{aligned} S^1_{\mathrm{B}} &= p(a)p(b|a) = p(a)p(b) \\ &= S^0_{\mathrm{M}}(\mu, \zeta_0, \gamma_0, L_0) S^0_{\mathrm{M}}(\mu_i, \zeta_0, \gamma_0, L_0), \end{aligned} \tag{1.109}$$

where $\mu = \cot\theta \geq 0$, $\mu_i = \cot\theta_i \leq 0$ and L_0 is the surface length.

In Figure 1.15(b), the transmitter and the receiver are on the same side of M_0 with respect to the zenith direction. As $\theta > \theta_i > 0$, the receiver is lower than the transmitter. Given that M_0 is seen by the receiver, it is sure that it would be also seen by the transmitter. Thus, $p(b|a) = 1$. The bistatic illumination function is given for this case by [4]

$$\begin{aligned} S^1_{\mathrm{B}} &= p(a)p(b|a) = p(a) \times 1 \\ &= S^0_{\mathrm{M}}(\mu, \zeta_0, \gamma_0, L_0). \end{aligned} \tag{1.110}$$

Figure 1.15(c) is very similar to (b), but with $0 < \theta < \theta_i$. Therefore, given that M_0 is seen by the transmitter, it is sure that it would be also seen by the receiver and $p(a|b) = 1$. The bistatic illumination function is expressed for this case by [4]

$$\begin{aligned} S^1_{\mathrm{B}} &= p(b)p(a|b) = p(b) \times 1 \\ &= S^0_{\mathrm{M}}(\mu_i, \zeta_0, \gamma_0, L_0). \end{aligned} \tag{1.111}$$

In conclusion, the bistatic illumination function is expressed from monostatic illumination functions.

1.6.2 Case of any uncorrelated process

In this section, the process is assumed to be uncorrelated.

1.6.2.1 Wagner and Smith formulations

From the Wagner (1.24) and Smith (1.27) models, the statistical bistatic shadowing function is expressed as

$$S_{\mathrm{B}}^{1}(\mu,\mu_i,\zeta_0,\gamma_0,L_0) = \Upsilon_t(\mu,\mu_i,\gamma_0) \times \begin{cases} \exp\{-\Lambda_t\left[P_\zeta(\zeta_0+\mu L_0)-P_\zeta(\zeta_0)\right]\}, & \text{Wagner} \\ \left[\dfrac{P_\zeta(\zeta_0)-P_\zeta(-\infty)}{P_\zeta(\zeta_0+\mu L_0)-P_\zeta(-\infty)}\right]^{\Lambda_t} & \text{Smith} \end{cases}, \tag{1.112}$$

where

$$\Lambda_t(\mu,\mu_i) = \begin{cases} \Lambda(\mu)+\Lambda_-(\mu_i) & \text{Case (a)} \\ \Lambda(\mu) & \text{Case (b)} \\ \Lambda(\mu_i) & \text{Case (c)} \end{cases}, \tag{1.113}$$

and

$$\Upsilon_t(\mu,\mu_i,\gamma_0) = \begin{cases} \Upsilon(\mu-\gamma_0)\,\Upsilon(\gamma_0-\mu_i) & \text{Case (a)} \\ \Upsilon(\mu-\gamma_0) & \text{Case (b)} \\ \Upsilon(\mu_i-\gamma_0) & \text{Case (c)} \end{cases}. \tag{1.114}$$

For the case (a), it is important to underline that the slope μ_i is negative and the surface points are illuminated if $\gamma_0-\mu_i \geq 0$ ($\gamma_0 \geq -|\mu_i|$). Cases (b) and (c) correspond to monostatic cases.

In addition, Λ is expressed from (1.22) and Λ_- is defined as

$$\begin{aligned} \Lambda_-(\mu_i) &= \frac{1}{\mu_i}\int_{-\infty}^{\mu_i}(\gamma-\mu_i)p_\gamma(\gamma)\mathrm{d}\gamma \\ &= -\frac{1}{|\mu_i|}\int_{-\infty}^{-|\mu_i|}(\gamma+|\mu_i|)p_\gamma(\gamma)\mathrm{d}\gamma \\ &= \frac{1}{|\mu_i|}\int_{\infty}^{|\mu_i|}(-\gamma+|\mu_i|)p_\gamma(-\gamma)\mathrm{d}\gamma \\ &= \frac{1}{|\mu_i|}\int_{|\mu_i|}^{\infty}(\gamma-|\mu_i|)p_\gamma(-\gamma)\mathrm{d}\gamma, \end{aligned} \tag{1.115}$$

where $|\mu_i| = -\mu_i$. If the slope PDF p_γ is assumed to be even, then $p_\gamma(-\gamma) = p_\gamma(\gamma)$, which implies that $\Lambda_-(\mu_i) = \Lambda(|\mu_i|)$. Moreover, $f(-\gamma_0) = \Upsilon(-\gamma_0+|\mu_i|) = \Upsilon(|\mu_i|-\gamma_0)$ where $f(\gamma_0) = \Upsilon(\gamma_0-\mu_i)$. As expected, if the surface slope is even, the statistical monostatic shadowing function of a receiver of incidence angle θ (rays coming from the right) equals that of a receiver of incidence angle $-\theta$ (rays coming from the left).

1.6.2.2 Average shadowing functions

As the monostatic case, expected values over γ_0, ζ_0 and both (γ_0,ζ_0) can be derived.

The average over the surface slope γ_0 needs to derive

$$\bar{\Upsilon}_t(\mu, \mu_i) = \int_{-\infty}^{+\infty} \Upsilon_t(\mu, \mu_i, \gamma_0) p_\gamma(\gamma_0) \mathrm{d}\gamma_0. \tag{1.116}$$

From (1.114), this leads for any slope PDF p_γ to

$$\bar{\Upsilon}_t(\mu, \mu_i) = \begin{cases} \displaystyle\int_{-|\mu_i|}^{\mu} p_\gamma(\gamma_0)\mathrm{d}\gamma_0 = \bar{\Upsilon}(\mu) - 1 + \int_{-\infty}^{|\mu_i|} p_\gamma(-\gamma_0)\mathrm{d}\gamma_0 & \text{Case (a)} \\ \displaystyle\int_{-\infty}^{\mu} p_\gamma(\gamma_0)\mathrm{d}\gamma_0 = \bar{\Upsilon}(\mu) & \text{Case (b)} \\ \displaystyle\int_{-\infty}^{\mu_i} p_\gamma(\gamma_0)\mathrm{d}\gamma_0 = \bar{\Upsilon}(\mu_i) & \text{Case (c)} \end{cases} . \tag{1.117}$$

Using the same way as the monostatic case, for an infinite surface length L_0, from (1.41) and (1.42), the average over the surface heights ζ_0 leads to

$$\int_{-\infty}^{+\infty} S_{\mathrm{B}}^{1}(\mu, \mu_i, \zeta_0, \gamma_0, \infty) p_\zeta(\zeta_0) \mathrm{d}\zeta_0 = \Upsilon_t \begin{cases} \dfrac{1 - e^{-\Lambda_t(\mu, \mu_i)}}{\Lambda_t(\mu, \mu_i)} & \text{Wagner} \\ \dfrac{1}{1 + \Lambda_t(\mu, \mu_i)} & \text{Smith} \end{cases} . \tag{1.118}$$

The average over γ_0 and ζ_0 leads to

$$\int_{-\infty}^{+\infty}\int_{-\infty}^{+\infty} S_{\mathrm{B}}^{1}(\mu, \mu_i, \zeta_0, \gamma_0, \infty) p_\gamma(\gamma_0) p_\zeta(\zeta_0) \mathrm{d}\gamma_0 \mathrm{d}\zeta_0$$

$$= \bar{\Upsilon}_t(\mu, \mu_i) \begin{cases} \dfrac{1 - e^{-\Lambda_t(\mu, \mu_i)}}{\Lambda_t(\mu, \mu_i)} & \text{Wagner} \\ \dfrac{1}{1 + \Lambda_t(\mu, \mu_i)} & \text{Smith} \end{cases}, \tag{1.119}$$

where the joint PDF $p_{\gamma,\zeta}(\gamma_0, \zeta_0) = p_\gamma(\gamma_0) p_\zeta(\zeta_0)$.

1.6.2.3 Case of a Gaussian process

For a Gaussian process with zero mean value, the slope PDF is even. Then, from (1.22) and (1.115), $\Lambda = \Lambda_-$ and it is expressed from (1.47). In addition, $\bar{\Upsilon}$ defined from (1.117) equals

$$\bar{\Upsilon}(v) = \frac{1 + \mathrm{erf}(v)}{2}, \tag{1.120}$$

where erf is the error function and $v = \cot\theta / (\sqrt{2}\sigma_\gamma)$.

1.6.3 Case of a correlated Gaussian process

From (1.5) and (1.6), the monostatic statistical shadowing function is expressed as

$$S_{\mathrm{M}}^{0}(\mu,\zeta_0,\gamma_0,L_0) = \Upsilon(\mu-\gamma_0)\exp\left[-\int_0^{L_0} g(\mu|\zeta_0,\gamma_0;\tau)\mathrm{d}\tau\right], \tag{1.121}$$

where g is defined from the Wagner (1.11) and Smith (1.15) models, respectively, as

$$g(\mu|\zeta_0,\gamma_0;\tau) = \begin{cases} \displaystyle\int_{\mu}^{+\infty}(\gamma-\mu)p(\zeta=\zeta_0+\mu\tau,\gamma|\zeta_0,\gamma_0;\tau)\mathrm{d}\gamma & \text{Wagner} \\ \dfrac{g_{\mathrm{W}}(\mu|\zeta_0,\gamma_0;\tau)}{\displaystyle\int_{-\infty}^{+\infty}\int_{-\infty}^{\zeta_0+\mu\tau} p(\zeta,\gamma|\zeta_0,\gamma_0;\tau)\mathrm{d}\gamma\,\mathrm{d}\zeta} & \text{Smith} \end{cases}. \tag{1.122}$$

Figure 1.15(a) requires a special attention, and it is expressed from (1.109), in which $S_{\mathrm{M}}^{0}(\mu_i,\zeta_0,\gamma_0,L_0) = S_{\mathrm{M}}^{0}(-|\mu_i|,\zeta_0,\gamma_0,L_0)$ must be derived. Changing (γ_0,γ,τ) by $(-\gamma_0,-\gamma,-\tau)$, the joint PDF $p(\zeta,-\gamma|z_0,-\gamma_0;-\tau)$ is the same as $p(\zeta,\gamma|z_0,\gamma_0;\tau)$ expressed from (1.58). Indeed as $C_{\zeta,0}(-\tau)=C_{\zeta,0}(\tau)$ (even function by definition), $C_{\zeta,1}(-\tau)=-C_{\zeta,1}(\tau)$ (derivative of an even function), $C_{\zeta,2}(-\tau)=C_{\zeta,2}(\tau)$ (derivation of an odd function), the functions C_{ij} expressed from (1.59) are odd for $\{i,j\}=\{1,3\}$ and $\{i,j\}=\{1,4\}$ and are even otherwise.

From the Wagner formulation, the function $g_{-}(\mu_i|\zeta_0,\gamma_0;\tau)$ with $\mu_i<0$ is then expressed as

$$\begin{aligned} g_{-}(\mu_i|\zeta_0,\gamma_0;\tau) &= \int_{-\infty}^{-|\mu_i|}(\gamma+|\mu_i|)p(\zeta=\zeta_0-|\mu_i|\tau,\gamma|\zeta_0,\gamma_0;\tau)\mathrm{d}\gamma \\ &= -\int_{\infty}^{|\mu_i|}(-\gamma+|\mu_i|)p(\zeta=\zeta_0-|\mu_i|\tau,-\gamma|\zeta_0,\gamma_0;\tau)\mathrm{d}\gamma \\ &= -\int_{|\mu_i|}^{\infty}(\gamma-|\mu_i|)p(\zeta=\zeta_0-|\mu_i|\tau,-\gamma|\zeta_0,\gamma_0;\tau)\mathrm{d}\gamma. \end{aligned} \tag{1.123}$$

For $\mu_i=-|\mu_i|\leq 0$, the integration over τ is $\tau\in[0;-L_0]$. Making the variable transformation $\tau\to-\tau$ and $\gamma_0\to-\gamma_0$, from (1.121) and (1.123), we can show

$$\begin{aligned} S_{\mathrm{M\text{-}}}^{0}(\mu_i,\zeta_0,-\gamma_0,L_0) &= \Upsilon(\gamma_0+|\mu_i|)\exp\left[-\int_0^{-L_0} g_{-}(\mu_i|\zeta_0,-\gamma_0;\tau)\mathrm{d}\tau\right] \\ &= \Upsilon(\gamma_0+|\mu_i|)\exp\left[\int_0^{L_0} g_{-}(\mu_i|\zeta_0,-\gamma_0;-\tau)\mathrm{d}\tau\right] \\ &= \Upsilon(\gamma_0+|\mu_i|)\exp\left[-\int_0^{L_0} g(|\mu_i||\zeta_0,\gamma_0;\tau)\mathrm{d}\tau\right] \\ &= S_{\mathrm{M}}^{0}(|\mu_i|,\zeta_0,-\gamma_0,L_0), \end{aligned} \tag{1.124}$$

because $p(\zeta, -\gamma|z_0, -\gamma_0; -\tau) = p(\zeta, \gamma|z_0, \gamma_0; \tau)$. In addition, as a Gaussian slope PDF p_γ is an even function, for the average over, the surface slope γ_0, $-\gamma_0$ can be replaced by $+\gamma_0$. From the Smith formulation, using the same way, we can show that (1.124) is also satisfied.

As the uncorrelated case, the statistical monostatic shadowing function of a receiver of incidence angle θ (rays coming from the right) equals that of a receiver of incidence angle $-\theta$ (rays coming from the left).

In conclusion, in Figure 1.15(a), from (1.109) and (1.124), the bistatic statistical shadowing function is

$$S_{\mathrm{B}}^{1}(\mu, \mu_i, \zeta_0, \gamma_0, L_0) = \Pi(\gamma_0) \exp\left\{-\int_0^{L_0} [g(|\mu_i||\zeta_0, \gamma_0; \tau) + g(\mu|\zeta_0, \gamma_0; \tau)]\,\mathrm{d}\tau\right\}, \tag{1.125}$$

where

$$\Pi(\gamma_0) = \begin{cases} 1 \text{ if } \gamma_0 \in [-|\mu_i|; \mu] \\ 0 \text{ otherwise} \end{cases}. \tag{1.126}$$

Using the variable transformations given from (1.63) and (1.68), the function S_{B}^{1} depends on (v, v_i, h_0, s_0, y_0) and g is expressed in Table 1.2.

1.6.4 Ray-tracing algorithm

In this section, the ray-tracing algorithm or MC process is generalized to the bistatic case.

In Figure 1.15(b) and (c), the algorithm presented in Figure 1.11 is applied. In Figure 1.15(a), this algorithm is applied twice with respect to the slope μ and $\mu_i = -|\mu_i|$. For the slope $\mu_i = -|\mu_i|$, the rays come from the left, and then minor changes are done. Nevertheless, as the surface slopes γ are even for a Gaussian process, algorithm of Figure 1.11 can be applied changing γ by $-\gamma$.

From the shadowed points, the average bistatic shadowing function over the surface heights and slopes is expressed as

$$\bar{\bar{S}}_{\mathrm{B}}^{1}(v, v_i) = \frac{1}{N} \sum_{i=1}^{i=N} b_i^{(\mathrm{T})} b_i^{(\mathrm{R})}, \tag{1.127}$$

where the Boolean $b_i^{(\mathrm{T})}$ and $b_i^{(\mathrm{R}))}$ are defined as

$$b_i^{(\mathrm{T})} = \begin{cases} 0 \text{ if the point is shadowed from the transmitter} \\ 1 \text{ otherwise} \end{cases}, \tag{1.128}$$

and

$$b_i^{(\mathrm{R})} = \begin{cases} 0 \text{ if the point is shadowed from the receiver} \\ 1 \text{ otherwise} \end{cases}. \tag{1.129}$$

As shown in [3], any expected value (or ensemble average) can be performed, and the marginal slope and height PDFs of the illuminated points can also be computed.

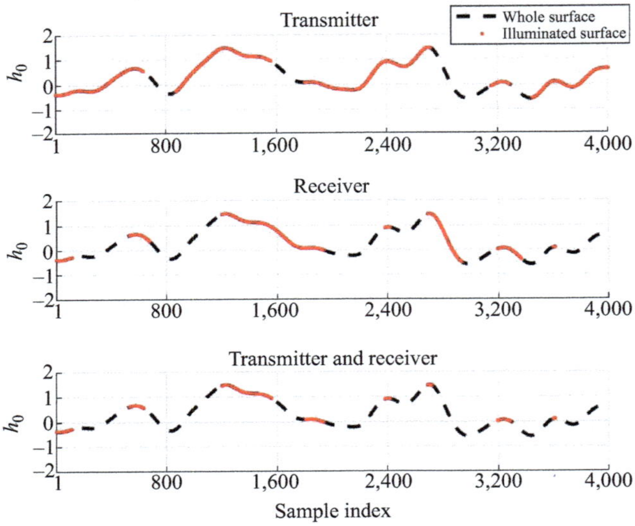

Figure 1.16 *Surface normalized heights h_0 of the whole and illuminated points versus the sample index. Top: From the transmitter only. Middle: From the receiver only. Bottom: From both the transmitter and receiver. The surface height autocorrelation function is Gaussian with $L_c = 200$, $\sigma_\varsigma = 1$ and the parameters $v = \mu/(\sigma_Y\sqrt{2}) = 0.3$ and $v_i = \mu_i/(\sigma_Y\sqrt{2}) = -0.5$ (Figure 1.15(a))*

Figure 1.16 presents an illustration of the algorithm. The top shows the surface points illuminated from rays propagating from the left (transmitter), whereas the middle shows the surface points illuminated from rays propagating from the right (receiver). The bottom shows the surface points illuminated both from the transmitter and the receiver. As we can see, the illuminated points are mostly located at the top of the surface (high elevations).

In the following, for the MC process, the surface length will be $L = 5,000L_c$.

1.6.5 Numerical results

In this section, for a Gaussian process, the Wagner and Smith models with and without correlation are compared with the benchmark method obtained from a ray-tracing algorithm or MC process. The surface length L_0 (or y_0) is assumed to be infinite.

1.6.5.1 Average over the surface heights

The average over the surface heights is expressed from (1.118). For the correlated case and for Figure 1.15(b) and (c), it is computed from (1.84), in which the integration over s_0 is not performed, and it is replaced by the unit step function $\Upsilon(v - s_0)$.

In addition, for the case (c) for which $0 < \theta < \theta_1$ (Figure 1.15(c)), v is changed by $|v_i|$. For the case (a), for which $\theta > 0$ and $\theta_i < 0$ (Figure 1.15(a)), from (1.125), it is expressed as

$$\begin{aligned}\bar{S}^1_{\mathrm{B}}(v, v_i, s_0, y_0) &= \int_{-\infty}^{+\infty} S^1_{\mathrm{B}}(v, v_i, h_0, s_0, y_0) p_\zeta(h_0) \mathrm{d}h_0 \\ &= \frac{\Pi(s_0)}{\sqrt{\pi}} \int_{-\infty}^{\infty} \exp\left\{-L_c \int_0^{y_t} [g(v, h_0, s_0, y) + g(|v_i|, h_0, s_0, y)] \, \mathrm{d}y \right. \\ &\quad \left. + G(v, h_0, y_t, y_0) + G(|v_i|, h_0, y_t, y_0)\right\} \exp(-h_0^2) \mathrm{d}h_0, \qquad (1.130)\end{aligned}$$

where

$$\Pi(s_0) = \begin{cases} 1 & \text{if } s_0 \in [-|v_i|; v] \\ 0 & \text{otherwise} \end{cases}, \qquad (1.131)$$

and $L_c g$ and G are expressed in Table 1.2. For the numerical implementation of the function g, see Section 1.4.4.1.

Figure 1.17 plots the average shadowing function over the heights ζ_0 versus s_0 computed from the Wagner (a) and the Smith (b) models, where $v = 0.3$ and $v_i = -0.5$.

As we can see, both Wagner's and Smith's models with correlation bend down significantly as s_0 approaches v_i or v, or equally as γ_0 approaches μ_i or μ, whereas the uncorrelated ones remains constant for $s_0 \in [-|v_i|; v]$. Next, the strengths with and without correlation jump to 0 at $s_0 > v$ and $s_0 < -|v_i|$ due to the Π function. Then, the correlation implies that all points with slope $s_0 \in [-|v_i|; v]$ are not equally shadowed. In comparison to the MC results, Figure 1.17 also shows that the models overpredict the occurrence of the surface slopes near v_i, whereas they match well near v. The strengths depend slightly of the surface height autocorrelation function (Gaussian or Lonrentzian; see Table 1.1 for their definition).

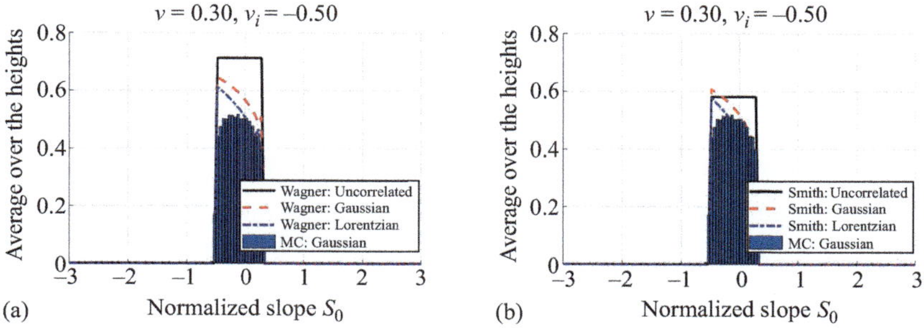

Figure 1.17 Average shadowing function over the heights ζ_0 versus s_0 computed from the Wagner (a) and the Smith (b) models, where $v = 0.3$ and $v_i = -0.5$

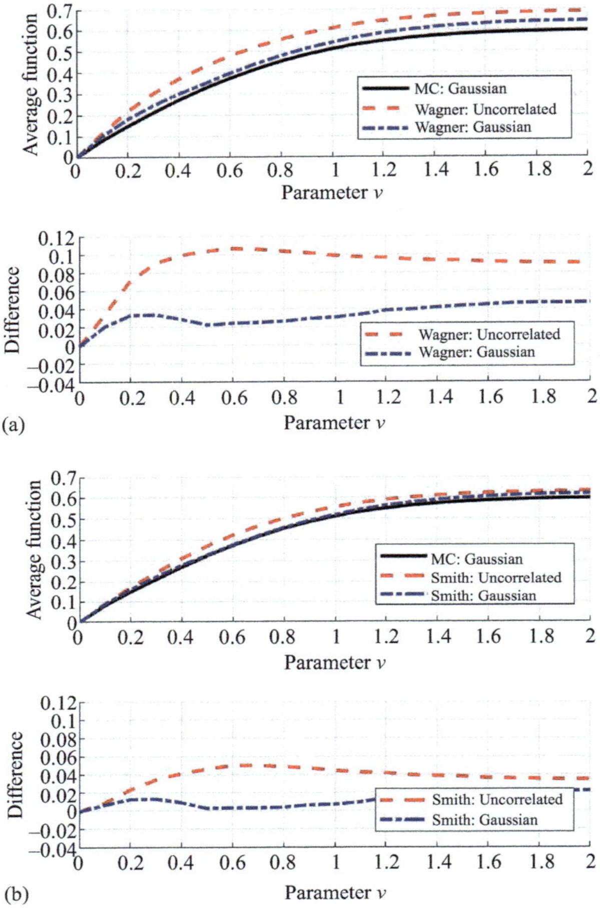

Figure 1.18 Top: Average shadowing function over the normalized slope s_0 and height h_0 versus v computed from the Wagner (a) and the Smith (b) models. Bottom: Difference against the Monte Carlo results. $v_i = -0.5$

1.6.5.2 Average over the surface heights and slopes

Figure 1.18 plots the average shadowing function $\bar{\bar{S}}_{\mathrm{B}}^1$ over the normalized slope s_0 and height h_0 versus v computed from the Wagner (a) and the Smith (b) models. At the bottom, the difference against the MC results is plotted. $\bar{\bar{S}}_{\mathrm{B}}^1$ is defined as

$$\bar{\bar{S}}_{\mathrm{B}}^1(v, v_i, y_0) = \frac{1}{\sqrt{\pi}} \int_{-\infty}^{\infty} \bar{S}_{\mathrm{B}}^1(v, v_i, s_0, y_0) \exp(-s_0^2) \mathrm{d}s_0, \tag{1.132}$$

where $\bar{S}_{\mathrm{B}}^1(v, v_i, y_0)$ is expressed from (1.130).

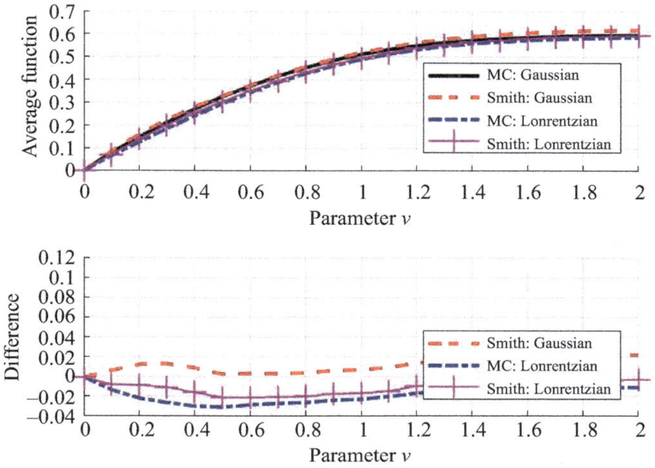

Figure 1.19 *Top: Average shadowing function over the normalized slope s_0 and height h_0 versus v computed from an MC process and the Smith model with correlation. Bottom: Difference against the MC results by considering a Gaussian autocorrelation function. $v_i = -0.5$*

As we can see, the correlation improves the results, and the Smith model is better than that of Wagner. In addition, for high values of v, the strengths become constant because the receiver viewed the whole surface, whereas some surface points illuminated by the transmitter are in shadow. Indeed, since $|v_i| = 0.5$, from Figure 1.6, the monostatic average-shadowing function approximately equals 0.7.

Figure 1.19 plots the average shadowing function over the normalized slope s_0 and height h_0 versus v computed from an MC process and the Smith model with correlation. At the bottom, the difference against the MC results is plotted by considering a Gaussian autocorrelation function.

As we can see, for a given autocorrelation function, the difference between the results is small, and the strengths are not very sensitive to the correlation function.

1.7 Generalization to the bistatic case in transmission

In this section, the concept of the shadowing function is extended to the case, for which the surface is illuminated by a transmitter (located above the surface) and observed by a receiver (located below the surface). In other words, as shown in Figure 1.20, this function gives the probability that an arbitrary point of the surface M_0 is seen by both the transmitter and the receiver of incidence angles θ_i and θ_t, respectively.

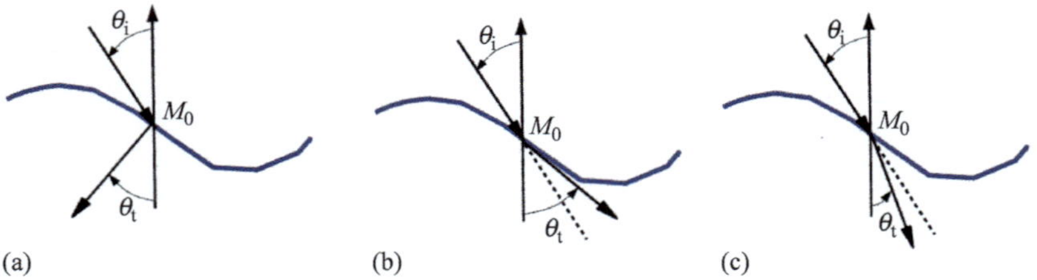

Figure 1.20 Bistatic case in transmission. (a): $\theta_t \geq 0$. (b): $\theta_t \in [-\pi/2;\theta_i]$. (c): $\theta_t \in [\theta_i; 0]$. $\theta_i \in [-\pi/2; 0]$ and $\theta_t \in [-\pi/2; \pi/2]$

1.7.1 Definition

The same way as in Section 1.6.1 is applied (bistatic case in reflection). Figure 1.20 shows three cases.

For Figure 1.20(a), the transmitter and the receiver are on the same side (with respect to the zenith direction) of the point of interest M_0 (note that θ_i is negative here as it is in the opposite direction of θ_t). Then

$$\begin{aligned} T_{\mathrm{B}}^{1} &= S_{\mathrm{M}}^{0}(\mu_i, \zeta_0, \gamma_0, L_0) T_{\mathrm{M}}^{0}(\mu_t, \zeta_0, \gamma_0, L_0) \\ &= S_{\mathrm{M}}^{0}(-|\mu_i|, \zeta_0, \gamma_0, L_0) T_{\mathrm{M}}^{0}(|\mu_t|, \zeta_0, \gamma_0, L_0), \end{aligned} \tag{1.133}$$

where $\mu_i = \cot\theta_i \leq 0$, $\mu_t = \cot\theta_t \geq 0$ and L_0 is the surface length. In addition, S_{M}^{0} is the monostatic statistical shadowing function in reflection, whereas T_{M}^{0} is that defined in transmission.

For Figure 1.20(b), the transmitter and the receiver are on opposite side with $\pi/2 \leq \theta_t \leq \theta_i < 0$ (with respect to the zenith direction) of the point of interest M_0 (note that θ_i and θ_t are negative). Then

$$\begin{aligned} T_{\mathrm{B}}^{1} &= S_{\mathrm{M}}^{0}(\mu_i, \zeta_0, \gamma_0, L_0) T_{\mathrm{M}}^{0}(\mu_t, \zeta_0, \gamma_0, L_0) \\ &= S_{\mathrm{M}}^{0}(-|\mu_i|, \zeta_0, \gamma_0, L_0) T_{\mathrm{M}}^{0}(-|\mu_t|, \zeta_0, \gamma_0, L_0), \end{aligned} \tag{1.134}$$

where $\mu_t = -\cot|\theta_t| \leq 0$.

For Figure 1.20(c), the transmitter and the receiver are on the opposite side with $\pi/2 \leq \theta_i < \theta_t < 0$ (with respect to the zenith direction) of the point of interest M_0 (note that θ_i and θ_t are negative). Then

$$\begin{aligned} T_{\mathrm{B}}^{1} &= S_{\mathrm{M}}^{0}(\mu_i, \zeta_0, \gamma_0, L_0) T_{\mathrm{M}}^{0}(\mu_t, \zeta_0, \gamma_0, L_0) \\ &= S_{\mathrm{M}}^{0}(-|\mu_i|, \zeta_0, \gamma_0, L_0) T_{\mathrm{M}}^{0}(-|\mu_t|, \zeta_0, \gamma_0, L_0). \end{aligned} \tag{1.135}$$

1.7.2 Case of any uncorrelated process

In this section, the process is assumed to be uncorrelated.

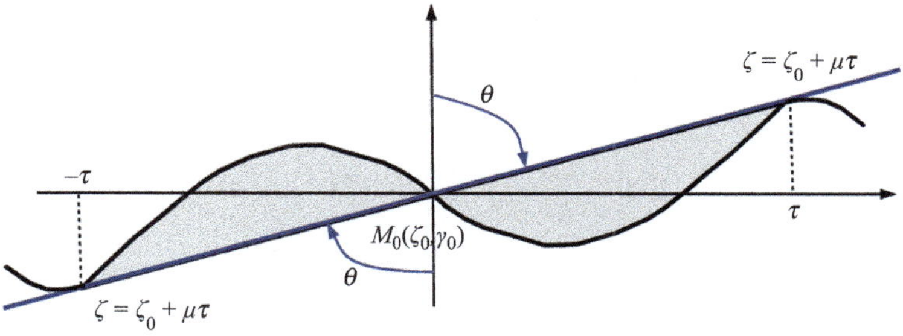

Figure 1.21 Illustration of the shadow both in reflection ($\tau \geq 0$) and in transmission ($\tau \leq 0$)

1.7.2.1 Wagner and Smith formulations

For $\mu_i = -|\mu_i| \leq 0$, from the Wagner (1.24) and Smith (1.27) models (see also (1.112), (1.113) and (1.114)), the statistical monostatic shadowing function is expressed as

$$S^0_{\mathrm{M}}(-|\mu_i|, \zeta_0, \gamma_0, L_0) = \Upsilon(\gamma_0 + |\mu_i|) \times \begin{cases} \exp\left\{-\Lambda_-(\mu_i)\left[P_\zeta(\zeta_0 + |\mu_i|L_0) - P_\zeta(\zeta_0)\right]\right\} & \text{Wagner} \\ \left[\dfrac{P_\zeta(\zeta_0) - P_\zeta(-\infty)}{P_\zeta(\zeta_0 + |\mu_i|L_0) - P_\zeta(-\infty)}\right]^{\Lambda_-(\mu_i)} & \text{Smith} \end{cases}, \tag{1.136}$$

where Λ_- is defined from (1.113). In the above equation, the term $P_\zeta(\zeta_0 + |\mu_i|L_0) - P_\zeta(\zeta_0)$ corresponds to the integration over τ, for which $\zeta = \zeta_0 + |\mu_i|L_0 \geq \zeta_0$. For $\tau \geq 0$ (reflection case), this is consistent with Figure 1.21 (in which, $|\mu| = \mu = \mu_i \geq 0$). For $\tau \leq 0$ (transmission case), $\zeta = \zeta_0 - |\mu_i|L_0 \leq \zeta_0$ and the integration over τ gives $P_\zeta(\zeta_0) - P_\zeta(\zeta_0 - |\mu_i|L_0)$.

For an infinite surface length L_0, $P_\zeta(\zeta_0) - P_\zeta(\zeta_0 - |\mu_i|L_0) = P_\zeta(\zeta_0) - P_\zeta(-\infty) = 1 - [P_\zeta(\infty) - P_\zeta(\zeta_0)]$, where $P_\zeta(\infty) - P_\zeta(-\infty) = 1$.

Then, for $\mu_t \geq 0$, from this simple physical consideration and (1.136), the statistical monostatic shadowing function in transmission is

$$T^0_{\mathrm{M}}(|\mu_t|, \zeta_0, \gamma_0, L_0) = \Upsilon(\mu_t - \gamma_0) \times \begin{cases} \exp\left\{-\Lambda(\mu_t)\left[P_\zeta(\zeta_0) - P_\zeta(\zeta_0 - \mu_t L_0)\right]\right\} & \text{Wagner} \\ \left[\dfrac{P_\zeta(\infty) - P_\zeta(\zeta_0)}{P_\zeta(\infty) - P_\zeta(\zeta_0 - |\mu_t|L_0)}\right]^{\Lambda(\mu_t)} & \text{Smith} \end{cases}. \tag{1.137}$$

The bistatic case also requires the derivation of T_{M}^{0} for $\theta = \theta_t < 0$. From symmetrical considerations, from (1.137), we have

$$T_{\mathrm{M}}^{0}(-|\mu_t|, \zeta_0, \gamma_0, L_0) = \Upsilon(|\mu_t| + \gamma_0) \times \begin{cases} \exp\{-\Lambda_{-}(\mu_t)\left[P_{\zeta}(\zeta_0) - P_{\zeta}(\zeta_0 - |\mu_t|L_0)\right]\} & \text{Wagner} \\ \left[\dfrac{P_{\zeta}(\infty) - P_{\zeta}(\zeta_0)}{P_{\zeta}(\infty) - P_{\zeta}(\zeta_0 - |\mu_t|L_0)}\right]^{\Lambda_{-}(\mu_t)} & \text{Smith} \end{cases}. \tag{1.138}$$

Assuming an infinite surface length L_0, the substitution of (1.138), (1.137) and (1.136) into (1.133)–(1.135) leads to

$$T_{\mathrm{B}}^{1}(\mu_i, \mu_t, \zeta_0, \gamma_0, \infty) = \Upsilon_t(\mu_i, \mu_t, \gamma_0) \times \begin{cases} \exp\left[-\Lambda_{-}(\mu_i)F_{\zeta}(\zeta_0) - L_t\left[1 - F_{\zeta}(\zeta_0)\right]\right\} & \text{Wagner} \\ \left[F_{\zeta}(\zeta_0)\right]^{\Lambda_{-}(\mu_i)}\left[1 - F_{\zeta}(\zeta_0)\right]^{L_t} & \text{Smith} \end{cases}, \tag{1.139}$$

where

$$L_t = \begin{cases} \Lambda(\mu_t) & \text{Case (a) } (\theta_t \geq 0) \\ \Lambda_{-}(\mu), \mu = \min(|\mu_i|, |\mu_t|) & \text{Cases (b) and (c) } (\theta_t < 0) \end{cases}, \tag{1.140}$$

$$\Upsilon_t(\mu_i, \mu_t, \gamma_0) = \begin{cases} \Upsilon(\gamma_0 + |\mu_i|)\,\Upsilon(\mu_t - \gamma_0) & \text{Case (a)} \\ \Upsilon(\mu + \gamma_0), \mu = \min(|\mu_i|, |\mu_t|) & \text{Cases (b) and (c)} \end{cases}, \tag{1.141}$$

and $F_{\zeta}(\zeta_0) = P_{\zeta}(\infty) - P_{\zeta}(\zeta_0)$.

If the surface slope PDF is even, from (1.115), $\Lambda_{-} = \Lambda$.

1.7.2.2 Average shadowing functions

Expected values over γ_0, ζ_0 and both (γ_0, ζ_0) can be derived.

The average over the surface slope γ_0 is given by

$$\bar{\Upsilon}_t(\mu_i, \mu_t) = \int_{-\infty}^{+\infty} \Upsilon_t(\mu_i, \mu_t, \gamma_0) p_{\gamma}(\gamma_0) \mathrm{d}\gamma_0. \tag{1.142}$$

From (1.141), for any slope PDF p_{γ}, this leads to

$$\bar{\Upsilon}_t(\mu_i, \mu_t) = \begin{cases} \displaystyle\int_{-|\mu_i|}^{\mu_t} p_{\gamma}(\gamma_0)\mathrm{d}\gamma_0 = \bar{\Upsilon}(|\mu_i|) - 1 + \int_{-\mu_t}^{\infty} p_{\gamma}(-\gamma_0)\mathrm{d}\gamma_0 \\ \text{Case (a)} \\ \displaystyle\int_{-\mu}^{\infty} p_{\gamma}(\gamma_0)\mathrm{d}\gamma_0 = \int_{-\infty}^{\mu} p_{\gamma}(-\gamma_0)\mathrm{d}\gamma_0 = \bar{\Upsilon}(\mu) \\ \text{Cases (b) and (c), } \mu = \min(|\mu_i|, |\mu_t|) \end{cases}. \tag{1.143}$$

From (1.139), the average over the surface heights ζ_0 is

$$\int_{-\infty}^{+\infty} T_{\mathrm{B}}^{1}(\mu_i, \mu_t, \zeta_0, \gamma_0, \infty) p_\zeta(\zeta_0) \mathrm{d}\zeta_0 = \Upsilon_t(\mu_i, \mu_t, \gamma_0)$$

$$\times \begin{cases} \dfrac{e^{-L_t}\left[1 - e^{-(\Lambda_- - L_t)}\right]}{\Lambda_- - L_t} = \dfrac{e^{-L_+} \sinh(L_-)}{2L_-} & \text{Wagner} \\ \mathrm{B}(1 + \Lambda_-, 1 + L_t) & \text{Smith} \end{cases}, \tag{1.144}$$

where $L_\pm = (\Lambda_- \pm L_t)/2$ and $\lim_{L_- \to 0} \sinh(L_-)/L_- = 1$. In addition, B stands for the Eulerian (or beta) function of first kind, defined as [12]

$$\mathrm{B}(1 + x, 1 + y) = \int_0^1 t^x (1 - t)^y \, \mathrm{d}t. \tag{1.145}$$

The average over γ_0 and ζ_0 is given by

$$\int_{-\infty}^{+\infty} \int_{-\infty}^{+\infty} T_{\mathrm{B}}^{1}(\mu_i, \mu_t, \zeta_0, \gamma_0, \infty) p_\gamma(\gamma_0) p_\zeta(\zeta_0) \mathrm{d}\gamma_0 \mathrm{d}\zeta_0$$

$$= \bar{\Upsilon}_t(\mu_i, \mu_t) \begin{cases} \dfrac{e^{-L_+} \sinh(L_-)}{2L_-} & \text{Wagner} \\ \mathrm{B}(1 + \Lambda_-, 1 + L_t) & \text{Smith} \end{cases}, \tag{1.146}$$

where the joint PDF $p_{\gamma,\zeta}(\gamma_0, \zeta_0) = p_\gamma(\gamma_0) p_\zeta(\zeta_0)$.

1.7.3 Case of a Gaussian-correlated process

From (1.124), we have shown for a Gaussian-correlated process that

$$S_{\mathrm{M}}^{0}(-|\mu_i|, \zeta_0, \gamma_0, L_0) = S_{\mathrm{M}}^{0}(|\mu_i|, \zeta_0, -\gamma_0, L_0). \tag{1.147}$$

In (1.133), the function $T_{\mathrm{M}}^{0}(|\mu_t|, \zeta_0, \gamma_0, L_0)$ ($|\mu_t| = \mu_t \geq 0$) can be derived from the statistical monostatic shadowing function defined in reflection from (1.121). In Figure 1.21, this case corresponds to $\tau \geq 0$. By doing a central symmetry, the function in transmission is obtained and corresponds to the case $\tau \leq 0$.

1.7.3.1 Monostatic statistical shadowing function

From physical considerations, the case in transmission is obtained from that in reflection replacing (τ, L_0) by $(-\tau, -L_0)$. Then, from (1.5) and (1.6), we have

$$\begin{aligned} T_{\mathrm{M}}^{0}(\mu_t, \zeta_0, \gamma_0, L_0) &= \Upsilon(\mu_t + \gamma_0) \exp\left[-\int_0^{-L_0} g(\mu_t|\zeta_0, \gamma_0; \tau) \mathrm{d}\tau \right] \\ &= \Upsilon(\mu_t + \gamma_0) \exp\left[+\int_0^{L_0} g(\mu_t|\zeta_0, \gamma_0; -\tau) \mathrm{d}\tau \right]. \end{aligned} \tag{1.148}$$

In addition, from the Wagner formulation, the integration range over γ of g_{W} becomes $]-\infty, -\mu_t]$, which implies that

$$\begin{aligned} g_{\mathrm{W}}(\mu_t|\zeta_0, -\gamma_0; -\tau) &= \int_{-\infty}^{-\mu_t} (\gamma + \mu_t) p(\zeta = \zeta_0 - \mu_t\tau, \gamma|\zeta_0, -\gamma_0; -\tau) \mathrm{d}\gamma \\ &= -\int_{\infty}^{\mu_t} (-\gamma + \mu_t) p(\zeta = \zeta_0 - \mu_t\tau, -\gamma|\zeta_0, -\gamma_0; -\tau) \mathrm{d}\gamma \\ &= -\int_{\mu_t}^{\infty} (\gamma - \mu_t) p(\zeta = \zeta_0 - \mu_t\tau, -\gamma|\zeta_0, -\gamma_0; -\tau) \mathrm{d}\gamma. \end{aligned} \tag{1.149}$$

In Section 1.6.3, for a Gaussian process, we showed that $p(\zeta, -\gamma|\zeta_0, -\gamma_0; -\tau) = p(\zeta, \gamma|\zeta_0, \gamma_0; \tau)$. Making the variable transformation $\gamma_0 \to -\gamma_0$, from the above equation, (1.148) becomes

$$\begin{aligned} T^0_{\mathrm{M}}(\mu_t, \zeta_0, -\gamma_0, L_0) &= \Upsilon(\mu_t - \gamma_0) \exp\left[-\int_0^{L_0} g(\mu_t|\zeta_0, -\gamma_0; -\tau) \mathrm{d}\tau\right] \\ &= \Upsilon(\mu_t - \gamma_0) \exp\left[-\int_0^{L_0} g(\mu_t|\zeta_0, \gamma_0; \tau) \mathrm{d}\tau\right], \end{aligned} \tag{1.150}$$

for a Gaussian (more generally, even) process. From the Smith model, following the same way, we can show that the above equation is also satisfied. In addition, in (1.15), the denominator is modified since in transmission $\zeta \in [\zeta_0 - \mu_t\tau, \infty]$ instead of $\zeta \in [-\infty; \zeta_0 + \mu\tau]$ in reflection. It becomes from (1.71)

$$\begin{aligned} &\int_{-\infty}^{+\infty} \int_{\zeta_0 - \mu_t\tau}^{\infty} p(\zeta, \gamma|\zeta_0, \gamma_0; \tau) \mathrm{d}\gamma \mathrm{d}\zeta = \int_{\zeta_0 - \mu_t\tau}^{\infty} \frac{p_3(\zeta_0, \zeta, \gamma_0; \tau)}{p_{\zeta,\gamma}(\zeta_0, \gamma_0)} \mathrm{d}\zeta \\ &= \int_{-\infty}^{+\infty} \frac{p_3(\zeta_0, \zeta, \gamma_0; \tau)}{p_{\zeta,\gamma}(\zeta_0, \gamma_0)} \mathrm{d}\zeta - \int_{-\infty}^{\zeta_0 - \mu_t\tau} \frac{p_3(\zeta_0, \zeta, \gamma_0; \tau)}{p_{\zeta,\gamma}(\zeta_0, \gamma_0)} \mathrm{d}\zeta \\ &= 1 - \left.(1.77)\right|_{\mu = -\mu_t}. \end{aligned} \tag{1.151}$$

In conclusion, the monostatic statistical shadowing function in transmission (1.148) is obtained from that of a reflection (1.121) changing $(\gamma_0, \zeta = \zeta_0 + \mu\tau)$ by $(-\gamma_0, \zeta = \zeta_0 - \mu_t\tau)$ and using the above equation.

1.7.3.2 Bistatic statistical shadowing function

The substitution of (1.150) and (1.147) (with (1.121)) into (1.133) leads for the case (a) to

$$\begin{aligned} T^1_{\mathrm{B}}(\mu_i, \mu_t, \zeta_0, -\gamma_0, L_0) &= \Upsilon_t(\mu_i, \mu_t, \gamma_0) \\ &\times \exp\left\{-\int_0^{L_0} [g(|\mu_i||\zeta_0, \gamma_0; \tau) + g(\mu_t|\zeta_0, \gamma_0; \tau)] \mathrm{d}\tau\right\}, \end{aligned} \tag{1.152}$$

where Υ_t is expressed from (1.141), $\zeta = \zeta_0 + |\mu_i|\tau$ and $\zeta = \zeta_0 - \mu_t\tau$.

In Figure 1.20, (b) and (c) require to derive the monostatic statistical shadowing function $T^0_{\rm M}(-|\mu_t|, \zeta_0, \gamma_0, L_0)$ for $\mu_t = -|\mu_t| \leq 0$. Using the same way as the bistatic case in reflection, we can show that

$$T^0_{\rm M}(-|\mu_t|, \zeta_0, \gamma_0, L_0) = T^0_{\rm M}(|\mu_t|, \zeta_0, -\gamma_0, L_0). \tag{1.153}$$

For Figure 1.20(b) and (c), the substitution of the above equation and (1.147) (with (1.121)) into (1.134) and (1.135) yields $T^1_{\rm B}$ is expressed as

$$\begin{aligned} T^1_{\rm B}(\mu_i, \mu_t, \zeta_0, -\gamma_0, L_0) &= \Upsilon_t(\mu_i, \mu_t, \gamma_0) \\ &\quad \times \exp\left\{-\int_0^{L_0} [g(|\mu_i||\zeta_0, \gamma_0; \tau) + g(|\mu_t||\zeta_0, \gamma_0; \tau)]\, {\rm d}\tau\right\}, \end{aligned} \tag{1.154}$$

where $\mu = \min(|\mu_i|, |\mu_t|)$, Υ_t is expressed from (1.141), $\zeta = \zeta_0 + |\mu_i|\tau$ and $\zeta = \zeta_0 - |\mu_t|\tau$.

1.7.3.3 Uncorrelated case

For the uncorrelated case and from the Wagner model, from (1.23), $g_{\rm W}(\mu_t|\zeta_0, \gamma_0; \tau) = \mu_t\Lambda(\mu_t)p_\zeta(\zeta)$. In (1.150) , the integration of $-g_{\rm W}$ over τ leads to

$$\Lambda(\mu_t)[P_\zeta(\zeta_0 - \mu_t L_0) - P_\zeta(\zeta_0)] = -\Lambda(\mu_t)[P_\zeta(\zeta_0) - P_\zeta(\zeta_0 - \mu_t L_0)]. \tag{1.155}$$

Equation (1.137) is then found. Introducing the variable transformations of (1.68), the function G defined from (1.67) is for $y_0 > y_t$

$$\begin{aligned} G_{\rm W}(v_t, h_0, y_t, y_0) &= -\frac{\eta\Lambda(v_t)v_t}{\sqrt{\pi}} \int_{y_t}^{y_0} e^{-(h_0 - y v_t \eta)^2}\, {\rm d}y \\ &= \frac{\Lambda(v_t)}{\sqrt{\pi}} \int_{h_0 - y_t v_t \eta}^{h_0 - y_0 v_t \eta} e^{-h^2}\, {\rm d}h \\ &= \frac{\Lambda(v_t)}{2} [{\rm erf}(h_0 - y_0 v_t \eta) - {\rm erf}(h_0 - y_t v_t \eta)], \end{aligned} \tag{1.156}$$

where $v_t = |\mu_t|/(\sigma_\gamma\sqrt{2})$.

For the Smith model, from (1.26), $g_{\rm S}$ becomes in transmission

$$g_{\rm S}(\mu_t|\zeta_0, \gamma_0; \tau) = \frac{\mu_t\Lambda(\mu_t)p_\zeta(\zeta_0 - \mu_t\tau)}{\displaystyle\int_{\zeta_0 - \mu_t\tau}^{\infty} p_\zeta(\zeta) d\zeta} = \frac{\mu_t\Lambda(\mu_t)p_\zeta(\zeta_0 - \mu_t\tau)}{P_\zeta(\infty) - P_\zeta(\zeta_0 - \mu_t\tau)}. \tag{1.157}$$

The integration of $-g_{\rm S}$ over τ in (1.150) leads to

$$\begin{aligned} &-\Lambda(\mu_t)\left\{\ln\left[P_\zeta(\infty) - P_\zeta(\zeta_0 - \mu_t L_0)\right] - \ln\left[P_\zeta(\infty) - P_\zeta(\zeta_0)\right]\right\} \\ &= \ln\left[\frac{P_\zeta(\infty) - P_\zeta(\zeta_0)}{P_\zeta(\infty) - P_\zeta(\zeta_0 - \mu_t L_0)}\right]^{\Lambda(\mu_t)}. \end{aligned} \tag{1.158}$$

Equation (1.137) is then found.

Introducing the variable transformations of (1.68), the function G defined from (1.67) is for $y_0 > y_t$

$$G_S(v_t, h_0, y_t, y_0) = -\frac{2\eta\Lambda(v_t)v_t}{\sqrt{\pi}}\int_{y_t}^{y_0}\frac{e^{-(h_0 - yv_t\eta)^2}\mathrm{d}y}{1-\mathrm{erf}(h_0 - yv_t\eta)}$$

$$= \frac{2\Lambda(v_t)}{\sqrt{\pi}}\int_{h_0 - y_t v_t\eta}^{h_0 - y_0 v_t\eta}\frac{e^{-h^2}\mathrm{d}h}{1-\mathrm{erf}(h)} = -\Lambda(v_t)\left[\ln\{1-\mathrm{erf}(h)\}\right]_{h_0 - y_t v_t\eta}^{h_0 - y_0 v_t\eta}$$

$$= \ln\left\{\left[\frac{1-\mathrm{erf}(h_0 - y_t v_t\eta)}{1-\mathrm{erf}(h_0 - y_0 v_t\eta)}\right]^{\Lambda(v_t)}\right\}. \quad (1.159)$$

1.7.4 Ray-tracing algorithm

In this section, the ray-tracing algorithm or MC process is generalized to the bistatic case.

In Figure 1.20(a), the algorithm presented in Figure 1.11 is applied to determine the shadowed points from the transmitter. Moreover, this algorithm is applied again to determine the shadowed points from the receiver located below the surface, which requires minor changes in the algorithm. In Figure 1.20(b) and (c), this modified algorithm is then applied twice with respect to the transmitter and receiver. From the shadowed points, the average bistatic shadowing function in transmission over the surface heights and slopes is expressed as

$$\bar{\bar{S}}_B^1(v_i, v_t) = \frac{1}{N}\sum_{i=1}^{i=N} b_i^{(T)} b_i^{(R)}, \quad (1.160)$$

where the Boolean $b_i^{(T)}$ and $b_i^{(R))}$ are defined as

$$b_i^{(T)} = \begin{cases} 0 \text{ if the point is shadowed from the transmitter} \\ 1 \text{ otherwise} \end{cases}, \quad (1.161)$$

and

$$b_i^{(R)} = \begin{cases} 0 \text{ if the point is shadowed from the receiver} \\ 1 \text{ otherwise} \end{cases}. \quad (1.162)$$

As shown in [3], any expected value (or ensemble average) can be performed and the marginal slope and height PDFs of the illuminated points can also be computed.

Figure 1.22 presents an illustration of the algorithm. The top shows the surface points illuminated from rays propagating from the left (transmitter), whereas the middle shows the surface points illuminated from rays propagating from the left but below the surface (receiver; Figure 1.20(a)). The bottom shows the surface points illuminated both from the transmitter and the receiver. As we can see, the illuminated points are more numerous from the receiver than the transmitter because $|v_t| < v_i$.

In the following, for the MC process, the surface length will be $L = 5,000L_c$.

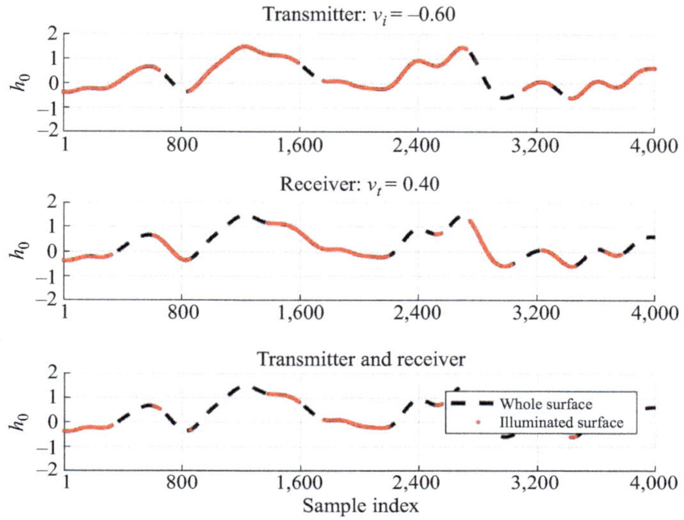

Figure 1.22 Surface normalized heights h_0 of the whole and illuminated points versus the sample index. Top: From the transmitter only. Middle: From the receiver only. Bottom: From both the transmitter and receiver. The surface height autocorrelation function is Gaussian with $L_c = 200$, $\sigma_\zeta = 1$ and the parameters $v_i = \mu_i/(\sigma_\gamma\sqrt{2}) = -0.6$ and $v_t = \mu_t/(\sigma_\gamma\sqrt{2}) = 0.4$ (Figure 1.20(a))

1.7.5 Numerical results for a Gaussian process

In this section, the Wagner and Smith models with and without correlation are compared with an MC process to access to the validity of these approaches. The surface length L_0 (or y_0) is assumed to be infinite.

As claimed by Pinel *et al.* [17], in the literature, we found very few papers dealing with the shadowing function in transmission. Tsang *et al.* [18] dealt with this case, but only very briefly. The latter gave without justification the following expression for the shadowing function average over the surface slopes

$$\int_{-\infty}^{+\infty} T_{\mathrm{B}}^{1}(\mu_i, \mu_t, \zeta_0, \gamma_0, \infty) p_\zeta(\zeta_0) d\zeta_0 = \Upsilon_t(\mu_i, \mu_t, \gamma_0) \times \frac{1}{1 + \Lambda(\mu_i) + \Lambda(|\mu_t|)}, \tag{1.163}$$

which differs from (1.144). In addition, the above is applied for any $\theta_i \in [-\pi/2; 0]$ and $\theta_t \in [-\pi/2; \pi/2]$, whereas we consider three cases. In fact, these authors used the same shadowing function as that derived in bistatic (a) case ((1.118), in which $\Lambda_t = \Lambda(\mu_i) + \Lambda(|\mu_t|)$) in reflection and from the Smith formulation.

1.7.5.1 Average over the surface heights

The average over the surface heights is expressed from (1.144). For the correlated case, from (1.152) and (1.154), it is expressed as

$$\bar{T}^1_{\mathrm{B}}(v_i, v_t, s_0, y_0) = \int_{-\infty}^{+\infty} T^1_{\mathrm{B}}(v_i, v_t, h_0, s_0, y_0) p_\zeta(h_0)\mathrm{d}h_0 = \frac{\Upsilon_t(v_i, v_t, \gamma_0)}{\sqrt{\pi}}$$
$$\times \int_{-\infty}^{+\infty} \exp\Big\{ - L_c \int_0^{y_t} [g(|v_i|, h_0, s_0, y) + g(|v_t|, h_0, s_0, y)]\,\mathrm{d}y$$
$$+ G(|v_i|, h_0, y_t, y_0) + G(|v_t|, h_0, y_t, y_0)\Big\} \exp(-h_0^2)\mathrm{d}h_0, \qquad (1.164)$$

where $L_c g$ and G are expressed in Table 1.2 with minor changes. In transmission, G is expressed from (1.156) and (1.159). From (1.151), for the Smith model in Table 1.2, $1 + \mathrm{erf}(\bullet)$ at the denominator becomes $1 + \mathrm{erf}(-\bullet)$ since $\mathrm{erf}(-x) = -\mathrm{erf}(x)$. For the numerical implementation of the function g, see Section 1.4.4.1.

Figure 1.23 plots the average shadowing function over the heights ζ_0 versus s_0 computed from the Wagner (a) and the Smith (b) models, where $v_t = 0.4$ and $v_i = -0.6$.

As we can see, both Wagner's and Smith's models with correlation bend down significantly as s_0 approaches v_t (for increasing s_0), or equally as γ_0 approaches μ_t, whereas the uncorrelated ones remains constant for $s_0 \in [-|v_i|; v_t]$. Next, the strengths with and without correlation jump to 0 at $s_0 > v_t$ and $s_0 < -|v_i|$ due to the Υ_t function. Then, the correlation implies that all points with slope $s_0 \in [-|v_i|; v_t]$ are not equally shadowed. In comparison to the MC results, Figure 1.23 also shows that the models overpredict the occurrence of the surface slopes over the range $s_0 \in [-|v_i|; v_t]$. The strengths depends slightly on the surface height autocorrelation function (Gaussian or Lonrentzian; see Table 1.1 for their definition).

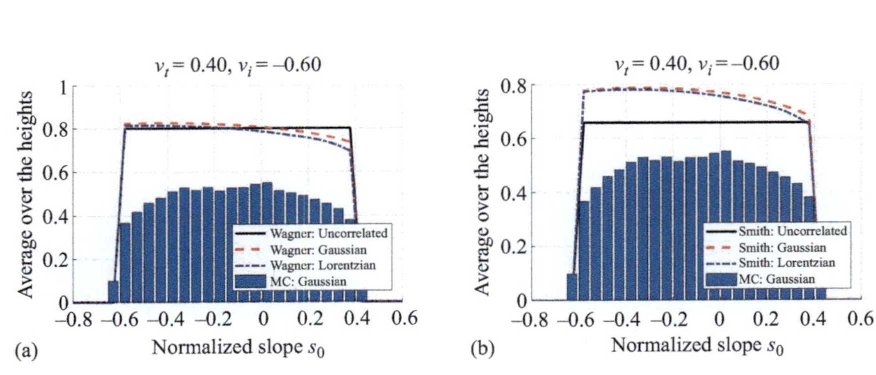

Figure 1.23 Average shadowing function over the heights ζ_0 versus s_0 computed from the Wagner (a) and the Smith (b) models, where $v_t = 0.4$ and $v_i = -0.6$

1.7.5.2 Average over the surface heights and slopes

Figure 1.24 plots the average shadowing function over the normalized slope s_0 and height h_0 versus v_t computed from the Wagner (a) and the Smith (b) models. At the bottom, the difference against the MC results is shown. In addition, the models of Tsang *et al.* [18] are plotted. $v_i = -0.6$.

As we can see, for $v_t > 0$, all the formulations overpredict the strengths. When the correlation is accounted for, the results match better with those obtained from

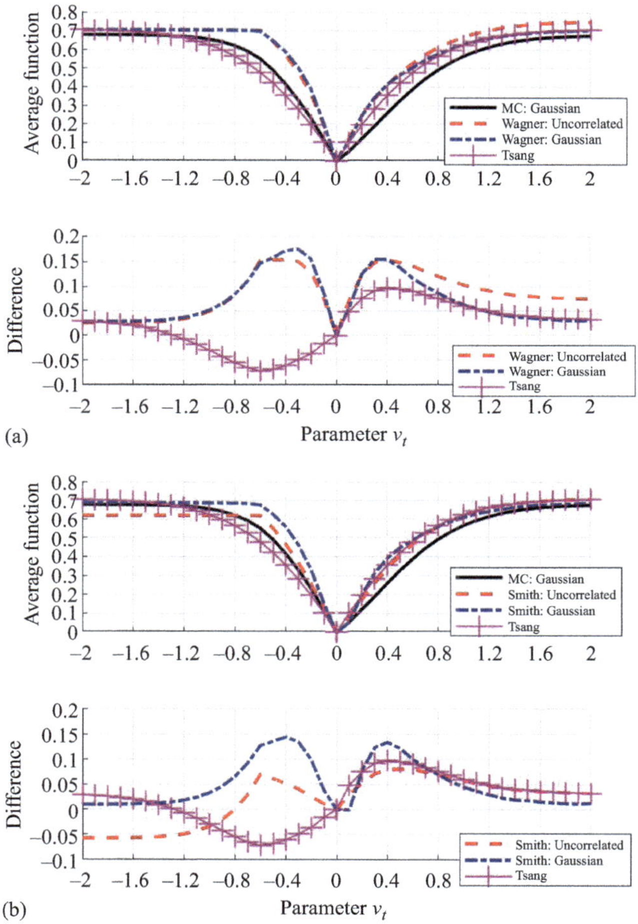

Figure 1.24 Top: Average shadowing function over the normalized slope s_0 and height h_0 versus v computed from the Wagner (a) and the Smith (b) models. Bottom: Difference against the Monte Carlo results. In addition, the models of Tsang et al. [18] is plotted. $v_i = -0.6$

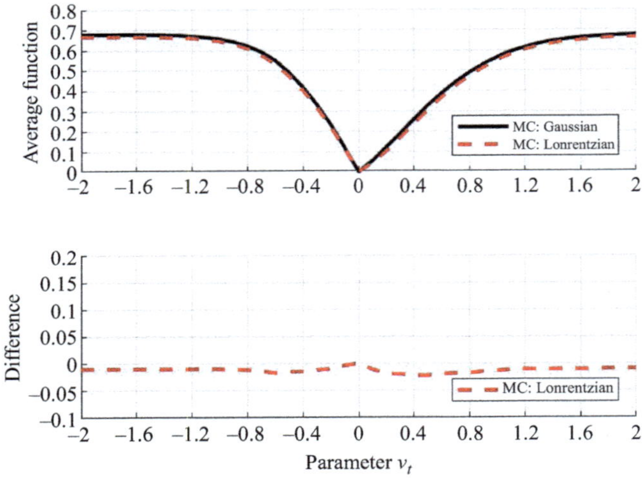

Figure 1.25 Top: Average shadowing function over the normalized slope s_0 and height h_0 versus v_t computed from an MC process and the Smith model with correlation. Bottom: Difference against the MC results by considering a Gaussian autocorrelation function. $v_i = -0.6$

MC, except for v_t close to 0 and by considering the Smith model. The model of Tsang gives satisfactory results.

For $v_t < 0$, Figure 1.24 shows that the correlation does not improve the results, unlike the case in reflection, except for $|v_t|$ close to 2 and from the Smith formulation. Indeed, the receiver viewed the whole surface, and the shadowing function is reduced to the monostatic case defined versus the transmitter. The model of Tsang gives satisfactory results and underestimates the strengths.

Figure 1.25 plots the average shadowing function over the normalized slope s_0 and height h_0 versus v_t computed from an MC process. At the bottom, the difference against the MC results by considering a Gaussian autocorrelation function is plotted. $v_i = -0.6$. As we can see, the strengths depends weakly on the choice of the correlation.

1.7.5.3 Conclusion

This section shows that the correlation does not improve the results in comparison to an MC process, unlike the bistatic shadowing function in reflection. Further investigations are then needed to improve its derivation.

1.8 Non-Gaussian distribution

In the previous sections, for the simulations, an uncorrelated Gaussian process is assumed. For practical applications, such as the modelling of the sea surface, the assumption of a Gaussian process is not fully satisfied. A simple means to model better the sea surface is to consider that the slopes follow a Gram–Charlier (GC) slope PDF [19–24] instead of a Gaussian PDF, whereas the height PDF follows a Gaussian process. When the correlation is neglected, the last assumption is not required since the shadowing function depends statistically only on the slope PDF.

1.8.1 Gram–Charlier distribution

Supposing that the distribution of the surface slopes γ_0 is approximately Gaussian and may be represented by the GC series

$$p_\gamma(\gamma_0) = \frac{1}{\sqrt{2\pi\kappa_2}} \exp\left(-\frac{f^2}{2}\right)\left[1 + \frac{\lambda_3\mathcal{H}_3}{6} + \left(\frac{\lambda_4\mathcal{H}_4}{24} + \frac{\lambda_3^2\mathcal{H}_6}{76}\right) + \cdots\right], \quad (1.165)$$

in which κ_n is the nth cumulant of p_γ, i.e. if

$$\mu_n = \int_{-\infty}^{+\infty} \gamma_0^n p_\gamma(\gamma_0)\mathrm{d}\gamma_0, \quad (1.166)$$

then

$$\begin{cases} \kappa_1 = \mu_1 \\ \kappa_2 = \mu_2 - \mu_1^2 \\ \kappa_3 = \mu_3 - 3\mu_1\mu_2 + 2\mu_1^3 \\ \ldots \end{cases} . \quad (1.167)$$

Also

$$f = \frac{\gamma_0 - \kappa_1}{\sqrt{\kappa_2}}, \quad (1.168)$$

$$\lambda_n = \frac{\kappa_n}{\kappa_2^{n/2}}, \quad (1.169)$$

and $\mathcal{H}_n$ is the nth Hermite polynomial defined as

$$\begin{cases} \mathcal{H}_3 = f^3 - 3f \\ \mathcal{H}_4 = f^4 - 6f^2 + 3 \\ \ldots \end{cases} . \quad (1.170)$$

Assuming that the mean value μ_1 is zero, $\kappa_n = \mu_n$ and $f = \gamma_0/\sqrt{\kappa_2} = \gamma_0/\sigma_\gamma$ where σ_γ is the slope standard deviation. In addition, neglecting the term λ_3^2 in (1.165), the slope PDF is

$$p_\gamma(\gamma_0) \approx \frac{1}{\sqrt{2\pi}\sigma_\gamma} \exp\left(-\frac{f^2}{2}\right)\left[1 + \frac{\lambda_3(f^3 - 3f)}{6} + \frac{\lambda_4(f^4 - 6f^2 + 3)}{24}\right], \quad (1.171)$$

where $f = \gamma_0/\sigma_\gamma$, $\lambda_3 = \kappa_3/\sigma_\gamma^3$ and $\lambda_4 = \kappa_4/\sigma_\gamma^4$. The numbers λ_3 and λ_4 are commonly named skewness and kurtosis coefficients, respectively.

We can see that λ_4 is related to the fourth-order statistics, for which only *even* powers of γ_0 are involved in (1.171). This means that for $\lambda_3 = 0$, the associated slope

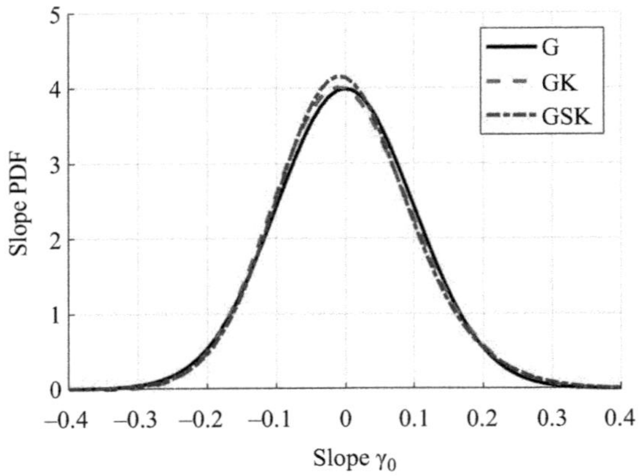

Figure 1.26 Slope PDF versus the slope γ_0. $\sigma_\gamma = 0.1$, $\lambda_3 = 0.2$ and $\lambda_4 = 0.3$

PDF remains symmetric ($p_\gamma(-\gamma_0) = p_\gamma(\gamma_0)$) since a Gaussian process is symmetric. Conversely, λ_3 is related to the third-order statistics, for which only *odd* powers of γ_0/σ_γ are involved in (1.171). This means that the associated slope PDF is no longer symmetric ($p_\gamma(-\gamma_0) \neq p_\gamma(\gamma_0)$).

To illustrate the above remarks, in Figure 1.26, the slope PDF, p_γ, is plotted versus the slope γ_0 for $\{\lambda_3 = 0, \lambda_4 = 0\}$ (Gaussian case with label 'G'), $\{\lambda_3 \neq 0, \lambda_4 = 0\}$ (label 'GS') and $\{\lambda_3 \neq 0, \lambda_4 \neq 0\}$ (label 'GSK'). $\sigma_\gamma = 0.1$, $\lambda_3 = 0.2$, $\lambda_4 = 0.3$ and, 'G', 'S' and 'K' refer to the Gaussian, skewness and kurtosis, respectively.

As expected, for 'G' case, the slope PDF is symmetric whereas for the two other cases, it is not symmetric due to the skewness effect. In addition, for 'GSK' case, the occurrence of small slopes close to zero is weakly greater than for the 'G' case, whereas the opposite effect occurs for moderate slopes.

1.8.2 Monostatic shadowing function

This section presents numerical results of the average monostatic shadowing function, $\bar{\bar{S}}^0_{\mathrm{M}}$, when the GC PDF is considered. The Smith model is used since it is the more accurate and the correlation is assumed to be negligible.

For any uncorrelated process, we showed in the previous section that

$$\bar{\bar{S}}^0_{\mathrm{M}}(v) = \frac{\bar{\Upsilon}(v)}{1 + L(v)}, \tag{1.172}$$

where

$$L(v) = \frac{1}{|\mu|} \int_{\mu}^{+\infty} (\gamma - |\mu|) p_\gamma\left(\mathrm{sign}(v)\gamma\right) \mathrm{d}\gamma, \tag{1.173}$$

$$\bar{\Upsilon}(v) = \int_{-\infty}^{|\mu|} p_\gamma\left(\mathrm{sign}(v)\gamma\right) \mathrm{d}\gamma, \tag{1.174}$$

and the symbol sign stands for the sign function. For $v \geq 0$, $\text{sign}(v) = 1$, otherwise -1. In addition, $v = \mu/(\sigma_\gamma\sqrt{2})$ and $\text{sign}(v) = \text{sign}(\mu)$. For the Gaussian case, p_γ is even, which implies that $L = \Lambda = \Lambda_-$ (expressed from (1.47) and (1.115)).

After simple but tedious calculations, the substitution of (1.171) into (1.173) yields

$$L(v) = L_G(|v|) + \lambda_3 \text{sign}(v) L_S(|v|) + \lambda_4 L_K(|v|), \tag{1.175}$$

where

$$\begin{cases} L_G(v) = \Lambda(v) = \dfrac{\exp(-v^2) - v\sqrt{\pi}\,\text{erfc}(v)}{2v\sqrt{\pi}} \\ L_S(v) = \dfrac{\exp(-v^2)}{6\sqrt{2\pi}} \\ L_K(v) = \dfrac{(2v^2 - 1)\exp(-v^2)}{48v\sqrt{\pi}} \end{cases}. \tag{1.176}$$

Moreover, the substitution of (1.171) into (1.174) yields

$$\bar{\Upsilon}(v) = \bar{\Upsilon}_G(|v|) + \lambda_3 \text{sign}(v) \bar{\Upsilon}_S(|v|) + \lambda_4 \bar{\Upsilon}_K(|v|), \tag{1.177}$$

where

$$\begin{cases} \bar{\Upsilon}_G(v) = \dfrac{1 + \text{erf}(v)}{2} \\ \bar{\Upsilon}_S(v) = \dfrac{(1 - 2v^2)\exp(-v^2)}{6\sqrt{2\pi}} \\ \bar{\Upsilon}_K(v) = \dfrac{v(3 - 2v^2)\exp(-v^2)}{24\sqrt{\pi}} \end{cases}. \tag{1.178}$$

By convention, for a ray illuminating the surface from the right, $\theta \geq 0$, $\mu \geq 0$, $v \geq 0$ and $\text{sign}(v) = +1$. For a ray illuminating the surface from the left, $\theta \leq 0$, $\mu \leq 0$, $v \leq 0$ and $\text{sign}(v) = -1$.

For $\lambda_3 = 0$, (1.175) and (1.177) show clearly that the average shadowing function is even with respect to θ because it depends on $|\theta|$ (or $|v|$). Then, the skewness is responsible for the asymmetry of the shadowing function.

Figure 1.27 plots the average shadowing function versus v. At the bottom, the difference against the Gaussian case is shown. The surface is illuminated from the right ($\theta \geq 0$). As we can see, the skewness produces a small deviation, and when the Kurtosis effect is added, the difference is larger. According to the value of v, this deviations can be positive or negative.

Figure 1.28 plots the average shadowing function versus $|v|$. At the bottom, the difference against the case $\theta \geq 0$ (or $v \geq 0$) is plotted. In the legend, the label '+' means that $\text{sign}(v) = +1$ (surface illuminated from the right, that is $\theta \geq 0$), whereas the label '$-$' means that $\text{sign}(v) = -1$ (surface illuminated from the left, that is $\theta \leq 0$). As expected, Figure 1.28 shows clearly that the skewness produces an asymmetry of

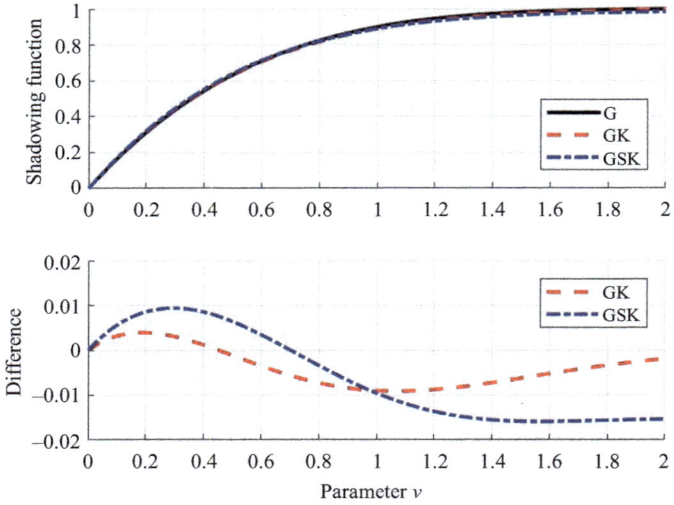

Figure 1.27 *Top: Average shadowing function versus v. Bottom: Difference against the Gaussian case. The surface is illuminated from the right ($\theta \geq 0$)*

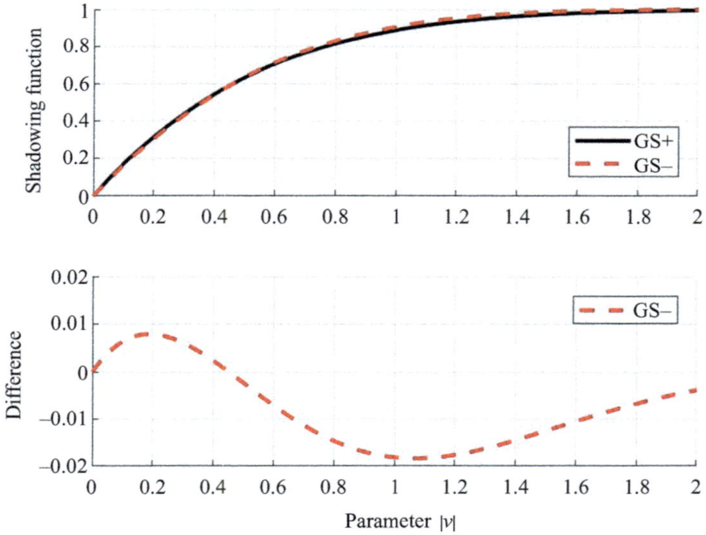

Figure 1.28 *Top: Average shadowing function versus |v|. Bottom: Difference against the case $\theta \geq 0$ (or $v \geq 0$). In the legend, the label '+' means that sign(v) = +1 (surface illuminated from the right, that is $\theta \geq 0$), whereas the label '−' means that sign(v) = −1 (surface illuminated from the left, that is $\theta \leq 0$)*

the average shadowing function. According to the value of v, the difference can be positive or negative.

1.8.3 Bistatic shadowing functions

If the correlation is neglected, we showed in the previous sections that the bistatic shadowing functions in reflection and transmission are expressed from the elementary functions $\bar{\Upsilon}$ and L. Thus, the extension to the bistatic case is easy.

Chapter 2
Shadowing function from a two-dimensional surface

Christophe Bourlier[1] and Hongkun Li[1]

This chapter extends the derivation of the shadowing function to a two-dimensional (2D) anisotropic random rough surface [7,25]. In polar coordinates ($x = R\cos\phi, y = R\sin\phi$) and the term 'anisotropic' means that the statistical properties of the surface depends on the azimuthal direction ϕ. Typically, for a sea surface, ϕ stands for the wind direction.

We will show that the 2D case is obtained from the one-dimensional (1D) case by making a rotation of an angle ϕ and next by calculating the marginal probability along this direction.

First, the monostatic shadowing function is derived, and next it is generalized to the bistatic case. Unlike the 1D case [8], the latter configuration requires a special attention and needs to account for the correlation with respect to the direction ϕ to avoid a discontinuity when the transmitter and the receiver are not in the same plane (the azimuthal directions differ).

2.1 Monostatic shadowing function

In this section, the 2D shadowing function is derived for a monostatic configuration. In order to apply the results obtained in Chapter 1, this function is expressed in polar coordinates. As shown in Figure 2.1, it is equivalent to make a cross section of the surface along the direction ϕ.

From the Smith and Wagner formulations, the shadowing function from a 1D surface depends on the joint PDF $p(\zeta, \gamma|\zeta_0, \gamma_0; \tau)$ of the slopes and heights of two separated surface points, named (ζ_0, γ_0) and (ζ, γ), of abscissa 0 and τ.

For a 2D surface, this joint PDF becomes $p(\zeta, \gamma_x, \gamma_y|\zeta_0, \gamma_{0x}, \gamma_{0y}; x, y)$, in which $(\zeta_0, \gamma_{0x}, \gamma_{0y})$ and $(\zeta, \gamma_x, \gamma_y)$ are the heights and slopes of the two points, respectively. In addition, (γ_x, γ_y) are the surface slopes with respect to the directions $((Ox), (Oy))$ defined in Cartesian coordinates (x, y).

[1]IETR Laboratory, CNRS, Nantes, France

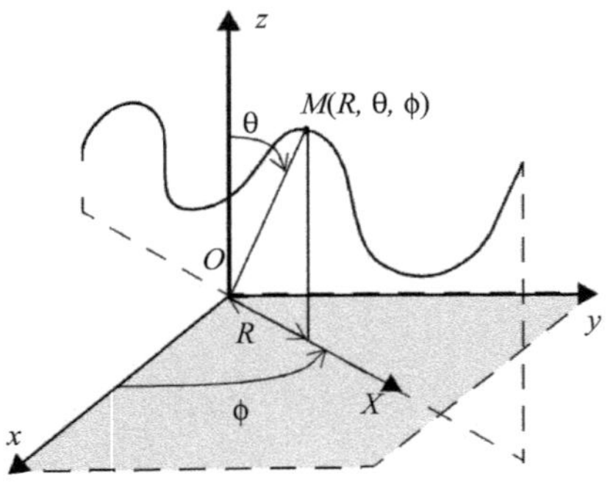

Figure 2.1 Illustration of the cross section of the surface with respect to the azimuthal direction ϕ

In order to express this joint probability along the azimuthal angle ϕ (or direction (OX)), the slopes (γ_X, γ_Y), defined with respect to ϕ and $\phi + \pi/2$ (or direction (OY)), are expressed from those (γ_x, γ_y) by making a rotation of an angle ϕ. This leads to

$$\begin{cases} \gamma_X = \ \ \gamma_x \cos\phi + \gamma_y \sin\phi \\ \gamma_Y = -\gamma_x \sin\phi + \gamma_y \cos\phi \end{cases} \Leftrightarrow \begin{cases} \gamma_x = \gamma_X \cos\phi - \gamma_Y \sin\phi \\ \gamma_y = \gamma_X \sin\phi + \gamma_Y \cos\phi \end{cases}. \tag{2.1}$$

The derivation of the PDF requires the introduction of the correlation functions between the surface slopes and heights.

2.1.1 Correlation functions

We introduce the following correlation functions:

$$\begin{cases} C_{\zeta_0,\zeta}(\vec{r}) & = \langle \zeta(\vec{r}_0)\zeta(\vec{r}_1)\rangle \\ C_{\zeta_0,\gamma_x}(\vec{r}) & = \langle \zeta(\vec{r}_0)\gamma_x(\vec{r}_1)\rangle = \dfrac{\partial}{\partial x_1}\langle \zeta(\vec{r}_0)\zeta(\vec{r}_1)\rangle \\ & = \dfrac{\partial C_{\zeta_0,\zeta}(\vec{r}_1-\vec{r}_0)}{\partial x_1} = \dfrac{\partial C_{\zeta_0,\zeta}(\vec{r})}{\partial x} \\ C_{\zeta_0,\gamma_y}(\vec{r}) & = \langle \zeta(\vec{r}_0)\gamma_y(\vec{r}_1)\rangle = \dfrac{\partial}{\partial y_1}\langle \zeta(\vec{r}_0)\zeta(\vec{r}_1)\rangle \\ & = \dfrac{\partial C_{\zeta_0,\zeta}(\vec{r}_1-\vec{r}_0)}{\partial y_1} = \dfrac{\partial C_{\zeta_0,\zeta}(\vec{r})}{\partial y} \\ C_{\gamma_{0y},\gamma_x}(\vec{r}) & = \langle \gamma_y(\vec{r}_0)\gamma_x(\vec{r}_1)\rangle = \dfrac{\partial^2}{\partial y_0 \partial x_1}\langle \zeta(\vec{r}_0)\zeta(\vec{r}_1)\rangle \\ & = \dfrac{\partial^2 C_{\zeta_0,\zeta}(\vec{r}_1-\vec{r}_0)}{\partial y_0 \partial x_1} = -\dfrac{\partial C_{\zeta_0,\zeta}(\vec{r})}{\partial y \partial x} \end{cases}, \tag{2.2}$$

where $\vec{r} = \vec{r}_1 - \vec{r}_0 = (x_1 - x_0, y_1 - y_0)$, $\vec{r}_0 = (x_0, y_0)$ and $\vec{r}_1 = (x_1, y_1)$. Since the process is assumed to be stationary (or homogeneous in space), the surface height autocorrelation function $C_{\zeta_0,\zeta}$ depends only on the difference $\vec{r} = \vec{r}_1 - \vec{r}_0$. In addition, the function $C_{a,b}$ stands for the surface cross-correlation function between two random variables a and b of abscissa $\vec{r}_0$ and $\vec{r}_1$, respectively.

The surface height autocorrelation function $C_{\zeta_0,\zeta}$ is expressed from the surface height spectrum (or power density function) $\hat{C}_{\zeta_0,\zeta}$ as

$$C_{\zeta_0,\zeta}(\vec{r}) = \frac{1}{(2\pi)^2} \int_{-\infty}^{+\infty} \int_{-\infty}^{+\infty} \hat{C}_{\zeta_0,\zeta}(k_x, k_y) e^{jk_x x + jk_y y} \mathrm{d}k_x \mathrm{d}k_y. \tag{2.3}$$

The integration domain $\Gamma = \mathbb{R}^2$ is decomposed into four non-overlapping subdomains as $\Gamma = \Gamma_1 \cup \Gamma_2 \cup \Gamma_3 \cup \Gamma_4$, in which $\Gamma_1 = \mathbb{R}^+ \times \mathbb{R}^+$, $\Gamma_2 = \mathbb{R}^+ \times \mathbb{R}^-$, $\Gamma_3 = \mathbb{R}^- \times \mathbb{R}^+$ and $\Gamma_4 = \mathbb{R}^- \times \mathbb{R}^-$. For the subdomain Γ_4, the variable transformations $(u = -k_x, v = -k_y)$ are applied, whereas for the subdomains Γ_2 and Γ_3, the variable transformations $(u = k_x, v = -k_y)$ and $(u = -k_x, v = k_y)$ are used, respectively. This leads to

$$\begin{aligned} C_{\zeta_0,\zeta}(\vec{r}) = \; & \frac{1}{(2\pi)^2} \int_0^{\infty} \int_0^{\infty} \Big\{ \hat{C}_{\zeta_0,\zeta}(u, v) \left[\cos(ux + vy) + j \sin(ux + vy)\right] \\ & + \hat{C}_{\zeta_0,\zeta}(-u, -v) \left[\cos(ux + vy) - j \sin(ux + vy)\right] \\ & + \hat{C}_{\zeta_0,\zeta}(-u, v) \left[\cos(ux - vy) - j \sin(ux - vy)\right] \\ & + \hat{C}_{\zeta_0,\zeta}(u, -v) \left[\cos(ux - vy) + j \sin(ux - vy)\right] \Big\} \mathrm{d}u \mathrm{d}v. \end{aligned} \tag{2.4}$$

Since $C_{\zeta_0,\zeta}$ is a real function, this implies that $\hat{C}_{\zeta_0,\zeta}(-u, -v) = \hat{C}_{\zeta_0,\zeta}(u, v)$ and $\hat{C}_{\zeta_0,\zeta}(-u, v) = \hat{C}_{\zeta_0,\zeta}(u, -v)$. In other words, the height spectrum $\hat{C}_{\zeta_0,\zeta}$ must be centrosymmetric, that is $\hat{C}_{\zeta_0,\zeta}(-\vec{k}) = \hat{C}_{\zeta_0,\zeta}(\vec{k})$, and

$$\begin{aligned} C_{\zeta_0,\zeta}(\vec{r}) &= \frac{2}{(2\pi)^2} \int_0^{\infty} \int_0^{\infty} \Big\{ \hat{C}_{\zeta_0,\zeta}(u, v) \cos(ux + vy) \\ &\quad + \hat{C}_{\zeta_0,\zeta}(-u, v) \cos(ux - vy) \Big\} \mathrm{d}u \mathrm{d}v \\ &= \frac{2}{(2\pi)^2} \int_0^{\infty} \mathrm{d}u \int_{-\infty}^{+\infty} \hat{C}_{\zeta_0,\zeta}(u, v) \cos(ux + vy) \mathrm{d}v \\ &= \frac{1}{(2\pi)^2} \int_{-\infty}^{+\infty} \int_{-\infty}^{+\infty} \hat{C}_{\zeta_0,\zeta}(u, v) \cos(ux + vy) \mathrm{d}u \mathrm{d}v. \end{aligned} \tag{2.5}$$

The above equation shows that $C_{\zeta_0,\zeta}(-x, -y) = C_{\zeta_0,\zeta}(\vec{r})$. In other words, the height autocorrelation function $\hat{C}_{\zeta_0,\zeta}$ is even both (or centrosymmetric) with respect to x and y.

From (2.2) and (2.5), the cross-correlation functions between ζ_0 and γ_x, ζ_0 and γ_y, γ_{0x} and γ_y are given by

$$\begin{cases} C_{\zeta_0,\gamma_x}(\vec{r}) = -\dfrac{1}{(2\pi)^2}\displaystyle\int_{-\infty}^{+\infty}\int_{-\infty}^{+\infty} u\hat{C}_{\zeta_0,\zeta}(u,v)\sin(ux+vy)\mathrm{d}u\mathrm{d}v \\ C_{\zeta_0,\gamma_y}(\vec{r}) = -\dfrac{1}{(2\pi)^2}\displaystyle\int_{-\infty}^{+\infty}\int_{-\infty}^{+\infty} v\hat{C}_{\zeta_0,\zeta}(u,v)\sin(ux+vy)\mathrm{d}u\mathrm{d}v \\ C_{\gamma_{0x},\gamma_y}(\vec{r}) = \dfrac{1}{(2\pi)^2}\displaystyle\int_{-\infty}^{+\infty}\int_{-\infty}^{+\infty} uv\hat{C}_{\zeta_0,\zeta}(u,v)\cos(ux+vy)\mathrm{d}u\mathrm{d}v \end{cases}. \tag{2.6}$$

Equation (2.6) shows that $C_{\zeta_0,\gamma_x}(-x,-y) = -C_{\zeta_0,\gamma_x}(\vec{r})$, $C_{\zeta_0,\gamma_y}(-x,-y) = -C_{\zeta_0,\gamma_y}(\vec{r})$ and $C_{\gamma_{0x},\gamma_y}(-x,-y) = C_{\gamma_{0x},\gamma_y}(\vec{r})$. In addition, $C_{\zeta_0,\gamma_x}(\vec{0}) = C_{\zeta_0,\gamma_y}(\vec{0}) = 0$. Since $C_{\zeta_0,\zeta}$ is a centrosymmetric function, it depends on the variables $X = x^2$ and $Y = y^2$. Then, the second partial derivative over x and y implies that the product $4xy$ occurs as a multiplicative term in the function C_{γ_{0x},γ_y}, which implies that $C_{\gamma_{0x},\gamma_y}(\vec{0}) = 0$. The fact that $C_{\zeta_0,\zeta}$ is even may be also obtained with $X = |x|$ and $Y = |y|$, but $C_{\zeta_0,\zeta}$ is not derivable at $\vec{r} = \vec{0}$.

2.1.2 Case of any uncorrelated process

This section derives the monostatic shadowing function by assuming that the joint PDF is uncorrelated but of any profile.

2.1.2.1 Definition of the marginal slope PDF

For any uncorrelated process, we have

$$\begin{aligned} p(\zeta,\gamma_x,\gamma_y|\zeta_0,\gamma_{0x},\gamma_{0y};,x,y) &= \frac{p_6(\zeta,\gamma_x,\gamma_y,\zeta_0,\gamma_{0x},\gamma_{0y};x,y)}{p_3(\zeta_0,\gamma_{0x},\gamma_{0y})} \\ &= p_3(\zeta,\gamma_x,\gamma_y). \end{aligned} \tag{2.7}$$

From (2.6), we have

$$\begin{cases} \langle\zeta\gamma_x\rangle = C_{\zeta_0,\gamma_x}(\vec{0}) = 0 \\ \langle\zeta\gamma_y\rangle = C_{\zeta_0,\gamma_y}(\vec{0}) = 0 \\ \langle\gamma_x\gamma_y\rangle = C_{\gamma_{0y},\gamma_x}(\vec{0}) = 0 \end{cases}. \tag{2.8}$$

In addition, from the above equation and (2.1), we show that

$$\begin{cases} \langle\zeta\rangle &= \langle\gamma_x\rangle = \langle\gamma_y\rangle = 0 \\ \langle\gamma_X\rangle &= \cos\phi\,\langle\gamma_x\rangle + \sin\phi\,\langle\gamma_y\rangle = 0 \\ \langle\gamma_X^2\rangle &= \cos^2\phi\,\langle\gamma_x^2\rangle + \sin\phi^2\,\langle\gamma_y^2\rangle + \sin(2\phi)\langle\gamma_x\gamma_y\rangle \\ &= \cos^2\phi\sigma_{\gamma_x}^2 + \sin^2\phi\sigma_{\gamma_y}^2 \end{cases}, \tag{2.9}$$

where the mean values over the surface heights and slopes are assumed to be zeros and $\langle \gamma_{x,y}^2 \rangle = \sigma_{\gamma_{x,y}}^2$ are the surface slope variances along the direction (Ox) and (Oy), respectively.

Then (2.7) becomes

$$p(\zeta, \gamma_x, \gamma_y | \zeta_0, \gamma_{0x}, \gamma_{0y}; , x, y) = p_\zeta(\zeta) p_\gamma(\gamma_x) p_\gamma(\gamma_y). \tag{2.10}$$

The marginal slope PDF along the direction ϕ is then obtained by integrating (2.10) over $\gamma_Y \in]-\infty; +\infty[$ and by using the variables transformations given by (2.1). Since the height ζ is invariant by rotation, the PDF p_ζ does not change.

2.1.2.2 Examples

Gaussian case

Assuming that the surface slopes follow a Gaussian process with a zero mean value, we have

$$p_\gamma(\gamma) = \frac{1}{\sigma_\gamma \sqrt{2\pi}} \exp\left(-\frac{\gamma^2}{2\sigma_\gamma^2}\right), \tag{2.11}$$

where σ_γ is the surface slope standard deviation.

Substituting (2.11) and (2.1) into (2.10) and integrating over $\gamma_Y \in]-\infty; +\infty[$, we can show that the surface slope probability of γ_X is

$$p_{\gamma_X}(\gamma_X) = \frac{1}{\sigma_{\gamma_X} \sqrt{2\pi}} \exp\left(-\frac{\gamma_X^2}{2\sigma_{\gamma_X}^2}\right), \tag{2.12}$$

where $\sigma_{\gamma_X}^2 = \cos^2\phi\,\sigma_{\gamma_x}^2 + \sin^2\phi\,\sigma_{\gamma_y}^2$. Then, the probability is also Gaussian as the 1D case but σ_γ is replaced by σ_{γ_X}, which is the slope standard deviation with respect to ϕ.

This result comes from the fact that the sum of two Gaussian variables is also a Gaussian variable of mean value $\langle \gamma_X \rangle$ and variance $\langle \gamma_X^2 \rangle$, which are consistent with (2.9).

Exponential case

Assuming that the surface slopes follow an exponential process with a zero mean value, we have

$$p_\gamma(\gamma) = \frac{1}{\sigma_\gamma \sqrt{2}} \exp\left(-\frac{|\gamma|\sqrt{2}}{\sigma_\gamma}\right). \tag{2.13}$$

Substituting (2.13) and (2.1) into (2.10) and integrating over $\gamma_Y \in]-\infty; +\infty[$, we can show that the surface slope probability of γ_X is

$$\begin{aligned} p_{\gamma_X}(\gamma_X) &= \frac{1}{2\sigma_{\gamma_x}\sigma_{\gamma_x}} \int_{-\infty}^{+\infty} e^{-\frac{|\gamma_X \cos\phi - \gamma_Y \sin\phi|\sqrt{2}}{\sigma_{\gamma_x}} - \frac{|\gamma_X \sin\phi + \gamma_Y \cos\phi|\sqrt{2}}{\sigma_{\gamma_y}}} \mathrm{d}\gamma_Y \\ &= \frac{1}{4\sigma_{\gamma_x}\sigma_{\gamma_x}} \sum_{p=1}^{p=3} e^{-\gamma_X(s_{1,p}\cos\phi + s_{2,p}\sin\phi)\sqrt{2}} \int_{\Gamma_p} e^{-\gamma_Y(-s_{1,p}\sin\phi + s_{2,p}\cos\phi)\sqrt{2}} \mathrm{d}\gamma_Y, \end{aligned} \tag{2.14}$$

where the integration domain $\Gamma = \mathbb{R}$ is decomposed into three non-overlapping subdomains as $\Gamma = \Gamma_1 \cup \Gamma_2 \cup \Gamma_3$, in which $\Gamma_1 =]-\infty; -\gamma_X \tan\phi]$, $\Gamma_2 = [-\gamma_X \tan\phi; \gamma_X/\tan\phi]$ and $\Gamma_3 = [\gamma_X/\tan\phi; +\infty]$. For $\gamma_X \in \Gamma_1 \cup \Gamma_2$, $|\gamma_X \cos\phi - \gamma_Y \sin\phi| = \gamma_X \cos\phi - \gamma_Y \sin\phi, s_{1,1} = s_{1,2} = 1$; otherwise, $|\gamma_X \cos\phi - \gamma_Y \sin\phi| = -(\gamma_X \cos\phi - \gamma_Y \sin\phi)$, $s_{1,3} = -1$. For $\gamma_X \in \Gamma_1$, $|\gamma_X \sin\phi + \gamma_Y \cos\phi| = -(\gamma_X \sin\phi + \gamma_Y \cos\phi)$, $s_{2,1} = -1$, otherwise $|\gamma_X \sin\phi + \gamma_Y \cos\phi| = \gamma_X \sin\phi + \gamma_Y \cos\phi, s_{2,2} = s_{2,3} = 1$. We assume that $\gamma_X \tan\phi \geq 0$.

The integration over γ_Y leads then to

$$p_{\gamma_X}(\gamma_X) = \frac{u_c \exp\left(-\gamma_X\sqrt{2}/u_c\right) - u_s \exp\left(-\gamma_X\sqrt{2}/u_c\right)}{\sqrt{2}\,(u_c^2 - u_s^2)}, \qquad \gamma_X \geq 0, \tag{2.15}$$

where $u_c = \sigma_{\gamma_x} \cos\phi$, $u_s = \sigma_{\gamma_y} \sin\phi$, $\cos\phi \geq 0$ and $\sin\phi \geq 0$, which means that $\gamma_X \geq 0$. For $\gamma_X \tan\phi \leq 0$, $\cos\phi \geq 0$ and $\sin\phi \geq 0$, (2.14) is obtained by replacing γ_X with $-\gamma_X = |\gamma_X|$. From symmetrical considerations, for $\phi \notin [0; \pi/2]$, (2.14) is again found changing $(\cos\phi, \sin\phi)$ by $(|\cos\phi|, |\sin\phi|)$.

In conclusion, for an exponential slope PDF, the marginal probability for any γ_X and ϕ is

$$p_{\gamma_X}(\gamma_X) = \frac{u_c \exp\left(-|\gamma_X|\sqrt{2}/u_c\right) - u_s \exp\left(-|\gamma_X|\sqrt{2}/u_c\right)}{\sqrt{2}\,(u_c^2 - u_s^2)}, \tag{2.16}$$

where $u_c = \sigma_{\gamma_x}|\cos\phi|$ and $u_s = \sigma_{\gamma_y}|\sin\phi|$.

We can verify that $\langle\gamma_X\rangle = 0$ and $\langle\gamma_X^2\rangle = \sigma_{\gamma_X}^2 = \cos^2\phi\sigma_{\gamma_x}^2 + \sin^2\phi\sigma_{\gamma_y}^2$, which are consistent with (2.9).

Unlike a Gaussian PDF, the marginal slope PDF p_{γ_X} has not the same shape as p_γ.

Gram–Charlier series

The previous examples assume that the surface slope PDF are even with respect to the directions (Ox) and (Oy). As shown in Chapter 1, a simple means to introduce the asymmetry is to consider that the PDF deviates slightly from a Gaussian one. Introducing only the skewness and the Kurtosis effects, the Gram–Charlier series up to the order four is written as

$$\begin{aligned} p(\gamma_x, \gamma_y) = {} & \frac{1}{2\pi\sigma_{\gamma_x}\sigma_{\gamma_y}} \exp\left(-\frac{\gamma_x^2}{2\sigma_{\gamma_x}^2} - \frac{\gamma_y^2}{2\sigma_{\gamma_y}^2}\right)\left[1 + \frac{c_{21}}{2}(\Gamma_y^2 - 1)\Gamma_x + \frac{c_{03}}{6}(\Gamma_x^2 - 3)\Gamma_x \right. \\ & \left. + \frac{c_{22}}{4}(\Gamma_x^2 - 1)(\Gamma_y^2 - 1) + \frac{c_{40}}{24}(\Gamma_y^4 - 6\Gamma_y^2 + 3) + \frac{c_{04}}{24}(\Gamma_x^4 - 6\Gamma_x^2 + 3)\right], \end{aligned} \tag{2.17}$$

where

$$\Gamma_{x,y} = \frac{\gamma_{x,y}}{\sigma_{\gamma_x,\gamma_y}}. \tag{2.18}$$

In this chapter, the formulation of Cox and Munk [21] (modelling of the slope PDF of the sea surface) is adopted.

In (2.17), $\{c_{21}, c_{03}\}$ are related to the third-order statistics or skewness, whereas $\{c_{22}, c_{04}, c_{40}\}$ are related to the fourth-order statistics or kurtosis. When these coefficients vanish, the slope PDF is Gaussian. Due to the introduction of the skewness, the slope PDF is no longer symmetric.

Substituting (2.1) into (2.17) and integrating over $\gamma_Y \in]-\infty; +\infty[$, after simple but tedious manipulations, we show that

$$\begin{aligned} p_{\gamma_X}(\gamma_X) = & \frac{1}{\sigma_{\gamma X}\sqrt{2\pi}} \exp\left(-\frac{\gamma_X^2}{2\sigma_{\gamma X}^2}\right)\left[1 + \frac{\alpha_K}{3}\left(3 - 6\frac{\gamma_X^2}{\sigma_{\gamma X}^2} + \frac{\gamma_X^4}{\sigma_{\gamma X}^4}\right)\right. \\ & \left. - \frac{\alpha_S}{3}\left(\frac{\gamma_X^3}{\sigma_{\gamma X}^3} - \frac{3\gamma_X}{\sigma_{\gamma X}}\right)\right], \end{aligned} \tag{2.19}$$

where

$$\alpha_S(\phi) = -\frac{\sigma_{\gamma_x}\cos\phi}{2\sigma_{\gamma X}^3}\left[c_{03}(\sigma_{\gamma_x}\cos\phi)^2 + 3c_{21}(\sigma_{\gamma_y}\sin\phi)^2\right], \tag{2.20}$$

and

$$\alpha_K(\phi) = \frac{1}{8\sigma_{\gamma X}^4}\left[c_{04}(\sigma_{\gamma_x}\cos\phi)^4 + c_{40}(\sigma_{\gamma_y}\sin\phi)^4 + \frac{3}{2}c_{22}\sigma_{\gamma_x}^2\sigma_{\gamma_y}^2\sin^2(2\phi)\right], \tag{2.21}$$

and $\sigma_{\gamma X}^2(\phi) = \langle\gamma_X^2\rangle$ is expressed from (2.8). The above equation shows that α_K is related to the fourth-order statistics, for which only *even* powers of $\gamma_X/\sigma_{\gamma X}$ are involved in (2.19). If $\alpha_S = 0$, the associated slope PDF remains symmetric ($p_{\gamma_X}(-\gamma_X) = p_{\gamma_X}(\gamma_X)$) since a Gaussian process is symmetric. Conversely, α_S is related to the third-order statistics, for which only *odd* powers of $\gamma_X/\sigma_{\gamma X}$ are involved in (2.19). This means that the associated slope PDF is no longer symmetric ($p_{\gamma_X}(-\gamma_X) \neq p_{\gamma_X}(\gamma_X)$).

For $\alpha_K = \alpha_S = 0$, the slope PDF is Gaussian, and since $\sigma_{\gamma X}(0) = \sigma_{\gamma X}(\pi)$, the PDF is equal in the up- and down-azimuthal directions.

For $\{\alpha_S \neq 0, \alpha_K = 0\}$, from (2.20) since $\alpha_S(\pi) = -\alpha_S(0)$ there is an asymmetry between the up- and down-azimuthal directions.

For $\{\alpha_S = 0, \alpha_K \neq 0\}$, from (2.21) since $\alpha_K(0) = \alpha_K(\pi)$, the slope PDF keeps the same symmetry properties as a Gaussian one.

For all cases, we have $\alpha_{S,K}(\pi + \phi) = \alpha_{S,K}(\pi - \phi)$ and $\sigma_{\gamma X}(\pi + \phi) = \sigma_{\gamma X}(\pi - \phi)$, which implies that the down-wind azimuthal direction $\phi = \pi$ is a symmetry axis.

Numerical results

Figure 2.2 plots the marginal slope PDF versus γ_X. $\phi = \pi/4$, $\sigma_{\gamma_x} = 0.2$, $\sigma_{\gamma_y} = 0.1$ and for the Gram–Charlier series, $c_{03} = 0.3$, $c_{21} = 0.6$, $c_{04} = 0.4$, $c_{40} = 0.4$ and $c_{22} = 0.8$.

As expected, the Gaussian and exponential probabilities are symmetric whereas the Gram–Charlier distribution is slightly asymmetric. In addition, for slopes near zero, the occurrence is larger for the exponential case in comparison to the Gaussian case, whereas for moderate slopes, the opposite effect appears.

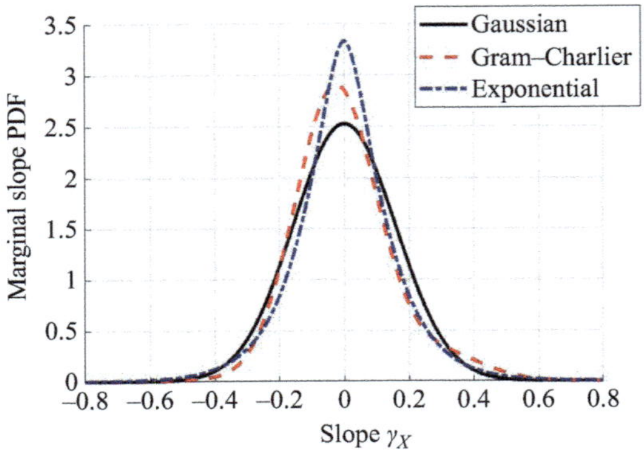

Figure 2.2 Marginal slope PDF versus γ_X. $\phi = \pi/4$, $\sigma_{\gamma_x} = 0.2$, $\sigma_{\gamma_y} = 0.1$ and for the Gram–Charlier series, $c_{03} = 0.3$, $c_{21} = 0.6$, $c_{04} = 0.4$, $c_{40} = 0.4$ and $c_{22} = 0.8$

2.1.2.3 Statistical shadowing function

Making a cross section of the surface along the azimuthal direction ϕ shown in Figure 2.1 and using the results of Chapter 1, the monostatic statistical shadowing function is

$$S^0_{\mathrm{M}}(\mu, \phi, \zeta_0, \gamma_{0X}, R_0) = \Upsilon(\mu - \gamma_{0X}) \exp\left[-\int_0^{R_0} g(\mu|\zeta_0, \gamma_{0X}; r, \phi)\mathrm{d}r\right], \quad (2.22)$$

where the function g equals

$$g_{\mathrm{W}}(\mu|\zeta_0, \gamma_{0X}; r, \phi) = \int_{\mu}^{+\infty} (\gamma_X - \mu)p(\zeta = \zeta_0 + \mu r, \gamma_X|\zeta_0, \gamma_{0X}; r, \phi)\mathrm{d}\gamma_X, \quad (2.23)$$

from the Wagner formulation and

$$g_{\mathrm{S}}(\mu|\zeta_0, \gamma_{0X}; r, \phi) = \frac{g_{\mathrm{W}}(\mu|\zeta_0, \gamma_{0X}; r, \phi)}{\int_{-\infty}^{+\infty}\int_{-\infty}^{\zeta_0+\mu r} p(\zeta, \gamma_X|\zeta_0, \gamma_{0X}; r, \phi)\mathrm{d}\gamma_X\mathrm{d}\zeta}, \quad (2.24)$$

from the Smith model.

The function S^0_{M} gives the probability that a point of the surface M of height ζ_0 and of slope γ_{0X} (defined with respect to the azimuthal direction ϕ) is seen by a receiver of incidence angle $\theta \geq 0$ (or of slope $\mu = \cot\theta \geq 0$) and of azimuthal direction ϕ.

For an uncorrelated process, we show in Chapter 1 that

$$S^0_{\mathrm{M}}(\mu, \phi, \zeta_0, \gamma_{0X}, R_0) = \Upsilon(\mu - \gamma_{0X}) \times \begin{cases} \exp\{-\Lambda\left[P_\zeta(\zeta_0 + \mu R_0) - P_\zeta(\zeta_0)\right]\} & \text{Wagner} \\ \left[\dfrac{P_\zeta(\zeta_0) - P_\zeta(-\infty)}{P_\zeta(\zeta_0 + \mu R_0) - P_\zeta(-\infty)}\right]^\Lambda & \text{Smith} \end{cases}, \quad (2.25)$$

where $P_\zeta(\zeta_0) - P_\zeta(-\infty)$ is the height CDF (cumulative density function) defined as

$$P_\zeta(\zeta_0) - P_\zeta(-\infty) = \int_{-\infty}^{\zeta_0} p_\zeta(\zeta)\mathrm{d}\zeta, \tag{2.26}$$

and

$$\Lambda(\mu, \phi) = \frac{1}{\mu}\int_{\mu}^{+\infty} (\gamma_X - \mu) p_{\gamma_X}(\gamma_X)\mathrm{d}\gamma_X. \tag{2.27}$$

2.1.2.4 Average shadowing function

Equation (2.25) shows that the statistical shadowing function depends on the random variables ζ_0 and γ_{0X}. Then, the averages over ζ_0 or/and γ_{0X} can be derived.

In Chapter 1, we show that the average over ζ_0 and γ_{0X} is expressed as

$$\begin{aligned}\bar{\bar{S}}^0_{\mathrm{M}}(\mu, \phi, R_0) &= \int_{-\infty}^{+\infty}\int_{-\infty}^{+\infty} S^0_{\mathrm{M}}(\mu, \zeta_0, \gamma_{0X}, R_0) p_{\gamma_X,\zeta}(\gamma_{0X}, \zeta_0)\mathrm{d}\gamma_{0X}\mathrm{d}\zeta_0 \\ &= \int_{-\infty}^{+\infty} \Upsilon(\mu - \gamma_{0X}) p_{\gamma_X}(\gamma_{0X})\mathrm{d}\gamma_{0X} \times \int_{-\infty}^{+\infty} \Psi(\mu, \zeta_0, R_0) p_\zeta(\zeta_0)\mathrm{d}\zeta_0,\end{aligned} \tag{2.28}$$

where Ψ equals the function after the symbol $\times$ in (2.25). Moreover, p_ζ stands for the surface height PDF and p_{γ_X} is the marginal surface slope PDF of γ_X.

For an infinite observation length R_0, in Chapter 1 we show that

$$\bar{\bar{S}}^0_{\mathrm{M}}(\mu, \phi, \infty) = \bar{\Upsilon}(\mu, \phi)\begin{cases}\dfrac{1 - e^{-\Lambda}}{\Lambda} & \text{Wagner} \\ \dfrac{1}{1 + \Lambda} & \text{Smith}\end{cases}, \tag{2.29}$$

where

$$\bar{\Upsilon}(\mu, \phi) = \int_{-\infty}^{\mu} p_{\gamma_X}(\gamma_{0X})\mathrm{d}\gamma_{0X}. \tag{2.30}$$

Equations (2.27), (2.29) and (2.30) show that the average shadowing function depends only on the slope PDF p_{γ_X}.

2.1.2.5 Numerical examples

Assuming a Gaussian slope PDF, p_{γ_X} is given by (2.12). Then, from (2.27) and (2.30), we show that

$$\begin{cases}\Lambda(v) = \dfrac{e^{-v^2} - v\sqrt{\pi}\,\mathrm{erfc}(v)}{2v\sqrt{\pi}} & \bar{\Upsilon}(v) = \dfrac{1 + \mathrm{erf}(v)}{2} \\ v(\theta, \phi) = \dfrac{\cot\theta}{\sigma_{\gamma_X}\sqrt{2}} & \sigma_{\gamma_X} = \sqrt{\cos^2\phi\sigma^2_{\gamma_x} + \sin^2\phi\sigma^2_{\gamma_y}}\end{cases}. \tag{2.31}$$

Assuming a Gram–Charlier slope PDF, p_{γ_X} is given by (2.19). Then, from (2.27) and (2.30), after simple but tedious calculations, we show that

$$\Lambda_{GC}(v) = \Lambda(v) - \frac{\alpha_s}{3}\Lambda_S(v) + \frac{\alpha_K}{3}\Lambda_K(v), \tag{2.32}$$

where

$$\begin{cases} \Lambda_S(v) = -\dfrac{\exp(-v^2)}{3\sqrt{2\pi}} \\ \Lambda_K(v) = \dfrac{(2v^2-1)\exp(-v^2)}{6v\sqrt{\pi}} \end{cases}. \tag{2.33}$$

and

$$\bar{\Upsilon}_{GC}(v) = \bar{\Upsilon}_G(v) - \frac{\alpha_s}{3}\bar{\Upsilon}_S(v) + \frac{\alpha_K}{3}\bar{\Upsilon}_K(v), \tag{2.34}$$

where

$$\begin{cases} \bar{\Upsilon}_G(v) = \dfrac{1+\mathrm{erf}(v)}{2} \\ \bar{\Upsilon}_S(v) = -\dfrac{(1-2v^2)\exp(-v^2)}{3\sqrt{2\pi}} \\ \bar{\Upsilon}_K(v) = \dfrac{v(3-2v^2)\exp(-v^2)}{3\sqrt{\pi}} \end{cases}. \tag{2.35}$$

Assuming an exponential slope PDF, p_{γ_X} is given by (2.15). Then, from (2.27) and (2.30), we show that

$$\begin{cases} \Lambda(\mu,\phi) = \dfrac{u_c^3\exp\left(-\mu\sqrt{2}/u_c\right) - u_s^3\exp\left(-\mu\sqrt{2}/u_c\right)}{2\sqrt{2}\,(u_c^2-u_s^2)\,\mu} \\ \bar{\Upsilon}(\mu,\phi) = \dfrac{u_c^2\left[1-(1/2)\exp\left(-\mu\sqrt{2}/u_c\right)\right] - u_s^2\left[1-(1/2)\exp\left(-\mu\sqrt{2}/u_s\right)\right]}{u_c^2-u_s^2} \end{cases}, \tag{2.36}$$

where $u_c = \sigma_{\gamma_x}|\cos\phi|$ and $u_s = \sigma_{\gamma_y}|\sin\phi|$.

Figure 2.3 plots the average shadowing function versus θ whereas at the bottom, the difference against the results computed from a Gaussian PDF is plotted. The simulation parameters are $\phi = \pi/4$, $\sigma_{\gamma_x} = 0.2$ and $\sigma_{\gamma_y} = 0.1$ and for the Gram–Charlier series, $c_{03} = 0.3$, $c_{21} = 0.6$, $c_{04} = 0.4$, $c_{40} = 0.4$ and $c_{22} = 0.8$.

As we can see, the surface is entirely illuminated for $\theta \leq 60^\circ$ and below this angle, the shadowing effect can be neglected. This is consistent with the criterion established in Chapter 1, which claims that for $v \geq v_0 = 2$, the shadow can be omitted. This implies that $\theta_0 = \mathrm{arccot}(v_0\sqrt{2}\sigma_{\gamma_X}) \approx 66^\circ$.

As expected, as θ tends towards 90°, $\bar{\bar{S}}^0_{\mathrm{M}}$ tends to zero (and for any ϕ). For angles θ ranging from 70° to 90°, exponential and Gram–Charlier slope PDFs predict larger strengths than those obtained from a Gaussian PDF and the difference is larger for the exponential PDF.

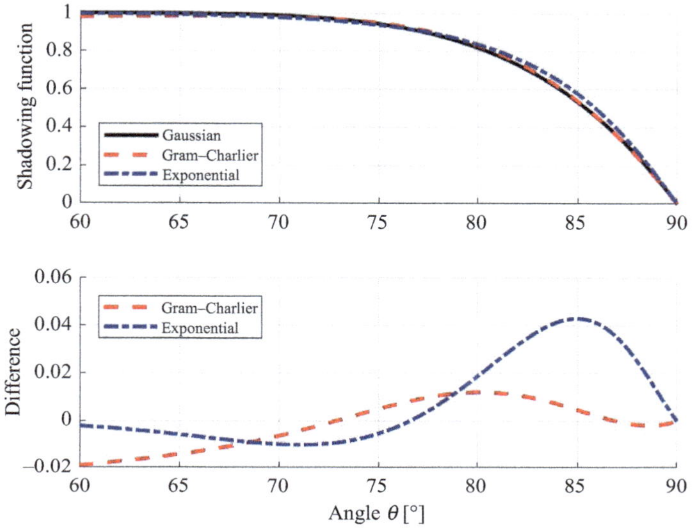

Figure 2.3 *Top: Average shadowing function versus* θ. *Bottom: Difference against the results computed from a Gaussian PDF.* $\phi = \pi/4$, $\sigma_{\gamma_x} = 0.2$, $\sigma_{\gamma_y} = 0.1$ *and for the Gram–Charlier series,* $c_{03} = 0.3$, $c_{21} = 0.6$, $c_{04} = 0.4$, $c_{40} = 0.4$ *and* $c_{22} = 0.8$

Figure 2.4 plots the average shadowing function versus ϕ whereas at the bottom, the difference against the results computed from a Gaussian PDF is plotted. The simulation parameters are $\theta = 85^\circ$, $\sigma_{\gamma_x} = 0.2$, $\sigma_{\gamma_y} = 0.1$ and for the Gram–Charlier series, $c_{03} = 0.3$, $c_{21} = 0.6$, $c_{04} = 0.4$, $c_{40} = 0.4$ and $c_{22} = 0.8$.

As we can see, the direction $\phi = \pi$ is a symmetric axis and the strengths obtained for $\phi = 0$ and $\phi = \pi$ differ only from the Gram–Charlier PDF due to the skewness. For the other PDFs, they are equal because the slope distributions are symmetric. For $\phi = \pi/2$, $\bar{\bar{S}}_{\text{M}}^{0}$ is larger than that obtained for $\phi = 0$ because the slope standard deviation is smaller. Indeed, $\sigma_{\gamma_X}(0) = \sigma_{\gamma_x} > \sigma_{\gamma_X}(\pi/2) = \sigma_{\gamma_y}$.

Figure 2.5 plots the same variations as in Figure 2.4 but for $\theta = 75^\circ$. As we can see, the variations along ϕ are smaller than those observed for $\theta = 85^\circ$.

2.1.3 Case of a correlated Gaussian process

When the correlation is omitted, the statistical shadowing function does not depend on the surface height autocorrelation function. The purpose of this section is to introduce the correlation by considering a Gaussian process.

2.1.3.1 Conditional PDF

Equations (2.23) and (2.24) show that the joint conditional PDF $p(\zeta, \gamma_X|\zeta_0, \gamma_{0X}; r, \phi)$ of the surface heights (ζ, ζ_0) and slopes (γ, γ_0) of two points separated by the distance

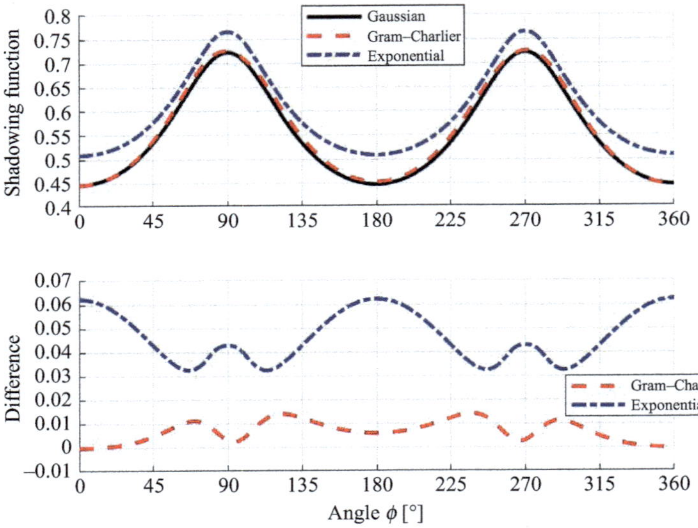

Figure 2.4 *Top: Average shadowing function versus* ϕ*. Bottom: Difference against the results computed from a Gaussian PDF.* $\theta = 85^o$, $\sigma_{\gamma_x} = 0.2$, $\sigma_{\gamma_y} = 0.1$ *and for the Gram–Charlier series,* $c_{03} = 0.3$, $c_{21} = 0.6$, $c_{04} = 0.4$, $c_{40} = 0.4$ *and* $c_{22} = 0.8$

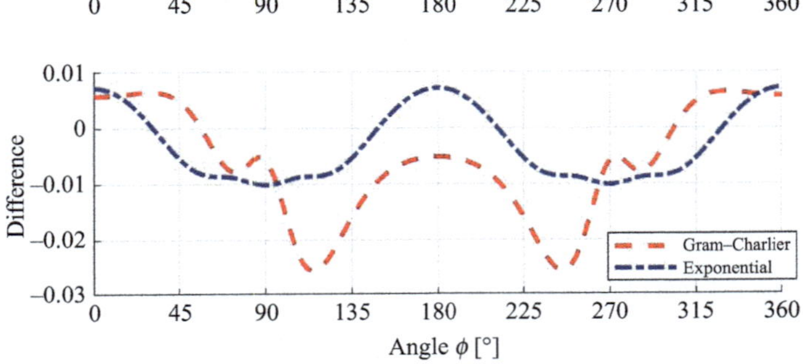

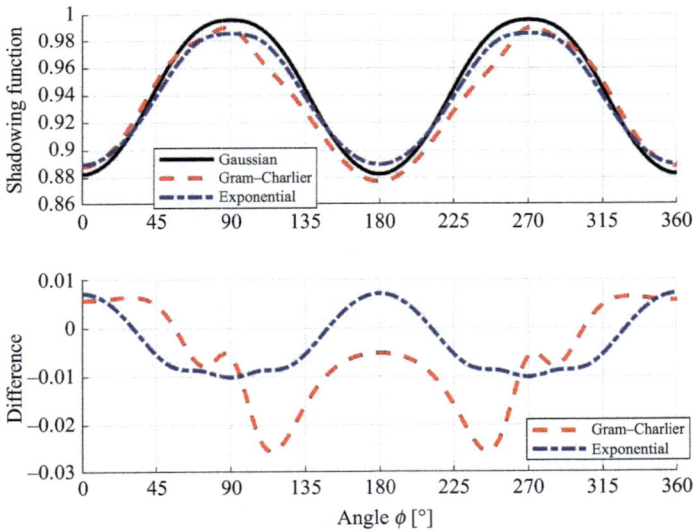

Figure 2.5 *Same variations as in Figure 2.4 but* $\theta = 75^o$

r must be derived in polar coordinates (r, ϕ). For a Gaussian process and from the Bayes theorem, it is expressed as

$$\begin{aligned} p(\zeta, \gamma_X | \zeta_0, \gamma_{0X}; r, \phi) &= \frac{p_4(\zeta, \gamma_X, \zeta_0, \gamma_{0X}; r, \phi)}{p_2(\zeta_0, \gamma_{0X})} \\ &= \frac{\sigma_{\gamma_X}\sigma_\zeta}{2\pi\sqrt{\det(C_4)}} \exp\left[-\frac{1}{2}\vec{V}_4^T \bar{C}_4^{-1} \vec{V}_4 + \frac{\zeta_0^2}{2\sigma_\zeta^2} + \frac{\gamma_{0X}^2}{2\sigma_{\gamma_X}^2}\right], \end{aligned} \quad (2.37)$$

where $\vec{V}_4$ is the vector of components $[\zeta_0 \ \zeta \ \gamma_{0X} \ \gamma_X]$, $[C_4]$ is the covariance matrix of elements $C_{4,ij} = \langle V_{4,i} V_{4,j} \rangle$ ($i = \{1, 4\}$ and $j = \{1, 4\}$ and $V_{4,i}$ denotes the component i of the vector $\vec{V}_4$), where the symbol $\langle \ldots \rangle$ stands for the ensemble average operator. In addition, the symbol det stands for the determinant of the matrix. The covariance matrix is expressed as

$$\bar{C}_4 = \begin{bmatrix} \langle \zeta_0^2 \rangle & \langle \zeta_0 \zeta \rangle & \langle \zeta_0 \gamma_{0X} \rangle & \langle \zeta_0 \gamma_X \rangle \\ \langle \zeta \zeta_0 \rangle & \langle \zeta^2 \rangle & \langle \zeta \gamma_{0X} \rangle & \langle \zeta \gamma_X \rangle \\ \langle \gamma_{0X} \zeta_0 \rangle & \langle \gamma_{0X} \zeta \rangle & \langle \gamma_{0X}^2 \rangle & \langle \gamma_{0X} \gamma_X \rangle \\ \langle \gamma_X \zeta_0 \rangle & \langle \gamma_X \zeta \rangle & \langle \gamma_X \gamma_{0X} \rangle & \langle \gamma_X^2 \rangle \end{bmatrix}. \quad (2.38)$$

Since this matrix is symmetric, only 10 elements must be derived. From (2.1), we have

$$\begin{cases} \langle \zeta_0^2 \rangle & = \sigma_\zeta^2 = C_{\zeta_0,\zeta}(\vec{0}) \\ \langle \zeta_0 \zeta \rangle & = C_{\zeta_0,\zeta}(\vec{r}) \\ \langle \zeta_0 \gamma_{0X} \rangle & = \cos\phi \langle \zeta_0 \gamma_{0x} \rangle + \sin\phi \langle \zeta_0 \gamma_{0y} \rangle \\ & = \cos\phi C_{\zeta_0,\gamma_x}(\vec{0}) + \sin\phi C_{\zeta_0,\gamma_y}(\vec{0}) \\ \langle \zeta_0 \gamma_X \rangle & = \cos\phi \langle \zeta_0 \gamma_x \rangle + \sin\phi \langle \zeta_0 \gamma_y \rangle \\ & = \cos\phi C_{\zeta_0,\gamma_x}(\vec{r}) + \sin\phi C_{\zeta_0,\gamma_y}(\vec{r}) \\ \langle \zeta^2 \rangle & = \sigma_\zeta^2 = C_{\zeta_0,\zeta}(\vec{0}) \\ \langle \zeta \gamma_{0X} \rangle & = \cos\phi \langle \zeta \gamma_{0x} \rangle + \sin\phi \langle \zeta \gamma_{0y} \rangle \\ & = \cos\phi C_{\zeta_0,\gamma_x}(-\vec{r}) + \sin\phi C_{\zeta_0,\gamma_y}(-\vec{r}) \\ \langle \zeta \gamma_X \rangle & = \cos\phi \langle \zeta \gamma_x \rangle + \sin\phi \langle \zeta \gamma_y \rangle \\ & = \cos\phi C_{\zeta_0,\gamma_x}(\vec{0}) + \sin\phi C_{\zeta_0,\gamma_y}(\vec{0}) \\ \langle \gamma_{0X}^2 \rangle & = \cos^2\phi \langle \gamma_{0x}^2 \rangle + \sin^2\phi \langle \gamma_{0y}^2 \rangle + \sin(2\phi) \langle \gamma_{0x} \gamma_{0y} \rangle \\ & = \cos^2\phi C_{\gamma_{0x},\gamma_x}(\vec{0}) + \sin^2\phi C_{\gamma_{0y},\gamma_y}(\vec{0}) + \sin(2\phi) C_{\gamma_{0x},\gamma_y}(\vec{0}) \\ \langle \gamma_{0X} \gamma_X \rangle & = \cos^2\phi \langle \gamma_{0x} \gamma_x \rangle + \sin^2\phi \langle \gamma_{0y} \gamma_y \rangle \\ & \quad + \cos\phi \sin\phi \left(\langle \gamma_{0x} \gamma_y \rangle + \langle \gamma_{0y} \gamma_x \rangle \right) \\ & = \cos^2\phi C_{\gamma_{0x},\gamma_x}(\vec{r}) + \sin^2\phi C_{\gamma_{0y},\gamma_y}(\vec{r}) \\ & \quad + \cos\phi \sin\phi \left[C_{\gamma_{0x},\gamma_y}(\vec{r}) + C_{\gamma_{0x},\gamma_y}(-\vec{r}) \right] \\ \langle \gamma_X^2 \rangle & = \cos^2\phi \langle \gamma_x^2 \rangle + \sin^2\phi \langle \gamma_y^2 \rangle + \sin(2\phi) \langle \gamma_x \gamma_y \rangle \\ & = \cos^2\phi C_{\gamma_{0x},\gamma_x}(\vec{0}) + \sin^2\phi C_{\gamma_{0y},\gamma_y}(\vec{0}) + \sin(2\phi) C_{\gamma_{0x},\gamma_y}(\vec{0}) \end{cases}, \quad (2.39)$$

where $C_{a,b}$ is the cross correlation between the random variables a and b, $C_{b,a}(\vec{r}) = C_{a,b}(-\vec{r})$, $\vec{r} = (x, y)$ and $\vec{0} = (0, 0)$. Then

$$\begin{cases} \langle \zeta_0^2 \rangle & = \langle \zeta^2 \rangle = \sigma_\zeta^2 = C_{\zeta_0,\zeta}(\vec{0}) \\ \langle \zeta_0 \zeta \rangle & = C_{\zeta_0,\zeta}(\vec{r}) \\ \langle \zeta_0 \gamma_{0X} \rangle & = \langle \zeta_0 \gamma_X \rangle \|_{\vec{r}=\vec{0}} \\ \langle \zeta_0 \gamma_X \rangle & = \cos\phi C_{\zeta_0,\gamma_x}(\vec{r}) + \sin\phi C_{\zeta_0,\gamma_y}(\vec{r}) \\ \langle \zeta \gamma_{0X} \rangle & = \langle \zeta_0 \gamma_X \rangle \|_{\vec{r}=-\vec{r}} \\ \langle \gamma_{0X}^2 \rangle & = \langle \gamma_X^2 \rangle = \langle \gamma_{0X} \gamma_X \rangle \|_{\vec{r}=\vec{0}} \\ \langle \gamma_{0X} \gamma_X \rangle & = \cos^2\phi C_{\gamma_{0x},\gamma_x}(\vec{r}) + \sin^2\phi C_{\gamma_{0y},\gamma_y}(\vec{r}) + \sin(2\phi) C_{\gamma_{0x},\gamma_y}(\vec{r}) \end{cases}, \quad (2.40)$$

where $C_{\gamma_{0x},\gamma_y}(-\vec{r}) = C_{\gamma_{0x},\gamma_y}(\vec{r})$ from (2.6).

From (2.3), the surface height autocorrelation function is given from the sea height spectrum $\hat{S}(k, \psi) = k\hat{C}_{\zeta_0,\zeta}(k_x, k_y)$ (factor k comes from the Jacobian of the variable transformations) expressed in polar coordinates ($k_x = k\cos\psi$, $k_y = k\sin(\psi)$) as

$$C_{\zeta_0,\zeta}(r, \phi) = \frac{1}{(2\pi)^2} \int_0^{2\pi} \int_0^{\infty} \hat{S}(k, \psi) e^{jkr\cos(\psi-\phi)} \mathrm{d}k \mathrm{d}\psi, \quad (2.41)$$

where $k = \sqrt{k_x^2 + k_y^2} \geq 0$, $\psi \in [0; 2\pi[$, $\cos\psi = k_x/k$, $\sin\psi = k_y/k$, $r = \sqrt{x^2 + y^2} \geq 0$, $\phi \in [0; 2\pi[$, $\cos\phi = x/r$ and $\sin\phi = y/r$.

In (2.40), the functions C_{ζ_0,γ_x} and C_{ζ_0,γ_y} are expressed from (2.2). From (2.3), $\langle \zeta_0 \gamma_X \rangle$ is then

$$\begin{aligned} \langle \zeta_0 \gamma_X \rangle &= \frac{j}{(2\pi)^2} \int_{-\infty}^{+\infty} \int_{-\infty}^{+\infty} \hat{C}_{\zeta_0,\zeta}(k_x, k_y) \left(k_x \cos\phi + k_y \sin\phi\right) e^{jk_x x + jk_y y} \mathrm{d}k_x \mathrm{d}k_y \\ &= \frac{j}{(2\pi)^2} \int_0^{2\pi} \int_0^{\infty} \hat{S}(k, \psi) \cos(\psi - \phi) k e^{jkr\cos(\psi-\phi)} \mathrm{d}k \mathrm{d}\psi \\ &= \frac{\partial C_{\zeta_0,\zeta}(r, \phi)}{\partial r}. \end{aligned} \quad (2.42)$$

In addition, since $\hat{S}(k, \psi)$ is a periodic function over ψ, $\langle \zeta_0 \gamma_X \rangle = 0$ for $\vec{r} = \vec{0}$. It is consistent with (2.6) (second line, because $\sin(0) = 0$).

In (2.40), the functions $C_{\zeta 0,\gamma_x}$, $C_{\zeta 0,\gamma_y}$ and $C_{\gamma_{0x},\gamma_y}(\vec{r})$ are expressed from (2.2). From (2.3), $\langle \gamma_{0X}\gamma_X \rangle$ is then

$$\begin{aligned}\langle \gamma_{0X}\gamma_X \rangle &= \frac{1}{(2\pi)^2}\int_{-\infty}^{+\infty}\int_{-\infty}^{+\infty}\hat{C}_{\zeta 0,\zeta}(k_x,k_y)\\ &\quad\times\left(k_x^2\cos^2\phi + k_y^2\sin^2\phi + 2k_xk_y\cos\phi\sin\phi\right)e^{jk_xx+jk_yy}\mathrm{d}k_x\mathrm{d}k_y\\ &= \frac{1}{(2\pi)^2}\int_0^{2\pi}\int_0^{\infty}\hat{S}(k,\psi)\\ &\quad\times(\cos\phi\cos\psi+\sin\phi\sin\psi)^2k^2e^{jkr\cos(\psi-\phi)}\mathrm{d}k\mathrm{d}\psi\\ &= \frac{1}{(2\pi)^2}\int_0^{2\pi}\int_0^{\infty}\hat{S}(k,\psi)\cos^2(\psi-\phi)k^2e^{jkr\cos(\psi-\phi)}\mathrm{d}k\mathrm{d}\psi\\ &= -\frac{\partial^2 C_{\zeta 0,\zeta}(r,\phi)}{\partial r^2}.\end{aligned} \tag{2.43}$$

In addition for $\vec{r}=\vec{0}$, (2.43) becomes $\langle \gamma_{0X}\gamma_X \rangle = \sigma_{\gamma X}^2$ and

$$\begin{aligned}\sigma_{\gamma X}^2 &= \frac{1}{(2\pi)^2}\int_{-\infty}^{+\infty}\int_{-\infty}^{+\infty}\hat{C}_{\zeta 0,\zeta}(k_x,k_y)\\ &\quad\times\left[k_x^2\cos^2\phi + k_y^2\sin^2\phi + k_xk_y\sin(2\phi)\right]\mathrm{d}k_x\mathrm{d}k_y\\ &= \cos^2\phi\,\sigma_{\gamma_x}^2 + \sin^2\phi\,\sigma_{\gamma_y}^2,\end{aligned} \tag{2.44}$$

where

$$\begin{cases}\sigma_{\gamma_x}^2 = \dfrac{1}{(2\pi)^2}\displaystyle\int_{-\infty}^{+\infty}\int_{-\infty}^{+\infty}\hat{C}_{\zeta 0,\zeta}(k_x,k_y)k_x^2\mathrm{d}k_x\mathrm{d}k_y\\ \quad\;\; = \dfrac{1}{(2\pi)^2}\displaystyle\int_0^{2\pi}\int_0^{\infty}\hat{S}(k,\psi)k^2\cos^2\psi\,\mathrm{d}k\mathrm{d}\psi\\ \sigma_{\gamma_y}^2 = \dfrac{1}{(2\pi)^2}\displaystyle\int_{-\infty}^{+\infty}\int_{-\infty}^{+\infty}\hat{C}_{\zeta 0,\zeta}(k_x,k_y)k_y^2\mathrm{d}k_x\mathrm{d}k_y\\ \quad\;\; = \dfrac{1}{(2\pi)^2}\displaystyle\int_0^{2\pi}\int_0^{\infty}\hat{S}(k,\psi)k^2\sin^2\psi\,\mathrm{d}k\mathrm{d}\psi\end{cases}. \tag{2.45}$$

From the above equation, the calculation of the slope variance can be performed in the directions (Ox) and (Oy) from the height spectrum either in Cartesian coordinates, $\hat{C}_{\zeta 0,\zeta}$, or in polar coordinates $\hat{S}(k,\psi) = k\hat{C}_{\zeta 0,\zeta}(k_x,k_y)$.

To sum up, the covariance matrix (2.46) is simplified as

$$\bar{C}_4 = \begin{bmatrix} \sigma_\zeta^2 & C_{\zeta 0,\zeta} & 0 & \dfrac{\partial C_{\zeta 0,\zeta}}{\partial r} \\ C_{\zeta 0,\zeta} & \sigma_\zeta^2 & -\dfrac{\partial C_{\zeta 0,\zeta}}{\partial r} & 0 \\ 0 & -\dfrac{\partial C_{\zeta 0,\zeta}}{\partial r} & \sigma_{\gamma X}^2 & \dfrac{\partial^2 C_{\zeta 0,\zeta}}{\partial r^2} \\ \dfrac{\partial C_{\zeta 0,\zeta}}{\partial r} & 0 & \dfrac{\partial^2 C_{\zeta 0,\zeta}}{\partial r^2} & \sigma_{\gamma X}^2 \end{bmatrix}, \tag{2.46}$$

where $C_{\zeta 0,\zeta}$ is expressed in polar coordinates (r, ϕ).

2.1.3.2 Wagner and Smith formulations

Comparing (2.46) with (1.56) of Chapter 1, we obtain exactly the same covariance matrix with the following changes:

$$\begin{cases} r & \to \tau \\ C_{\zeta 0,\zeta} & \to C_{\zeta 0,0} \\ \sigma_{\gamma X} & \to \sigma_\gamma \end{cases}. \tag{2.47}$$

This means that all the results established for a 1D surface can be applied to a 2D surface with the changes expressed from (2.47). Then, the monostatic statistical shadowing functions with a Gaussian correlation of Wagner and Smith are expressed in Table 1.2.

For Gaussian and Lorentzian profiles, $C_{\zeta 0,\zeta}$ is defined as

$$C_{\zeta 0,\zeta}(x,y) = \sigma_\zeta^2 \begin{cases} \exp\left(-\dfrac{x^2}{L_{cx}^2} - \dfrac{y^2}{L_{cy}^2}\right) = \exp\left(-\dfrac{r^2}{L_c^2}\right) \\ \dfrac{1}{1 + (x^2/L_{cx}^2) + (y^2/L_{cy}^2)} = \dfrac{1}{1 + (r^2/L_c^2)} \end{cases}, \tag{2.48}$$

where $1/L_c^2 = \cos^2\phi/L_{cx}^2 + \sin^2\phi/L_{cy}^2$. In polar coordinates, the same surface height autocorrelation functions as defined for a 1D surface are retrieved by introducing an equivalent correlation length L_c, which depends on ϕ. They are expressed in Table 1.1 of Chapter 1. Bourlier *et al.* [7] investigated also the case of a modified Lorentzian profile defined as $C_{\zeta 0,\zeta}(x,y) = [(1 + x^2/L_{cx}^2)(1 + y^2/L_{cy}^2)]^{-1}$ and the case of a sea spectrum.

Comparing with the Monte Carlo procedure presented in Chapter 1, they obtained the same conclusions drawn in Chapter 1, that is the Smith formulation is more accurate and the correlation improves slightly the results.

2.2 Bistatic shadowing function

2.2.1 *Introduction*

As reported in Chapter 1, the bistatic shadowing function is derived by multiplying two independent 1D monostatic illumination functions: the first one associated to the transmitter and the second one to the receiver. As pointed out by Heitz *et al.* [8], this assumption is not valid for many 2D configurations. Indeed, if the transmitter and receiver azimuthal directions are close, the illumination probabilities associated to these directions become strongly correlated and then they cannot be multiplied. Numerically, a discontinuity can occur when the transmitter and receiver are the same [7].

While this case is often neglected, it is not unusual in real-life situations. Typically, in long-range acquisition when the transmitter and the receiver are located on the same device, their directions are close comparatively to the faraway target.

2.2.2 *Derivation*

2.2.2.1 Smith monostatic model

The surface heights ζ and slopes (γ_x, γ_y) are assumed to be uncorrelated so that (2.10) can be applied. Then the marginal PDF p_{γ_X} of the surface slope $\gamma_X = \gamma_x \cos\phi + \gamma_y \sin\phi$ can be derived by integrating the slope PDF of (γ_x, γ_y) with respect to $\gamma_Y \in \mathbb{R} = -\gamma_x \sin\phi + \gamma_y \cos\phi$.

For an infinite observation length and by considering the Smith model (the more accurate) without correlation, we showed that the monostatic shadowing function is expressed as

$$S^0_{\mathrm{M}}(\mu, \phi, \zeta_0, \gamma_{0X}) = \Upsilon(\mu - \gamma_{0X}) \left[F_\zeta(\zeta_0)\right]^{\Lambda(\mu,\phi)}, \tag{2.49}$$

where $F_\zeta(\zeta_0) = P_\zeta(\zeta_0) - P_\zeta(-\infty)$ is the CDF of the surface heights and P_ζ is a primitive of the surface height PDF p_ζ. In addition, Λ is defined by (2.27).

2.2.2.2 Bistatic shadowing function

As shown in Figure 2.6, the receiver is defined from the point A of polar angles (θ_1, ϕ_1) and the transmitter from the point B of polar angles (θ_2, ϕ_2), in which $\theta_p \in [0; \pi/2]$ and the azimuthal angles $\phi_p \in [0; 2\pi[$ $(p = \{1, 2\})$. The associated slopes are $\mu_p = \cot\theta_p \geq 0$.

Two cases can be distinguished.

2.2.2.3 Case $\phi = 0$

The case where $\phi = |\phi_2 - \phi_1| = 0$ is equivalent to an in-plane configuration. The events A and B are correlated because when the two rays are above the surface, the probability that the surface point is viewed both by the transmitter and the receiver is

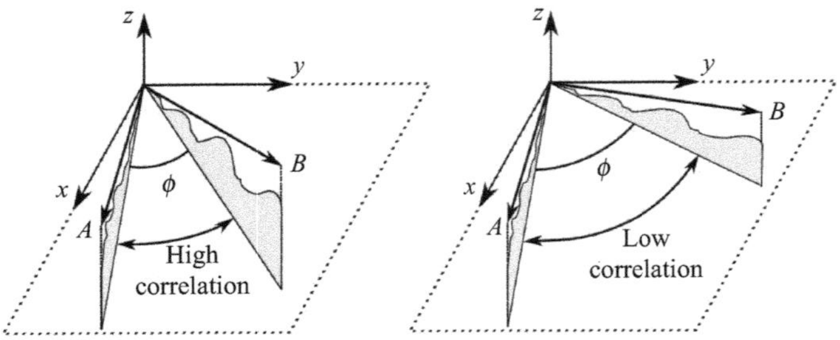

Figure 2.6 Illustration of the azimuthal correlation between two directions A and B in the bistatic configuration

determined only by the probability that the lowest ray is above the surface. Then, the statistical function is

$$S_{\mathrm{B}}^{1}(A,B,\zeta_0,\gamma_{0X_1},\gamma_{0X_2}) = \min\left[S_{\mathrm{M}}^{0}(A,\zeta_0,\gamma_{0X_1}),S_{\mathrm{M}}^{0}(B,\zeta_0,\gamma_{0X_2})\right]$$
$$= \begin{cases} S_{\mathrm{M}}^{0}(A,\zeta_0,\gamma_{0X_1}) \text{ if } \mu_2 \geq \mu_1 \\ S_{\mathrm{M}}^{0}(B,\zeta_0,\gamma_{0X_2}) \text{ if } \mu_2 < \mu_1 \end{cases}, \quad (2.50)$$

where $\gamma_{X_p} = \gamma_x \cos\phi_p + \gamma_y \sin\phi_p$.

2.2.2.4 Case $\phi \neq 0$ and $\mu_1 \leq \mu_2$

Bourlier *et al.* [7] assumed that the events associated to the points A and B are independent (the probabilities can be estimated separately and multiplied), leading to

$$S_{\mathrm{B}}^{1}(A,B,\zeta_0,\gamma_{0X_1},\gamma_{0X_2}) = S_{\mathrm{M}}^{0}(A,\zeta_0,\gamma_{0X_1})S_{\mathrm{M}}^{0}(B,\zeta_0,\gamma_{0X_2}). \quad (2.51)$$

However, if A and B are close (Figure 2.6), the events can be strongly correlated and the above equation does not hold. This occurs when the absolute azimuthal difference $\phi = |\phi_2 - \phi_1|$ is small.

For $\mu_1 \leq \mu_2$ ($\zeta_1 \leq \zeta_2$) and $\phi \in [0;\pi/2]$, by introducing the azimuthal correlation coefficient c_0, Heitz *et al.* [8] then showed

$$S_{\mathrm{B}}^{1}(A,B,\zeta_0,\gamma_{0X_1},\gamma_{0X_2}) = S_{\mathrm{M}}^{0}(A,\zeta_0,\gamma_{0X_1})S_c(A,B,\zeta_0,\gamma_{0X_2}), \quad (2.52)$$

where

$$S_c(A,B,\zeta_0,\gamma_{0X_2}) = \Upsilon\left(\mu_2 - \gamma_{0X_2}\right)\left[F_\zeta(\zeta_0)\right]^{c_0(\mu_1,\mu_2,\phi,\zeta_0)\Lambda(\mu_2,\phi_2)}, \quad (2.53)$$

in which

$$c_0(\mu_1,\mu_2,\phi,\zeta_0) = \frac{\displaystyle\int_0^\infty \frac{p_\zeta(\zeta_2)F_d(\zeta_2,\zeta_2)\mathrm{d}r_2}{\displaystyle\int_{-\infty}^{\zeta_2} p_\zeta(\zeta_2')F_d(\zeta_2,\zeta_2')\mathrm{d}\zeta_2'}}{\displaystyle\int_0^\infty \frac{p_\zeta(\zeta_2)\mathrm{d}r_2}{F_\zeta(\zeta_2)}}, \quad (2.54)$$

$$\begin{cases} F_\zeta(\zeta) = P_\zeta(\zeta) - P_\zeta(-\infty) = \int_{-\infty}^{\zeta} p_\zeta(\zeta_0)\mathrm{d}\zeta_0 \\ F_d(\zeta_2, \zeta_2') = \int_{-\infty}^{\zeta_1} p_d(\zeta)\mathrm{d}\zeta = \frac{1}{2}\left[1 + \mathrm{erf}\left(\frac{\zeta_1 - m_d}{\sqrt{2}\sigma_d}\right)\right] \\ p_d(\xi) = \frac{1}{\sqrt{2\pi}\sigma_d}\exp\left(-\left[\frac{\zeta - m_d}{\sqrt{2}\sigma_d}\right]^2\right) \\ m_d(\zeta_2') = \frac{\zeta_2'}{1 + d^2\sigma_\gamma^2/\sigma_\zeta^2} \quad \sigma_d^2(\zeta_2, \zeta_0) = \frac{d^2\sigma_\gamma^2}{1 + d^2\sigma_\gamma^2/\sigma_\zeta^2} \\ d(\zeta_2, \zeta_0) = r_2(\zeta_2, \zeta_0)\tan\phi \quad r_2(\zeta_2, \zeta_0) = \frac{\zeta_2 - \zeta_0}{\mu_2} \\ \zeta_1(\zeta_2, \zeta_0) = \zeta_0 + \mu_1 r_1(\zeta_2, \zeta_0) = \zeta_0 + \frac{\mu_1 r_2(\zeta_2, \zeta_0)}{\cos\phi} \end{cases}, \tag{2.55}$$

and $\zeta_2 = \zeta_0 + \mu_2 r_2$. The term c_0 can be interpreted as a correlation coefficient between the azimuthal directions of the transmitter and receiver.

If $\phi = \pi/2$, then $d \to +\infty$ ($r_2 \geq 0$), $m_d \to 0^+ \times \zeta_2'$, $\sigma_d \to \sigma_\zeta$ and $\zeta_1 \to +\infty$ where $\mu_1 \geq 0$. Then $(\zeta_1 - m_d)/(\sqrt{2}\sigma_d) \to +\infty$ and $F_d \to 1$, which implies from (2.54) that

$$c_0|_{\phi=\pi/2} = 1. \tag{2.56}$$

As expected, the azimuthal correlation coefficient is maximal for $\phi = \pi/2$.

If $\phi = 0$, then $d = 0^+$, $m_d(\zeta_2') = \zeta_2'$, $\sigma_d = 0^+$, $[\zeta_1 - m_d(\zeta_2)]/(\sqrt{2}\sigma_d) = (\zeta_1 - \zeta_2)/(\sqrt{2}\sigma_d) \to -\infty$ since $\zeta_1 - \zeta_2 \leq 0$, which implies that $F_d(\zeta_2, \zeta_2) \to 0$. In addition, the PDF $p_d(\zeta) = \delta(\zeta_1 - \zeta_2')$, in which δ is the Dirac delta distribution. An integration by parts leads to

$$\begin{aligned} \int_{-\infty}^{\zeta_2} p_\zeta(\zeta_2')F_d(\zeta_2, \zeta_2')\mathrm{d}\zeta_2' &= \left[F_d(\zeta_2, \zeta_2')P_\xi(\zeta_2')\right]_{\zeta_2'=-\infty}^{\zeta_2'=\zeta_2} + \int_{-\infty}^{\zeta_2} p_d(\zeta_2')P_\xi(\zeta_2')\mathrm{d}\zeta_2' \\ &= F_d(\zeta_2, \zeta_2)P_\xi(\zeta_2) - F_d(\zeta_2, -\infty)P_\xi(-\infty) \\ &\quad + P_\xi(\zeta_2')\delta(\zeta_1 - \zeta_2') \\ &= P_\xi(\zeta_2) - P_\xi(-\infty) + 0, \end{aligned} \tag{2.57}$$

where $\partial F_d(\zeta_2, \zeta_2')/\partial\zeta_2' = -\delta(\zeta_1 - \zeta_2')$, $F_d(\zeta_2, \zeta_2) = 1$ and $F_d(\zeta_2, -\infty) = 1$. In addition, the product with the Dirac delta function vanishes because ζ_1 differs from ζ_2'. For $\phi = 0$, the above equation gives a finite value and since $F_d(\zeta_2, \zeta_2) = 0$, from (2.54), we have

$$c_0|_{\phi=0} = 0. \tag{2.58}$$

As expected, the azimuthal correlation coefficient is minimal for $\phi = 0$. Moreover, from (2.53), $S_c = 1$, which implies from (2.52), $S_{\mathrm{B}}^1(A, B, \zeta_0, \gamma_{0X_1}, \gamma_{0X_2}) = S_{\mathrm{M}}^0(A, \zeta_0, \gamma_{0X_1})$. Equation (2.50) is retrieved. As pointed out by Heitz *et al.* [8], the introduction of c_0 allow us to have no discontinuity between the cases $\phi = 0$ and $\phi \neq 0$.

2.2.2.5 Case for any ϕ and μ_1

For $\mu_1 > \mu_2$ ($\zeta_2 \leq \zeta_1$) and $\phi \in [0;\pi/2]$, $S_{\rm B}^1$ is obtained from (2.51) by replacing (A,B) with (B,A). In addition for $\phi \in]\pi/2;\pi]$, $c_0 = 1$ and from symmetrical considerations, $S_{\rm B}^1|_{\phi=-\phi} = S_{\rm B}^1|_{\phi}$.

In conclusion for any ϕ and μ_1, from (2.49), (2.52) and (2.53), the bistatic statistical shadowing function is

$$\begin{aligned} S_{\rm B}^1(A,B,\zeta_0,\gamma_{0X_1},\gamma_{0X_2}) &= \Upsilon(\mu_1-\gamma_{0X_1})\,\Upsilon\left(\mu_2-\gamma_{0X_2}\right) \\ &\quad\times\left[F_\zeta(\zeta_0)\right]^{c_0(\mu_p,\mu_q,\phi,\zeta_0)\Lambda(\mu_q,\phi_q)+\Lambda(\mu_p,\phi_p)}, \end{aligned} \tag{2.59}$$

where $(p,q)=(1,2)$ if $0 \leq \mu_1 \leq \mu_2$ and $(p,q) = (2, 1)$ otherwise.

2.2.3 *Average shadowing functions*

Equation (2.59) shows that $S_{\rm B}^1$ depends statistically on the surface heights ζ_0 and surface slopes $(\gamma_{0X_1},\gamma_{0X_2})$.

The average shadowing function over ζ_0 and $(\gamma_{0X_1},\gamma_{0X_2})$ is defined as

$$\bar{\bar{S}}_{\rm B}^1(A,B) = \int_{-\infty}^{+\infty}\int_{-\infty}^{+\infty}\int_{-\infty}^{+\infty} S_{\rm B}^1(A,B,\zeta_0,\gamma_{0X_1},\gamma_{0X_2})p_3(\zeta_0,\gamma_{0X_1},\gamma_{0X_1})\mathrm{d}\zeta_0\mathrm{d}\gamma_{0X_1}\mathrm{d}\gamma_{0X_2}. \tag{2.60}$$

The cross and autocorrelations between the random variables ζ_0, γ_{0X_1} and γ_{0X_2} are expressed as

$$\begin{cases} \langle\zeta_0^2\rangle &= \sigma_\zeta^2 \\ \langle\zeta_0\gamma_{0X_1}\rangle &= \cos\phi_1\,\langle\zeta_0\gamma_{0x}\rangle + \sin\phi_1\,\langle\zeta_0\gamma_{0y}\rangle = 0 \\ \langle\zeta_0\gamma_{0X_2}\rangle &= \cos\phi_2\,\langle\zeta_0\gamma_{0x}\rangle + \sin\phi_2\,\langle\zeta_0\gamma_{0y}\rangle = 0 \\ \langle\gamma_{0X_1}^2\rangle &= \cos^2\phi_1\,\langle\gamma_{0x}^2\rangle + +\sin^2\phi_1\,\langle\gamma_{0y}^2\rangle + \sin(2\phi_1)\langle\gamma_{0x}\gamma_{0y}\rangle \\ &= \cos^2\phi_1\sigma_{\gamma_x}^2 + \sin^2\phi_1\sigma_{\gamma_y}^2 = \sigma_{\gamma X}^2(\phi_1) \\ \langle\gamma_{0X_2}\gamma_{0X_1}\rangle &= \langle(\cos\phi_1\gamma_{0x}+\sin\phi_1\gamma_{0y})\,(\cos\phi_2\gamma_{0x}+\sin\phi_2\gamma_{0y})\rangle \\ &= \rho_{12}\sigma_{\gamma X}(\phi_1)\sigma_{\gamma X}(\phi_2) \\ \langle\gamma_{0X_2}^2\rangle &= \sigma_{\gamma X}^2(\phi_2) \end{cases}, \tag{2.61}$$

where

$$\rho_{12} = \frac{\langle\gamma_{0X_2}\gamma_{0X_1}\rangle}{\sigma_{\gamma X}(\phi_1)\sigma_{\gamma X}(\phi_2)} = \frac{\sigma_x^2\cos\phi_1\cos\phi_2 + \sigma_y^2\sin\phi_1\sin\phi_2}{\sigma_{\gamma X}(\phi_1)\sigma_{\gamma X}(\phi_2)}, \tag{2.62}$$

and $\langle\zeta_0\gamma_{0x}\rangle = \langle\zeta_0\gamma_{0y}\rangle = \langle\gamma_{0x}\gamma_{0y}\rangle = 0$, $\langle ab\rangle = \langle ba\rangle$ (for more details see Section 2.1.3.1). Moreover, σ_ζ is the surface height standard deviation and $(\sigma_{\gamma_x},\sigma_{\gamma_y})$ those of the surface slopes with respect to the directions (Ox) and (Oy), respectively.

The associated covariance matrix is then

$$\bar{C}_3 = \begin{bmatrix} \sigma_\zeta^2 & 0 & 0 \\ 0 & \sigma_{\gamma X}^2(\phi_1) & \rho_{12}\sigma_{\gamma X}(\phi_1)\rho_{12}\sigma_{\gamma X}(\phi_2) \\ 0 & \rho_{12}\sigma_{\gamma X}(\phi_1)\rho_{12}\sigma_{\gamma X}(\phi_2) & \sigma_{\gamma X}^2(\phi_2) \end{bmatrix}. \tag{2.63}$$

Assuming a Gaussian process, from (2.59) and (2.63), (2.60) becomes

$$\begin{aligned} \bar{\bar{S}}_{\mathrm{B}}^1(A,B) = & \int_{-\infty}^{+\infty} \left[F_\zeta(\zeta_0)\right]^{L(\mu_1,\phi_1,\mu_2,\phi_2,\zeta_0)} p_\zeta(\zeta_0)\mathrm{d}\zeta_0 \\ & \times \int_{-\infty}^{+\infty}\int_{-\infty}^{+\infty} \Upsilon(\mu_1 - \gamma_{0X_1})\,\Upsilon(\mu_2 - \gamma_{0X_2})\,p_2(\gamma_{0X_1},\gamma_{0X_2})\mathrm{d}\gamma_{0X_1}\mathrm{d}\gamma_{0X_2}, \end{aligned} \tag{2.64}$$

where $L(\mu_1,\phi_1,\mu_2,\phi_2,\zeta_0) = c_0(\mu_p,\mu_q,\phi,\zeta_0)\Lambda(\mu_q,\phi_q) + \Lambda(\mu_p,\phi_p)$ and

$$p_\zeta(\zeta_0) = \frac{1}{\sqrt{2\pi}\sigma_\zeta}\exp\left(-\frac{\zeta_0^2}{2\sigma_\zeta^2}\right), \tag{2.65}$$

$$\begin{aligned} p_2(\gamma_{0X_1},\gamma_{0X_2}) = & \frac{1}{2\pi\sigma_{\gamma X1}\sigma_{\gamma X2}\sqrt{1-\rho_{12}^2}} \\ & \times \exp\left(-\frac{\gamma_{0X_1}^2\sigma_{\gamma X1}^2 + \gamma_{0X_2}^2\sigma_{\gamma X2}^2 - 2\rho_{12}\gamma_{0X_1}\gamma_{0X_2}\sigma_{\gamma X1}\sigma_{\gamma X2}}{2\sigma_{\gamma X1}^2\sigma_{\gamma X2}^2(1-\rho_{12}^2)}\right), \end{aligned} \tag{2.66}$$

and $\sigma_{\gamma X p} = \sigma_{\gamma X}(\phi_p)$.

In (2.64), if $c_0 \approx \tilde{c}_0(\mu_p,\mu_q,\phi)$ is assumed to be independent of the heights ζ_0, then the integration over ζ_0 leads to

$$\int_{-\infty}^{+\infty} \left[F_\zeta(\zeta_0)\right]^{L(\mu_1,\phi_1,\mu_2,\phi_2,\zeta_0)} p_\zeta(\zeta_0)\mathrm{d}\zeta_0 \approx \frac{1}{1+\tilde{L}(\mu_1,\phi_1,\mu_2,\phi_2)}, \tag{2.67}$$

where $\tilde{L}(\mu_1,\phi_1,\mu_2,\phi_2) = \tilde{c}_0(\mu_p,\mu_q,\phi)\Lambda(\mu_q,\phi_q) + \Lambda(\mu_p,\phi_p)$. If no assumption is applied, the integration over ζ_0 is numerical.

In (2.64), the use of the variable transformations $\gamma_{0X_2} = s_2\sigma_{\gamma X2}\sqrt{2(1-\rho_{12}^2)}$ and $\gamma_{0X_1} = s_1\sigma_{\gamma X1}\sqrt{2}$, and the integration over s_2 leads to

$$\frac{1}{2\sqrt{\pi}}\int_{-\infty}^{v_1} \exp(-s_1^2)\left[1 + \mathrm{erf}\left(\frac{v_2 - \rho_{12}s_1}{\sqrt{1-\rho_{12}^2}}\right)\right]\mathrm{d}s_1, \tag{2.68}$$

where $v_1 = \cot\theta_1/[\sqrt{2}\sigma_{\gamma X}(\phi_1)]$ and $v_2 = \cot\theta_2/[\sqrt{2}\sigma_{\gamma X}(\phi_2)]$.

The correlation coefficient $\rho_{12} \in [-1;1]$. For instance, if the surface is assumed to be isotropic, then $\sigma_{\gamma_x} = \sigma_{\gamma_y}$ and from (2.62), $\rho_{12} = \cos(\phi_2 - \phi_1) = \cos\phi$.

For $\phi_1 = \phi_2$, $\phi = 0$ and $\rho_{12} = 1$, which implies that the kernel in (2.68) has a discontinuity because $1/\sqrt{1-\rho_{12}^2} \Rightarrow +\infty$. The kernel contributes for $v_2 - \rho_{12}s_1 = v_2 - s_1 \geq 0 \Rightarrow s_1 \leq v_2$. Then

$$\text{Equation (2.68)} = \frac{1}{\sqrt{\pi}} \int_{-\infty}^{\min(v_1,v_2)} \exp\left(-s_1^2\right) \mathrm{d}s_1 = \frac{1 + \operatorname{erf}(\min(v_1, v_2))}{2}, \phi = 0. \tag{2.69}$$

For $\phi_2 - \phi_1 = \pi$, $\phi = \pi$ and $\rho_{12} = -1$, which implies that the kernel in (2.68) has a discontinuity because $1/\sqrt{1-\rho_{12}^2} \Rightarrow +\infty$. Since $[1 + \operatorname{erf}(\infty)]/2 = 1$ and $[1 + \operatorname{erf}(-\infty)]/2 = 0$, the kernel contributes for $v_2 - \rho_{12}s_1 = v_2 + s_1 \geq 0 \Rightarrow s_1 \geq -v_2$. Then

$$\text{Equation (2.68)} = \frac{1}{\sqrt{\pi}} \int_{-v_2}^{v_1} \exp\left(-s_1^2\right) \mathrm{d}s_1 = \frac{\operatorname{erf}(v_1) + \operatorname{erf}(v_2)}{2}, \phi = \pi. \tag{2.70}$$

Cases $\phi = 0$ and $\phi = \pi$ occurs when the transmitter and receiver are in the same plane, which is equivalent to a 2D problem and the surface can be considered 1D. Equations (2.69) and (2.70) are then consistent with those derived in Chapter 1 ((1.117) and (1.120) with $\mu = \mu_2$ and $\mu_i = \mu_1$).

2.2.4 Numerical results

2.2.4.1 Correlation coefficient c_0

In (2.54), the use of the variable transformations $\zeta_2 = \zeta_0 + \mu_2 r_2$ and next $h_2 = \zeta_2/(\sqrt{2}\sigma_\zeta)$, $h_2' = \zeta_2'/(\sqrt{2}\sigma_\zeta)$ leads to

$$c_0(v_1, v_2, \phi, h_0) = \frac{\displaystyle\int_{h_0}^{\infty} \frac{p_h(h_2)F_d(h_2,h_2)\mathrm{d}h_2}{\displaystyle\int_{-\infty}^{h_2} p_h(h_2')F_d(h_2,h_2')\mathrm{d}h_2'}}{\ln |F_h(h_2)|_{h_2=h_0}^{h_2=\infty}} = \frac{\displaystyle\int_{h_0}^{\infty} \frac{p_h(h_2)F_d(h_2,h_2)\mathrm{d}h_2}{\displaystyle\int_{-\infty}^{h_2} p_h(h_2')F_d(h_2,h_2')\mathrm{d}h_2'}}{\ln [F_h(\infty)] - \ln [F_h(h_0)]}, \tag{2.71}$$

where

$$\begin{cases} F_h(h) = \displaystyle\int_{-\infty}^{h} p_h(h_0)\mathrm{d}h_0 & \\ F_d(h_2, h_2') = \dfrac{1}{2}\left[1 + \operatorname{erf}\left(\dfrac{h_1 - m_d}{\sigma_d}\right)\right] & \\ m_d(h_2') = \dfrac{h_2'}{1+d^2} & \sigma_d^2(h_2, h_0) = \dfrac{d^2}{1+d^2} \\ d(h_2, h_0) = r_2(h_2, h_0)\tan\phi & r_2(h_2, h_0) = \dfrac{h_2 - h_0}{v_2} \\ h_1(h_2, h_0) = h_0 + \dfrac{v_1 r_2(h_2, h_0)}{\cos\phi} & \end{cases}. \tag{2.72}$$

Assuming a Gaussian surface height PDF, we have

$$p_h(h_0) = \frac{1}{\sqrt{\pi}} \exp\left(-h_0^2\right) \quad \ln\left[F_h(\infty)\right] - \ln\left[F_h(h_0)\right] = -\ln\left[\frac{1+\mathrm{erf}(h_0)}{2}\right]. \tag{2.73}$$

In (2.71), h_2 ranges from h_0 to $h_{2,\max}$ with a sampling step Δh_2 and h'_2 ranges from $h'_{2,\min}$ to h_2 with a sampling step $\Delta h'_2$. Typically, $h_{2,\max} = 3$ since $\exp\left(-h^2_{2,\max}\right) \approx 0$, $h'_{2,\min} = -3$ and $\Delta h_2 = \Delta h'_2 = 0.01$. The accuracy of the two numerical integrations increases as the sampling steps decreases and are done from a conventional trapezoidal rule.

Figure 2.7 plots the coefficient c_0 versus the angle ϕ and h_0 and for given values of v_1 (constant on each column) and v_2. The full black curve plots the mean value of c_0 defined as

$$\begin{aligned}\bar{c}_0(v_1, v_2, \phi) &= \int_{-\infty}^{+\infty} c_0(v_1, v_2, \phi, h_0) p_h(h_0) \mathrm{d}h_0 \\ &= \frac{1}{\sqrt{\pi}} \int_{-\infty}^{+\infty} c_0(v_1, v_2, \phi, h_0) e^{-h_0^2} \mathrm{d}h_0,\end{aligned} \tag{2.74}$$

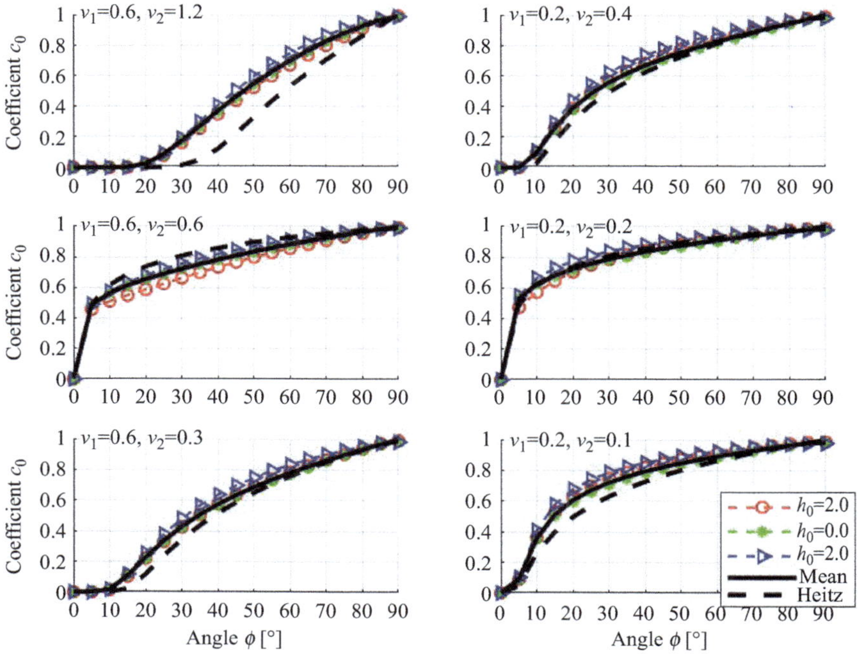

Figure 2.7 Coefficient c_0 versus the angle ϕ and h_0 and for given values of v_1 (constant on the left) and v_2 (constant on the right). The full black curve plots the mean value of c_0 and the dashed black curve the model of Heitz [8]

and the dashed black curve plots the empirical model of Heitz [8] expressed as

$$c_0 \approx \tilde{c}_0(\phi, v_1 - v_2) = \begin{cases} \dfrac{\ln\left(1 + \alpha\phi^{\beta}\right)}{\ln\left(1 + \alpha\left[\pi/2\right]^{\beta}\right)} & \phi \in [0; \pi/2[\\ 1 & \phi \in [\pi/2; \pi] \end{cases}, \begin{cases} \alpha = \dfrac{0.17}{|v_1 - v_2|^{10.49}} \\ \beta = 8.85 \end{cases}. \tag{2.75}$$

As we can see, the coefficient c_0 increases as ϕ increases and tends towards 1 for $\phi = \pi/2$. As v_1 or/and v_2 decreases, the coefficient increases, which means that the correlation is stronger. In other words, the less point is visible, the more correlation effect becomes significant. Figure 2.7 also reveals that the mean value $\bar{c}_0 \approx c_0(v_1, v_2, \phi, h_0 = 0)$, which accelerates the computation of $\bar{c}_0$.

2.2.4.2 Average shadowing functions

Figure 2.8 plots the bistatic average shadowing functions versus the angle $\phi = |\phi_2 - \phi_1|$ and for different values of the angles θ_1 and θ_2 given in the titles of the sub-figures. The simulation parameters are $\sigma_{\gamma_x} = 0.3$, $\sigma_{\gamma_y} = 0.2$ and $\phi_1 = 0$ ($\phi = |\phi_2|$). In the legend

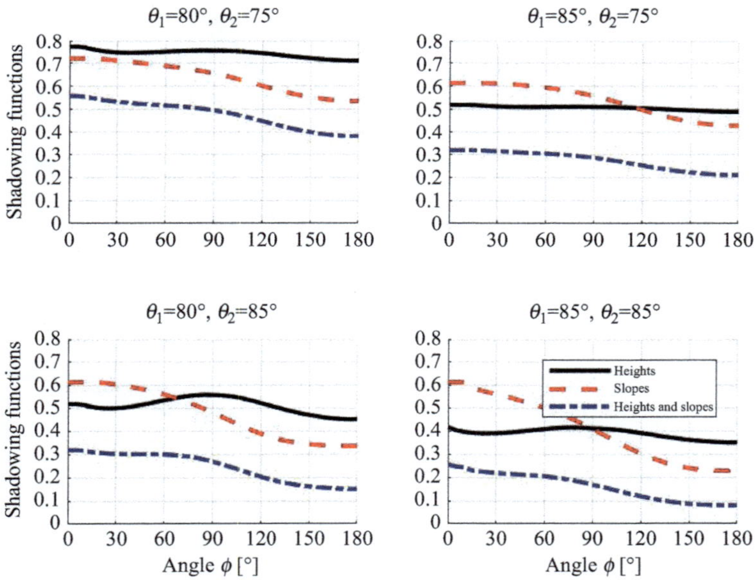

Figure 2.8 Bistatic average shadowing functions versus the angle $\phi = |\phi_2 - \phi_1|$ and for different values of the angles θ_1 and θ_2 given in the titles of the sub-figures. The simulation parameters are $\sigma_{\gamma_x} = 0.3$, $\sigma_{\gamma_y} = 0.2$ and $\phi_1 = 0$ ($\phi = |\phi_2|$)

- The label 'Heights' means that the statistical shadowing function is averaged over the surface heights expressed as

$$\int_{-\infty}^{+\infty} \left[\frac{1}{2}\text{erfc}(h_0)\right]^{L(v_1,v_2,\phi,h_0)} e^{-h_0^2}\text{d}h_0 = \bar{H}. \tag{2.76}$$

- The label 'Slopes' means that the statistical shadowing function is averaged over the surface slopes (2.68) leading to $\bar{\Upsilon}$.
- The label 'Heights and Slopes' means that the statistical shadowing function is averaged over the surface heights and slopes ((2.68) multiplied by (2.76)) ($\bar{\bar{S}}_{\text{B}}^1 = \bar{\Upsilon} \times \bar{H}$).

Figure 2.8 shows that $\bar{H}$ is not a monotonic function of ϕ, whereas $\bar{\Upsilon}$ decreases when ϕ grows. As shown previously, $\bar{\Upsilon}$ ranges from $[1 + \text{erf}(\min(v_1, v_2))]/2 = \bar{\Upsilon}_{\max}$ ($\phi = 0$) to $[\text{erf}(v_1) + \text{erf}(v_2)]/2 = \bar{\Upsilon}_{\min}$ ($\phi = \pi$).

For $\phi = 0$, the surface slope standard deviation $\sigma_{\gamma_X}(0) = \sigma_{\gamma_x}$, whereas for $\phi = \pi/2$, $\sigma_{\gamma_X}(\pi/2) = \sigma_{\gamma_y} < \sigma_{\gamma_x}$ and for $\phi = \pi$, $\sigma_{\gamma_X}(\pi) = \sigma_{\gamma_x}$. Then, $\sigma_{\gamma_X}(\phi)$ decreases as $\phi \in [0, \pi/2]$ decreases and next grows as $\phi \in]\pi/2; \pi]$ increases. The function $\Lambda(\mu_2, \phi_2) = \Lambda(v_2)$ follows the same behaviour.

For $\theta_1 > \theta_2$ and $\phi \in [0; \pi/2]$, as c_0 is an increasing function of ϕ and $\Lambda(v_2)$ is a decreasing function of ϕ, $L = c_0\Lambda_2 + \Lambda_1$ (Λ_1 is a constant), L can increase or decrease. For $\theta_2 < \theta_1$, this way also holds. For $\phi \in]\pi/2, \pi]$, $c_0 = 1\ \forall\phi$ and as Λ_2 decreases as ϕ grows, L also decreases. These statements explain the behaviour of $\bar{H}$ in Figure 2.8.

As expected, when θ_1 or/and θ_2 increase, the average shadowing function $\bar{\bar{S}}_{\text{B}}^1 = \bar{\Upsilon} \times \bar{H}$ decreases, and it is a decreasing function of ϕ.

Figure 2.9 plots the bistatic average shadowing functions over the heights versus the angle $\phi = |\phi_2 - \phi_1|$ and for different values of the angles θ_1 and θ_2 given in the titles. The simulation parameters are $\sigma_{\gamma_x} = 0.3$, $\sigma_{\gamma_y} = 0.2$ and $\phi_1 = 0$ ($\phi = |\phi_2|$). In addition, the relative difference against the 'Exact' results is plotted versus the angle $\phi = |\phi_2 - \phi_1|$. Figure 2.10 plots the same as in Figure 2.9 but θ_2 differs.

In the legend

- The label 'Exact' means that the shadowing is computed from (2.76) (no assumption).
- The label 'Mean' means that the shadowing is computed from the following equation (2.67):

$$\frac{1}{1 + \tilde{c}_0(v_p, v_q, \phi)\Lambda(v_q) + \Lambda(v_p)}, \tag{2.77}$$

  in which $\tilde{c}_0 = \bar{c}_0$ computed from (2.74).
- The label '$h_0 = 0$' means that the shadowing is computed from the above equation, in which $\tilde{c}_0 = c_0(v_1, v_2, \phi, 0)$ expressed by (2.71).

As shown in Figures 2.9 and 2.10, the two approximations give satisfactory results since the relative error does not exceed 0.15%. The advantage to use the '$h_0 = 0$'

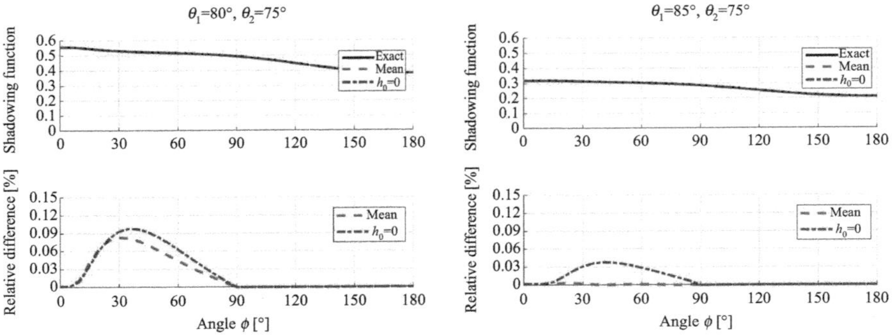

Figure 2.9 *Left panel and top: Bistatic average shadowing functions over the heights versus the angle $\phi = |\phi_2 - \phi_1|$ and for different values of the angles θ_1 and θ_2 given in the titles. The simulation parameters are: $\sigma_{\gamma_x} = 0.3$, $\sigma_{\gamma_y} = 0.2$ and $\phi_1 = 0$ ($\phi = |\phi_2|$). Left panel and bottom: Relative difference against the exact results versus the angle $\phi = |\phi_2 - \phi_1|$. Right panel: same as in the left panel but the angle θ_1 differs*

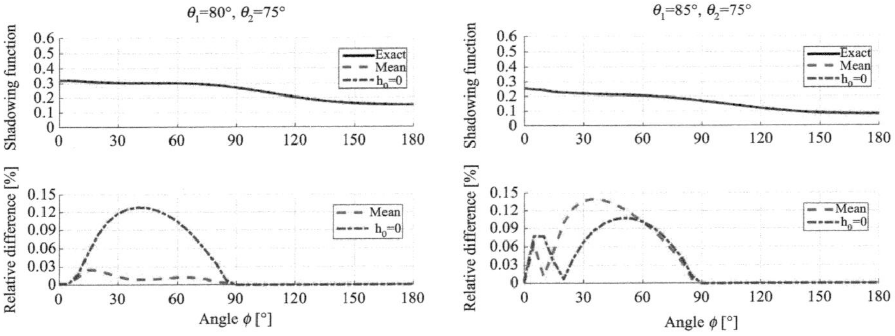

Figure 2.10 *Same as in Figure 2.9 but θ_2 differs*

approximation is that the evaluation of $\tilde{c}_0$ requires 2-fold numerical integrations, whereas the other one needs 3-fold numerical integrations.

2.2.4.3 Evaluation of a special function

In the numerator of (2.71), the following integral

$$G(x_0; a, m) = \int_{-\infty}^{x_0} g(x; a, m)\mathrm{d}x \text{ where } \begin{cases} p(x; a) = \sqrt{\dfrac{a}{\pi}} e^{-ax^2} \\ F(x; m) = \dfrac{1 + \mathrm{erf}(m - x)}{2} \\ g(x; a, m) = p(x; a)F(x; m) \end{cases}, \quad (2.78)$$

must be evaluated, in which $a \geq 0$ (real number).

Setting $u(x) = \int p(x;a)\mathrm{d}x = \mathrm{erf}(\sqrt{a}x)/2$ and $v(x) = F(x;a)$ and making an integration by parts, we can show that

$$\begin{aligned} G(x_0;a,m) &= [u(x)v(x)]_{-\infty}^{x_0} - \int_{-\infty}^{x_0} u(x)v'(x)\mathrm{d}x \\ &= \frac{1}{4}\left[\mathrm{erf}(\sqrt{a}x)\{1+\mathrm{erf}(m-x)\}\right]_{-\infty}^{x_0} + \frac{1}{2\sqrt{\pi}}\int_{-\infty}^{x_0} \mathrm{erf}(\sqrt{a}x)e^{-(m-x)^2}\mathrm{d}x \\ &= \frac{1}{4}\left\{\mathrm{erf}(\sqrt{a}x_0)\left[1+\mathrm{erf}(m-x_0)\right]+2\right\} \\ &\quad - \frac{1}{2\sqrt{a\pi}}\int_{-\infty}^{y_0} \mathrm{erf}(m\sqrt{a}-y)e^{-y^2/a}\mathrm{d}y \\ &= \frac{1}{4}\left\{\mathrm{erf}(\sqrt{a}x_0)\left[1+\mathrm{erf}(m-x_0)\right]+2\right\} - \frac{1}{2}G\left((m-x_0)\sqrt{a};\frac{1}{a},m\sqrt{a}\right), \end{aligned} \tag{2.79}$$

where $y = (m-x)\sqrt{a}$. First, particular cases are investigated.

Case 1: $a = 1$ and $m = 0$
For this case, using the variable transformation $u = F(x;0) = \mathrm{erf}(x)$, we obtain

$$G(x_0;1,0) = \frac{1}{2\sqrt{\pi}}\int_{-\infty}^{x_0} e^{-x^2}\frac{1-\mathrm{erf}(x)}{2}\mathrm{d}x = \frac{2\mathrm{erf}(x_0)-\mathrm{erf}^2(x_0)+3}{8}. \tag{2.80}$$

Case 2: $a \to \infty$
For this case $p(x;a) = \delta(x)$, where δ is Dirac delta distribution, which implies that

$$G(x_0;\infty,m) = \int_{-\infty}^{x_0} \delta(x)\frac{1+\mathrm{erf}(m-x)}{2}\mathrm{d}x = \begin{cases} \dfrac{1+\mathrm{erf}(m)}{2} & x_0 \geq 0 \\ 0 & x_0 < 0 \end{cases}. \tag{2.81}$$

Case 3: $m \to \pm\infty$
For this case $F(x;\pm\infty) = [1+\mathrm{sign}(m)]/2$, where the symbol sign stands for the sign function, which implies that

$$G(x_0;a,\pm\infty) = \frac{1+\mathrm{sign}(m)}{2}\int_{-\infty}^{x_0} p(x;a)\mathrm{d}x = \frac{1+\mathrm{sign}(m)}{4}\left[1+\mathrm{erf}(\sqrt{a}x_0)\right]. \tag{2.82}$$

General property
If f is a decreasing function over x and non-negative throughout the interval $[x_1;x_2]$ and g is integrable over $x \in [x_1;x_2]$, then the second mean-value theorem ([26] page 243) states that

$$\int_{x_1}^{x_2} f(x)g(x)\mathrm{d}x = f(x_1)\int_{x_1}^{\xi_1} g(x)\mathrm{d}x, \quad x_1 \leq \xi_1 \leq x_2. \tag{2.83}$$

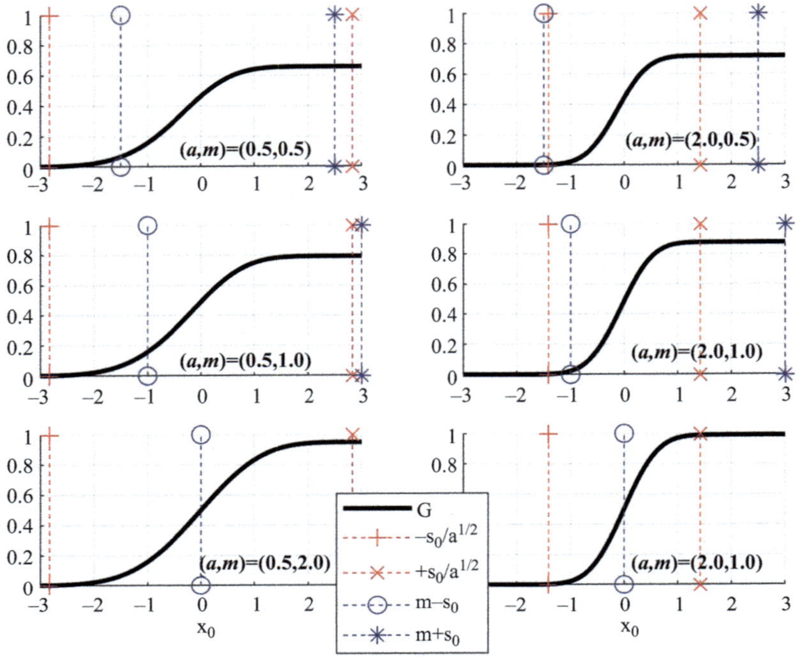

Figure 2.11 *In full black curve, G (defined by (2.78)) versus x_0 for different values of a and $m > 0$. In addition, the other curves show particular values*

Setting $g(x) = p(x; a), f(x) = F(x; m) \geq 0$, $x_2 = x_0$ and $x_1 = -\infty$, the application of the above theorem leads to

$$G(x_0; a, m) = \sqrt{\frac{a}{\pi}} F(-\infty; m) \int_{-\infty}^{\xi_0} e^{-ax^2} \mathrm{d}x = \frac{1 + \operatorname{erf}(\sqrt{a}\xi_0)}{2}, \tag{2.84}$$

where $\xi_0 = \xi_1 \in] - \infty; x_0]$ and depends on x_0, a and m.

Figures 2.11 and 2.12 plot G (defined by (2.78)) versus x_0 for different values of a and for $m > 0$ and $m < 0$, respectively. G is computed numerically from a conventional trapezoidal rule.

As we can see, G behaves as the erf function defined by (2.84) and ranges from 0 to 1. For $m > 0$ and, as m and/or a grow, the asymptotic value obtained for $x_0 \to \infty$ increases, whereas for $m < 0$, this remark does not hold. As expected, as $x_0 \to -\infty$, G tends towards zero. More precisely, for $x_0 < -s_0/\sqrt{a}$, G vanishes, whereas for $x_0 > s_0/\sqrt{a}$, the function G becomes a constant.

Figures 2.13 and 2.14 plot $w_0 = \xi_0\sqrt{a} = \operatorname{erfinv}(2G - 1)$ computed from G and (2.84) versus x_0 for different values of a and for $m > 0$ and $m < 0$, respectively. The symbol erfinv stands for the inverse error function ($\operatorname{erf}(\operatorname{erfinv}(x)) = x$).

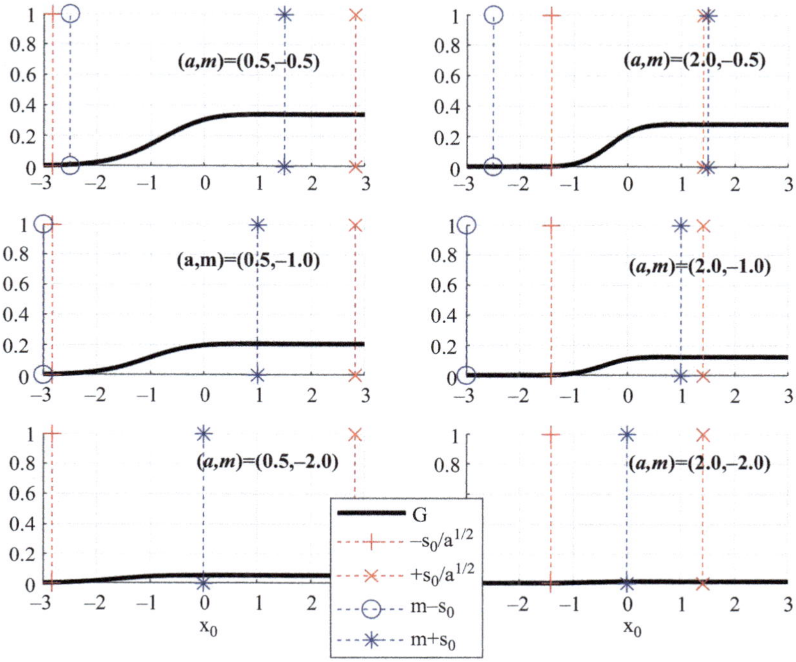

Figure 2.12 Same variations as in Figure 2.11 but for $m < 0$

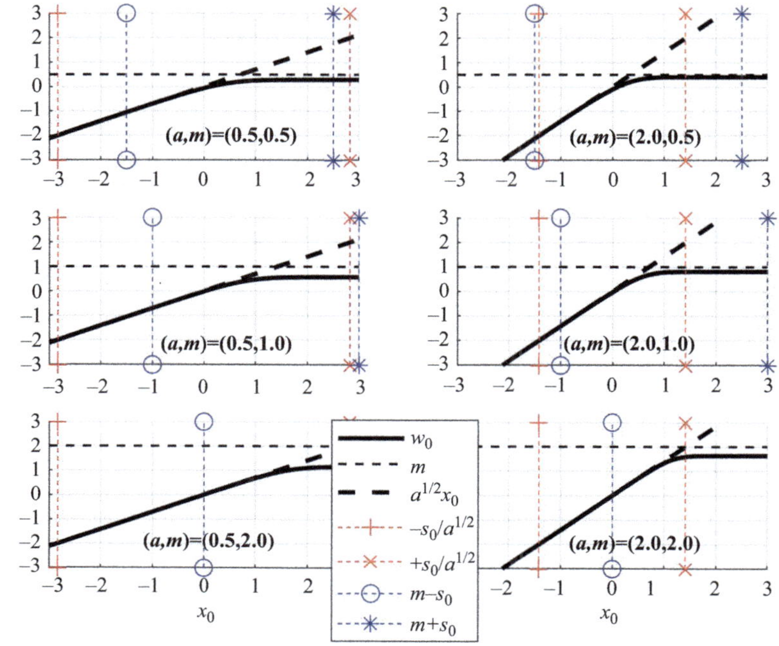

Figure 2.13 $w_0 = \xi_0\sqrt{a} = erfinv(2G\text{-}1)$ computed from G and (2.84) versus x_0 for different values of a and $m > 0$. In addition, the other curves show particular values

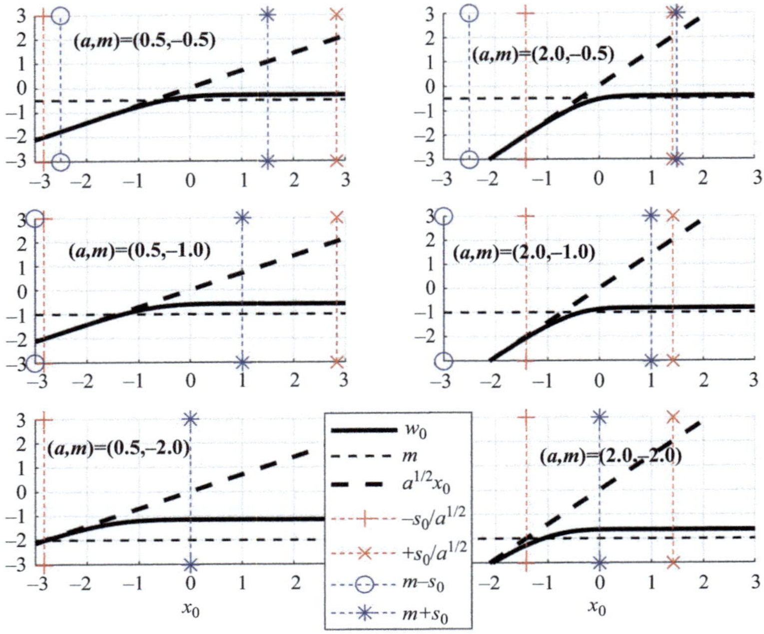

Figure 2.14 Same variations as in Figure 2.13 but for m < 0

For $m > 0$, Figure 2.13 shows that as m or/and a grow, w_0 behaves as $\sqrt{a}x_0$ over the range $x_0 \in]-\infty; x_{0,\max}]$, in which $x_{0,\max}$ increases. In addition as a or/and m increases, the asymptotic value of w_0 tends to m, which is consistent with (2.81) (and (2.82)). For $m < 0$, Figure 2.14 shows that as a grows, the asymptotic value of w_0 tends also to m and more rapidly as m increases.

Primitive of $G(x; a, m)$

From (2.84), since G is expressed from the erf function, the derivative of G, $\tilde{g}$, defined as $\tilde{g} = \partial G/\partial x_0$, can be written as

$$\tilde{g}(x; a, m) = \frac{1}{\sigma_1\sqrt{\pi}} \exp\left(-\left[\frac{x - m_1}{\sigma_1}\right]^2\right)$$
$$\Rightarrow \tilde{G}(x_0; a, m) = \int_{-\infty}^{x_0} \tilde{g}(x; a, m)\mathrm{d}x = \frac{1}{2}\left[1 + \mathrm{erf}\left(\frac{x - m_1}{\sigma_1}\right)\right], \quad (2.85)$$

where the mean value m_1 and σ_1 depends on a, m and x_0.

The identification of (2.85) with (2.84) leads to

$$\sqrt{a}\xi_0 = w_0 = \frac{x_0 - m_1}{\sigma_1} \quad \frac{1}{\sigma_1} = \frac{\partial w_0}{\partial x_0} = \sqrt{a}\frac{\partial \xi_0}{\partial x_0}, \quad (2.86)$$

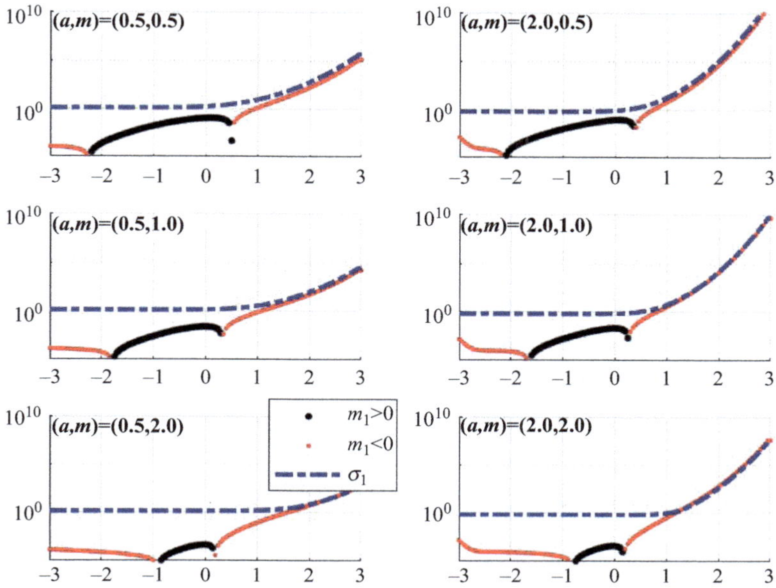

Figure 2.15 σ_1 *and* $m_1 = x_0 - w_0\sigma_1$ *(defined by (2.86)) versus* x_0 *for different values of a and* $m > 0$*. The y axis is log-scaled, and then the cases* $m_1 = |m_1| > 0$ *and* $m_1 = -|m_1| < 0$ *are plotted with different colours*

because the partial derivative of (2.84) gives

$$\sqrt{\frac{a}{\pi}}\frac{\partial \xi_0}{\partial x_0}e^{-a\xi_0^2} = \text{Equation (2.85)} = \frac{1}{\sigma_1\sqrt{\pi}}\exp\left(-\left[\frac{x-m_1}{\sigma_1}\right]^2\right). \tag{2.87}$$

Figure 2.15 plots σ_1 and $m_1 = x_0 - w_0\sigma_1$ (defined by (2.86)) versus x_0 for different values of a and $m > 0$. The y axis is log-scaled, and then the cases $m_1 = |m_1| > 0$ and $m_1 = -|m_1| < 0$ are plotted with different colours. As we can see, for negative values of x_0, m_1 is close to zero, whereas σ_1 is a constant (equals $1/\sqrt{a}$ from Figure 2.13) and as x_0 grows, both m_1 and σ_1 increase exponentially, for which their ratio is a constant. Indeed, $w_0 = (x_0 - m_1)/\sigma_1 = x_0/\sigma_1 - m_1/\sigma_1 \to -m_1/\sigma_1 = |m_1|/\sigma_1$, which is consistent with Figure 2.13.

Figure 2.16 plots g (defined by (2.78)), $\tilde{g}$ (defined by (2.85)) and $\tilde{g}_\infty$ for different values of a and $m > 0$. As we can see, the function g and $\tilde{g}$ perfectly match, which means that g can be modelled as a Gaussian function.

The function $\tilde{g}_\infty$ equals $\tilde{g}$, in which m_1 and σ_1 are substituted by constant values $m_{1,\infty}$ and $\sigma_{1,\infty}$ (independent of x_0), respectively. We can note that

$$\tilde{a}_1 = \int_{-\infty}^{+\infty} x\tilde{g}_\infty(x)\mathrm{d}x = m_{1,\infty} \quad \tilde{a}_2 = \int_{-\infty}^{+\infty} x^2\tilde{g}_\infty(x)\mathrm{d}x = m_{1,\infty}^2 + \frac{\sigma_{1,\infty}^2}{2}, \tag{2.88}$$

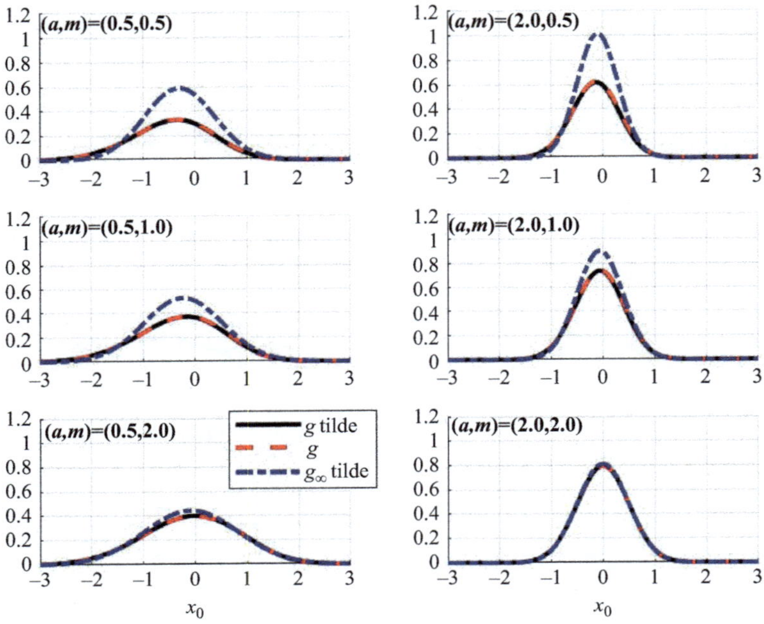

Figure 2.16 g, $\tilde{g}$ *and* $\tilde{g}_\infty$ *versus* x_0 *for different values of a and* $m > 0$

and

$$a_1 = \int_{-\infty}^{+\infty} xg(x)\mathrm{d}x \quad a_2 = \int_{-\infty}^{+\infty} x^2 g(x)\mathrm{d}x. \tag{2.89}$$

From (2.89), a_1 and a_2 are calculated numerically from the trapezoidal method and from (2.88), $m_{1,\infty} = \tilde{a}_1 = a_1$ and $\sigma_{1,\infty}^2 = 2(\tilde{a}_2 - \tilde{a}_1^2) = 2(a_2 - a_1^2)$.

Figure 2.16 shows that $\tilde{g}_\infty$ differs from $\tilde{g}$, which clearly shows that m_1 and σ_1 depend on x_0, which makes their derivation very complicated. To find a relation between the m_1 and σ_1, different ways have been investigated.

Modelling of m_1 and σ_1

The first way to model m_1 and σ_1 is to identify the function g (2.78) with $\tilde{g}$ (2.85), leading to

$$\sqrt{\frac{a}{\pi}} e^{-ax^2} \frac{1 + \mathrm{erf}(m - x)}{2} = \frac{1}{\sigma_1 \sqrt{\pi}} \exp\left(-\left[\frac{x - m_1}{\sigma_1}\right]^2\right), \tag{2.90}$$

where this equality must be satisfied for any x. Taking the logarithmic of the above equation, we have

$$-ax^2 + \ln\left[\frac{1 + \mathrm{erf}(m - x)}{2}\right] + \frac{1}{2}\ln(a) = -\left(\frac{x - m_1}{\sigma_1}\right)^2 - \ln(\sigma_1). \tag{2.91}$$

If $m - x \gg 1$, then $\mathrm{erf}(m - x) \approx 1$, and the above equation becomes

$$ax^2 - \frac{1}{2}\ln(a) \approx \left(\frac{x}{\sigma_1} - \frac{m_1}{\sigma_1}\right)^2 + \ln(\sigma_1) \quad \text{for} \quad m - x \gg 1. \tag{2.92}$$

Assuming that $x \gg m_1$, then the above equation is satisfied if $a = 1/\sigma_1^2$ or $\sigma_1 = 1/\sqrt{a}$. This is consistent with Figures 2.13 and 2.14.

As expected, (2.91) shows that the equality is satisfied for any x if the set (m_1, σ_1) depends on x. A possible set of functions is

$$\begin{cases} ax^2 = \left[\dfrac{x - m_1(x)}{\sigma_1(x)}\right]^2 \Rightarrow \sigma_1^2(x) = \dfrac{1}{a}\left[1 - \dfrac{m_1(x)}{x}\right]^2 = \dfrac{1}{a}\dfrac{4}{[1 + \mathrm{erf}(m - x)]^2} \\ \sqrt{a}\left[\dfrac{1 + \mathrm{erf}(m - x)}{2}\right] = \dfrac{1}{\sigma_1(x)} \Rightarrow \dfrac{2}{1 + \mathrm{erf}(m - x)} = \left|1 - \dfrac{m_1(x)}{x}\right| \end{cases}. \tag{2.93}$$

The first line of the above equation gives an expression of $\sigma_1(x)$ versus a, m and x. Unfortunately, the numerical results showed that this equation is only valid for $m - x \ll 1$. A second attempt is to make a second-order Taylor series expansion near $x = x_0$ by assuming that m_1 and σ_1 are constant near x_0 and match the terms of same powers on x, leading to three equations. Different values of x_0 (like $x_0 = m$ and $x_0 = m_1$) have been tested, but the numerical results are not conclusive.

The second way uses the property that the integral of the function $x^n g$ can be derived analytically for only odd positive n integers. In addition, the resulting integral is identified to that of $x^n \tilde{g}$ to derive m_1 $(n = 1)$ and σ_1 $(n = 3)$.

From (2.85), we have

$$\begin{cases} \tilde{a}_1(a, m) = \displaystyle\int_{-\infty}^{+\infty} x\tilde{g}(x; a, m)\mathrm{d}x = m_1 \\ \tilde{a}_3(a, m) = \displaystyle\int_{-\infty}^{+\infty} x^3\tilde{g}(x; a, m)\mathrm{d}x = m_1^3 + \dfrac{3}{2}m_1\sigma_1^2 \Rightarrow \sigma_1^2 = \dfrac{2(\tilde{a}_3 - m_1^3)}{3m_1} \end{cases}. \tag{2.94}$$

In addition, from (2.78) and doing an integration by parts, we can show

$$\begin{aligned} a_1(a, m) &= \int_{-\infty}^{+\infty} xg(x; a, m)\mathrm{d}x = \frac{1}{2}\sqrt{\frac{a}{\pi}}\int_{-\infty}^{+\infty} xe^{-ax^2}\left[1 + \mathrm{erf}(m - x)\right]\mathrm{d}x \\ &= -\sqrt{\frac{a}{\pi}}\int_{-\infty}^{+\infty} xe^{-ax^2}\mathrm{erf}(x - m)\mathrm{d}x \\ &= -\frac{1}{4}\sqrt{\frac{a}{\pi}}\left[\frac{e^{-(am^2/(a+1))}}{a\sqrt{a+1}}\mathrm{erf}\left(\frac{x(a+1) - m}{\sqrt{a+1}}\right) - \frac{e^{-ax^2}\mathrm{erf}(x - m)}{a}\right]_{x=-\infty}^{x=+\infty} \\ &= -\frac{e^{-\eta^2}}{2\sqrt{a\pi}\,(a+1)}, \quad \eta = m\sqrt{\frac{a}{a+1}}, \end{aligned} \tag{2.95}$$

and

$$\begin{aligned} a_3(a,m) &= \int_{-\infty}^{+\infty} x^3 g(x;a,m)\mathrm{d}x = -\sqrt{a}\frac{\partial}{\partial a}\left(\frac{a_1}{\sqrt{a}}\right) \\ &= -\frac{e^{-(am^2/(a+1))}}{4\sqrt{\pi}\,[a(a+1)]^{3/2}}\left(2+3a+2\eta^2\right). \end{aligned} \tag{2.96}$$

Since $a_{1,3} = \tilde{a}_{1,3}$, the identifications of (2.96) and (2.95) with (2.94) lead to

$$\left\{ \begin{aligned} m_1(m,a) &= -\frac{e^{-\eta^2}}{2\sqrt{a\pi}\,(a+1)},\ \eta = m\sqrt{\frac{a}{a+1}} \\ \sigma_1^2(m,a) &= \frac{1}{3a(a+1)}\left(2+3a+2\eta^2 - \frac{e^{-\eta^2}}{2\pi}\right) \end{aligned} \right. . \tag{2.97}$$

Numerical tests showed that m_1 and σ_1 do not coincide with those computed numerically.

This issue is under investigation.

Chapter 3
Shadowing function with multiple reflections

Hongkun Li[1] *and Christophe Bourlier*[1]

3.1 Monostatic case from a 1D surface

In this chapter, multiple reflections from a randomly rough surface is addressed. Multiple reflections occur when an incident ray coming from the upper space interacts with the surface or when an emitted ray originating from a facet of the rough surface reflects multiple times in the surface before it finally heads into the upper space.

For instance, multiple surface reflections are very important when studying the sea surface infrared emissivity and reflectivity at grazing incidence angles. It is reported that models of sea surface emissivity without considering multiple surface reflections underestimate the surface emissivity [27–29], and energy conservation is not well fulfilled [30] (see also Chapter 4). The key to the problem lies in deriving the shadowing function with surface reflections.

When the number of the reflections N equals 0 and 1 for the monostatic and bistatic cases, respectively, the shadowing function with multiple reflections corresponds to the case without reflection, which is thoroughly discussed in Chapters 1 and 2. Published models of shadowing function with multiple reflections are derived from the ones without reflection.

In this section, the monostatic shadowing function with multiple reflections is derived. A monostatic configuration implies that the ray of interest originates from a surface facet and only one receiver is involved. Several important models about shadowing function with single reflection are reviewed in Sections 3.1.3–3.1.5. Next, Section 3.1.6 extends the formulation to higher order reflections.

3.1.1 Introduction of the problem

The monostatic shadowing function with N reflections gives the probability that a ray emitted by a facet of the surface is reflected N times before it is captured by the receiver. Figure 3.1 shows the case when one surface reflection occurs.

A ray coming from the source point M_1 propagates along the direction $\hat{s}'(\theta_1)$ with a zenith angle $\theta_1' \in [0°; 180°]$ and then intersects the surface at another point M_0

[1]IETR Laboratory, CNRS, Nantes, France

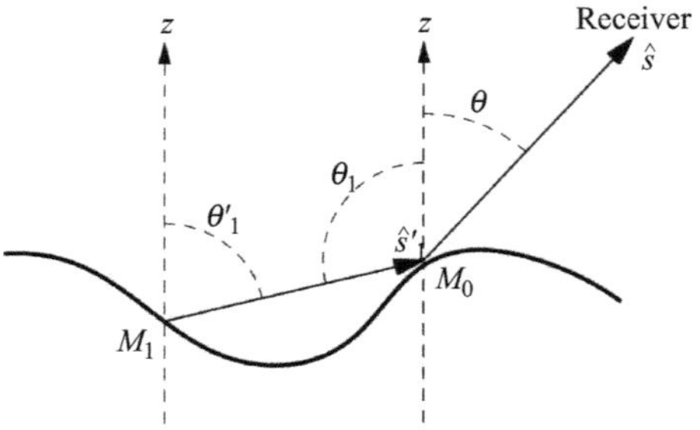

Figure 3.1 Illustration of one surface reflection: the emitted ray form the source point M_1 propagates along the direction $\hat{s}'$ and intersects the surface at M_0 where it is reflected into the receiver direction $\hat{s}$

where it is reflected to the receiver along the direction $\hat{s}(\theta)$ of angle $\theta \in [0°; 180°]$. Following the discussion in Chapter 1, the point M_1 lies in shadow and M_0 is viewed by the receiver.

Imagine that the task is to derive the energy measured by a receiver when looking at the rough surface with the angle θ. First, the energy emitted by the point M_0 is taken into account. However, the energy coming from the point M_1 which is reflected by the point M_0 towards the receiver must be also accounted for. Furthermore, additional reflections may also occur and the corresponding energy must be added. The key to the problem is to derive a shadowing function with N reflections to evaluate the probability of the occurrence of such reflections.

It is obvious that for any number of reflections, no obstacle must be present between any two adjacent intersections (first event) and next between M_0 and the receiver (second event) so that the ray is viewed. For the second event, the associated probability is expressed from the statistical monostatic shadowing function thoroughly investigated in Chapter 1. For the first event, since the physical process is similar to the second event, it is not surprising that the statistical shadowing function with multiple reflections is expressed from that without reflection.

For the sake of clarity, first, only one reflection is considered. For a monostatic configuration and adopting the notation in Figure 3.1, the shadowing function with one reflection can be defined as the probability that the ray $M_0(\hat{s})$ of starting point M_0 and of propagation direction $\hat{s}$ does not intersect the surface while the incident ray of direction $\hat{s}'$ comes from a surface point M_1. Mathematically, it is given by

$$\begin{aligned} S_M^1 &= p(\hat{s}(\theta) \text{ does not intersect} \cap \text{Incident ray from surface}) \\ &\approx p(\hat{s}(\theta) \text{ does not intersect}) \times p(\text{Incident ray from surface}) \\ &= S_M^0 \times P_1. \end{aligned} \tag{3.1}$$

For now, let us consider that the events '$M_0(\hat{s})$ does not intersect the surface' and 'the incident ray originates from the surface' are independent (the correlation will be addressed in Section 3.1.5.1). The probability that '$M_0(\hat{s})$ does not intersect the surface' is expressed from the monostatic statistical shadowing function without reflection S_M^0. The probability that 'the incident ray originates from the surface' is unknown and is noted as P_1 hereafter.

The key of the problem now lies in determining the probability P_1. Or inversely, imagine that the ray is emitted from the receiver and propagates along the direction $-\hat{s}$. What is the chance that after its first intersection with the surface point M_0, its reflected ray of direction $-\hat{s}'$ intersects the surface at an arbitrary point M_1? Answering this question is the main task of this section.

In the literature, there are several models which try to solve this issue. In the following subsections, several analytical models, from the simplest to the most complicated, are reported. First of all, a numerical ray-tracing algorithm is built to serve as a reference to evaluate the accuracy of the models.

3.1.2 Numerical ray-tracing algorithm

The derivation of the monostatic shadowing function with multiple reflections from a randomly rough surface is a difficult task. A Monte Carlo ray-tracing algorithm is a relevant tool to validate the closed expressions. All the models are based on the geometry optics approximation, but no additional assumptions are required with the ray-tracing algorithm. In addition, such an algorithm can help us to understand the shadowing mechanisms. In this section, a one-dimensional (1D) surface is assumed, which means that the rough surface depends only on the abscissa x.

This algorithm is an extension of the one introduced in Chapter 1. First, the rough surface is generated from a given surface height autocorrelation function (or its associated height spectrum) and by assuming a Gaussian process (of the surface heights and slopes). For the next simulations, the autocorrelation function is assumed to be Gaussian. Readers are free to use other autocorrelation functions or to assume that the surface points are uncorrelated [28]. From the ray-tracing algorithm, all surface points which are illuminated directly by the receiver are found following the same way as in Section 1.5.

The Monte Carlo process means that a large number N of randomly rough surfaces of same statistical features (only the seed of the Gaussian process changes) are generated, on which the ray-tracing algorithm is performed. Height and slope histograms of the points M_0 and M_1 and the associated mean values are computed by averaging over the results obtained from each surface. In this section, N equals 3,000 to reach the convergence.

3.1.2.1 Definition of the geometry

For the ray-tracing algorithm, the propagation reverse path is used for easy implementation. Figure 3.2 shows the case of one surface reflection. A ray emitted by the receiver propagates along the direction $\hat{s}_i$ and intersects the surface at the point M_0 where it is reflected in the specular direction $\hat{s}_r$. The subscript 'i' stands for incidence, and 'r' stands for reflection.

In Figure 3.2, the incident direction $\hat{s}_i$ is expressed from the zenith angle θ as

$$\hat{s}_i = -\begin{bmatrix} \sin\theta \\ \cos\theta \end{bmatrix}. \tag{3.2}$$

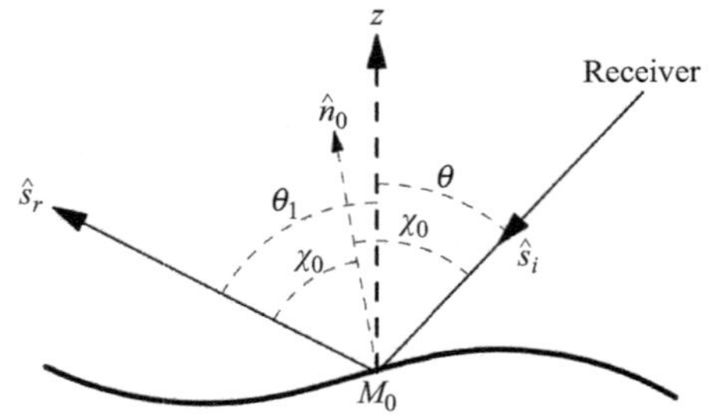

Figure 3.2 Reflection from the propagation reverse path: a ray emitted by the receiver propagates along the direction $\hat{s}_i$ and intersects the surface at the point M_0 where it is reflected in the specular direction $\hat{s}_r$

The unit normal vector to the point M_0 is defined from the surface slope γ_0 as

$$\hat{n}_0 = \frac{1}{\sqrt{1+\gamma_0^2}}\begin{bmatrix}-\gamma_0\\ 1\end{bmatrix}. \tag{3.3}$$

The local angle of incidence χ_0 is then given by

$$\cos\chi_0 = -\hat{s}_i \cdot \hat{n}_0 = \frac{\cos\theta - \gamma_0 \sin\theta}{\sqrt{1+\gamma_0^2}}. \tag{3.4}$$

Since physically the local angle of incidence χ_0 should not exceed 90° – otherwise the incident ray comes from the back of the surface – the condition $\cos\chi_0 \geq 0$ must be satisfied, or equivalently from (3.4)

$$\cos\theta - \gamma_0 \sin\theta \geq 0 \Rightarrow \gamma_0 \leq \cot\theta. \tag{3.5}$$

Equation (3.5) is a relevant criterion when deriving the shadowing function. A surface reflection may occur only if the slope of the surface facet γ_0 does not exceed the slope of the incident ray $\cot\theta$.

With the knowledge of the incident direction $\hat{s}_i$ and the surface normal vector $\hat{n}_0$, the specular direction is defined as

$$\hat{s}_r = 2\hat{n}_0\left(\hat{n}_0 \cdot \hat{s}_i\right) + \hat{s}_i = \begin{bmatrix}\sin\theta_1\\ \cos\theta_1\end{bmatrix}. \tag{3.6}$$

3.1.2.2 Implementation

The ray-tracing algorithm is a generalization of that developed without reflection (Section 1.5). As shown in Figure 3.2 and for sake of clarity, the receiver is located on the right-hand side of the rough surface. For an even surface, there is no difference whether the receiver is located on the left or on the right.

The first step is to generate a rough surface according to the desired statistical features, then apply the ray-tracing algorithm to locate all the surface points which are illuminated directly by the receiver, or which are not in shadow along the observation angle θ. An illustration of an illuminated point can be shown in Figure 3.2 M_0.

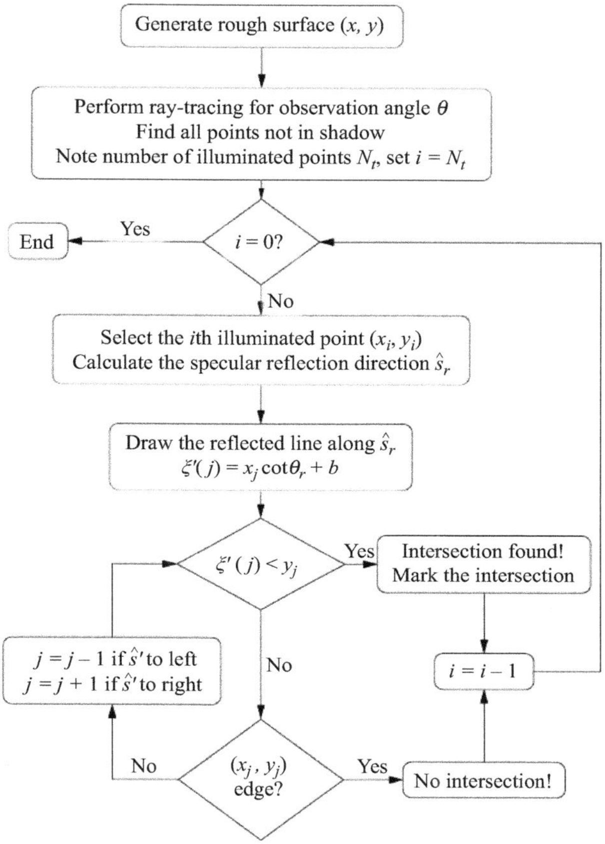

Figure 3.3 Illustration of the ray-tracing algorithm with one surface reflection

The second step is to determine if the reflected ray at any illuminated point obtained in the previous step intersects the surface again. The specular reflection direction $\hat{s}_r$ can be derived from (3.6). The reflected ray $M_0(\hat{s}_r)$ can then be drawn. Mathematically, the reflected ray is given by

$$\begin{cases} z = (x - x_0)\cot\theta_1 + z_0 \\ x < x_0 \text{ if } \hat{s}_r \text{ propagates to the left} \\ x > x_0 \text{ if } \hat{s}_r \text{ propagates to the right} \end{cases}, \tag{3.7}$$

where M_0 has Cartesian coordinates (x_0, z_0). Note that θ_1 is negative since it is measured anticlockwise from the zenith direction. Then this ray is traced to determine whether it intersects the surface at a point other than M_0. If so, each intersection is denoted as M_1. The points M_0 and M_1 forms a group. At the end of this step, all illuminated points M_0 for which the reflected ray does not intersect the surface elsewhere are removed. A detail illustration of the algorithm with one reflection is shown in Figure 3.3.

The statistical properties of the point M_0 can be studied, such as distributions of its height and slope, which point out what slopes or heights are more likely to lead to one surface reflection. This corresponds to the marginal height or slope histogram of the monostatic shadowing function with one reflection. The ratio of the number of groups N_1 of the points M_0 and M_1, or equally the number of remaining points M_0, over the total number of the surface points N_{total} corresponds to the average monostatic shadowing function with one reflection

$$\bar{\bar{S}}_M^1(\theta) = \frac{N_1}{N_{\text{total}}}. \tag{3.8}$$

The average monostatic shadowing function with one reflection gives the percentage of the surface where one surface reflection occurs.

If more reflections occur, step two is repeated with M_0 replaced by M_i where $i+1$ equals the number of reflections to be considered. For example, if $i=1$, then two reflections are considered. For N reflections, all intersections, $M_0, M_1, \ldots, M_i$ must be grouped. Similarly, the associated marginal height or slope histograms as well as the corresponding average shadowing function can be performed.

A ray-tracing algorithm allows us to determine the heights and slopes of the point M_i. In addition, it is easy to determine the direction of a given reflected ray and furthermore whether this ray intersects the surface again. If so, information about the new intersection is also available, which make it possible to determine if more reflections occur. For an analytical model, this information are generally unavailable, especially when two or more reflections are involved. A more detailed discussion is addressed in Section 3.1.6.

3.1.2.3 Numerical results

Illustration

Figure 3.4 shows a piece of a rough surface, in which the ray-tracing algorithm is applied. The blue lines are the incident rays coming from the receiver. These rays intersect the surface at points marked by the red symbol '+'. The reflected rays are marked in red.

In Figure 3.4, there are two types of reflected rays according to its propagation direction: the upwards ones and the downwards ones. It is clear that a ray propagating downwards can intersect the surface again. Attention should be paid to those propagating upwards. Published models are usually built by determining how likely a ray of such kind intersects the surface again.

In Figure 3.4, all the reflected rays propagate to the left. In fact, the number of the left-propagated reflected rays are large when the observation angle θ is large. To have a reflected ray propagating to the right, the slope of the first intersection should be very large, which is not likely to occur. When reducing θ, the number of reflected rays which propagate to the right increases, as the required slope becomes smaller.

Histograms of the first intersection

Figure 3.5 shows the height and slope histograms of the first intersection obtained by the ray-tracing algorithm. The incidence angle $\theta = 80°$, the surface slope root-mean-square (RMS) $\sigma_\gamma = 0.3$ and the number of surface realizations is $N = 3,000$.

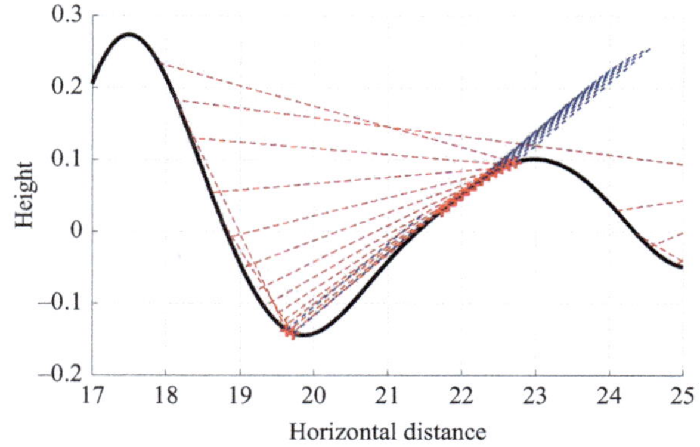

Figure 3.4 *Illustration of the ray-tracing algorithm with one surface reflection. $\theta = 85°$, the incident rays come from the receiver marked in blue, the reflected rays marked in red, and the first intersection marked by the red symbol '+'*

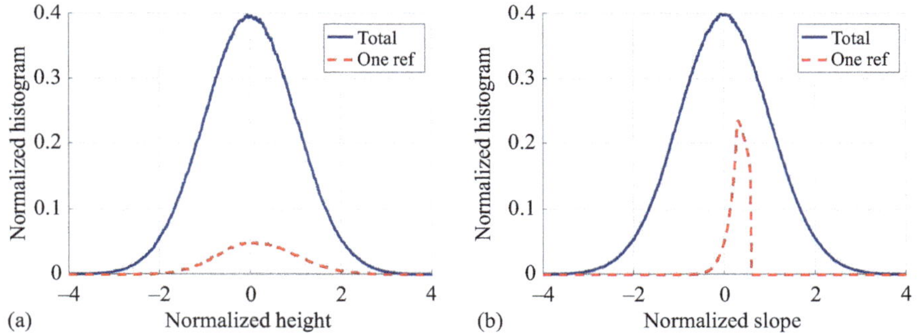

Figure 3.5 *Height (a) and slope (b) histograms of the first intersection obtained from a Monte Carlo ray-tracing algorithm. The incidence angle $\theta = 80°$, the surface RMS $\sigma_\gamma = 0.3$ and the number of surface realizations $N = 3{,}000$*

These points are marked with the red symbol '+' in Figure 3.4. The surface heights and slopes are normalized by their respective RMS value σ_ξ and σ_γ. In addition, the histograms are normalized so that their integration over the normalized heights or slopes of the whole surface equals one.

As expected, Figure 3.5(a) shows that the height histogram of the whole surface follows a Gaussian distribution. The height histogram of the first intersection still has a bell shape, but its magnitude is significantly reduced since the number of points leading to one surface reflection is smaller than that of the whole surface.

Figure 3.5(b) shows that he slope histogram of the whole surface follows a Gaussian distribution. However, the one for the first intersection differs from zero only on

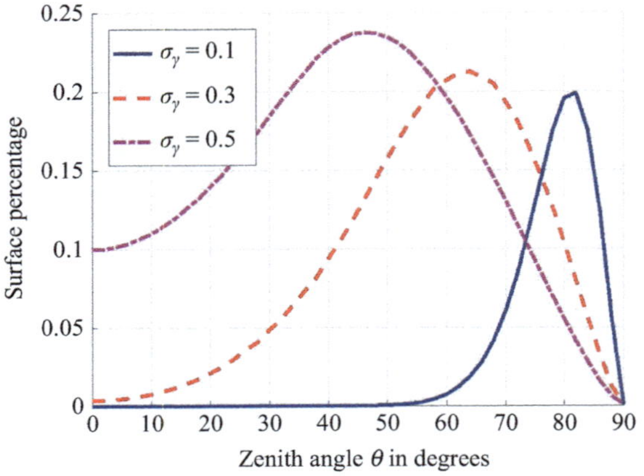

Figure 3.6 Area percentage of the surface leading to one surface reflection computed from a Monte Carlo ray-tracing algorithm. The surface RMS slopes are $\sigma_\gamma = 0.1$, 0.3 and 0.5 and the number of surface realizations is $N = 3{,}000$

a small range. In fact, in Figure 3.4, it is already shown that most of the first intersections marked by the red symbol '+' have slopes close to the incident ray slope. In other words, all slopes do not have an equal probability to give rise to a surface reflection (some are more promising than the others). This point is crucial to derive physical quantities, which depend on the facet slope, such as the sea surface infrared emissivity with one reflection.

Percentage of area leading to reflections

The percentage of the surface area where one reflection occurs can be calculated by dividing the number of the remained points M_0 by the total number of the surface points (Equation (3.8)). This value can be also computed by integrating the height or the slope histogram (shown in Figure 3.5) over the surface normalized heights or slopes.

Figure 3.6 shows the Monte Carlo ray-tracing results of the area percentage of rough surfaces leading to one surface reflection. The surface RMS slopes are $\sigma_\gamma = 0.1, 0.3$ and 0.5 and the number of surface realizations is $N = 3,000$.

As we can see, the percentage of the surface area where one reflection occurs is not as large as that for the directly illuminated area which is close to the one when θ is small. However, the value is still too significant to ignore. The maximum value increases slightly as σ_γ increases. For $\sigma_\gamma = 0.5$, the maximum is about 0.25, which means that one surface reflection occur on 25 per cent of the surface points.

It is also noticeable that one surface reflection tends to occur when the incident angle θ is large. For $\sigma_\gamma = 0.1$, one surface reflection is only found when θ is larger than 60°. The range of θ becomes larger when σ_γ increases. For high RMS slope, surface reflections occurs even for small θ.

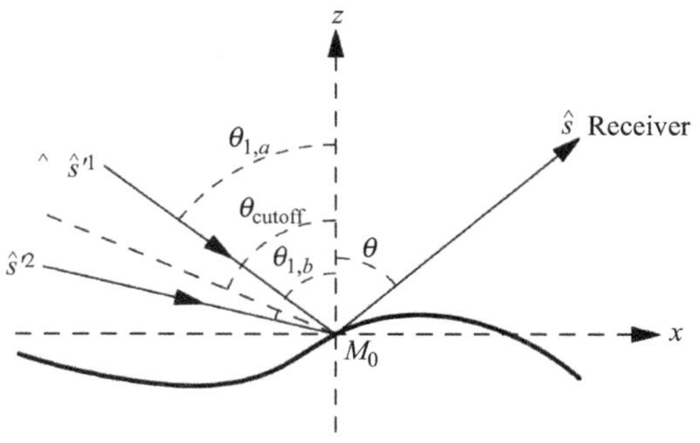

Figure 3.7 Surface reflection and cut-off angle

As discussed previously in Section 3.1.2.3, it is already clear that some surface heights and slopes are more likely to give rise to one surface reflection than the others. In other words, referring to the percentage of surface area where one reflection occurs as the percentage that a given surface point leads to one reflection will not be very precise, because the information on the height and slope of this surface point is ignored. Section 3.1.5 addresses this point.

3.1.3 Empirical models defined by a cut-off angle

3.1.3.1 Model of Watts *et al.*

Watts *et al.* [31] derived a shadowing function with one surface reflection (its name is different) for the derivation of the sea surface-emitted–surface-reflected infrared emissivity. This model is empirical with a very straight forward idea. It is a good example to begin with.

Watts *et al.* [31] pointed out that the key to the problem is to distinguish whether the incident ray (as shown in Figure 3.7) comes from the sky or from the rough surface, which corresponds to P_1 defined by (3.1). There is no doubt that when $\theta_1 > 90°$, the incident ray originates from the rough surface because the sky is located in the upper half medium of the space. However, for $\theta_1 \leq 90°$, it is difficult to tell at which point the incident ray comes from the sky and not from the surface.

Watts *et al.* [31] then defined a cut-off angle basing on the height to wavelength ratio of the surface wave as

$$\theta_{\text{cutoff}} = \frac{\pi}{2} - \tan^{-1}\left(\frac{2H}{L}\right), \tag{3.9}$$

where H is the wave height and L is the wavelength. If $\theta_1 \geq \theta_{\text{cutoff}}$, the incident ray comes from the sky, otherwise it comes from the surface.

In the particular case shown in Figure 3.7, the incident ray along the direction $\hat{s}'^1$ comes from the sky, while the one along the direction $\hat{s}'^2$ comes from the surface. Mathematically, the probability that the incident ray comes from the surface is

$$P_1 = \begin{cases} 1, & \text{if} \quad \theta_1 \geq \theta_{\text{cutoff}} \\ 0, & \text{if} \quad \theta_1 < \theta_{\text{cutoff}} \end{cases}. \tag{3.10}$$

Watts *et al.* [31] also explained that the definition of the cut-off angle also depends on the location of the facet, for example, in a trough or on a peak. However, it was only practical to obtain a rough estimation of the cut-off angle for the whole surface. The readers can adjust the height to wavelength ratio H/L (or the steepness of the surface) according to the surface roughness. They also pointed out that for a sea surface, the typical value of θ_{cutoff} lies somewhere between 79° and 89.1°.

3.1.3.2 Model of Wu and Smith

A very similar model of the shadowing function with one surface reflection is reported by Wu and Smith [27]. The idea is still to introduce an empirical cut-off angle. However, Wu and Smith pointed out that only a few reflections near the trough can benefit from the full advantage of the steepness of the surface. In addition, the definition of the cut-off angle in (3.9) is overestimated for other parts of the surface. Thus, the probability that the incident ray comes from the sea should monotonically decrease from one for $\theta_1 > 90°$ to zero for $\theta_1 < \theta_{\text{cutoff}}$.

Due to the lack of additional knowledge, Wu and Smith [27] defined the cut-off angle to be 85° for a sea surface. In other words, they assumed that the rays with $\theta_1 < 85°$ come from the sky and the ones with $\theta_1 > 90°$ come from the sea surface. For the rays with $85° \leq \theta_1 \leq 90°$, the probability that they originate from the sea surface was given empirically as

$$P_1 = \begin{cases} 1, & \text{if} \quad \theta_1 > 90° \\ 1 - (\theta_1 - 85°)^2/25, & \text{if} \quad 85° \leq \theta_1 \leq 90°. \\ 0, & \text{if} \quad \theta_1 < 85° \end{cases} \tag{3.11}$$

From the above discussion, we can see that it is difficult to well define the cut-off angle θ_{cutoff}, and even more difficult to determine the probability that a ray originates from the sea given $\theta_{\text{cutoff}} \leq \theta_1 \leq 90°$.

For $\theta_{\text{cutoff}} = 85°$ and 90°, it is reported in reference [31] that the sea surface emissivity with one surface reflection differs from each other by about 2.5×10^{-3} for moderate wind speeds. This difference increases as the wind speed increases, or equivalently, as the height to wavelength ratio H/L grows (see Figure 5 of [31]).

To better estimate the occurrence of one surface reflection, a model of higher accuracy is required.

3.1.3.3 Numerical results

References [27,31] did not calculate explicitly the shadowing function with one reflection. They aimed at determining the sea surface infrared emissivity and accounted for one surface reflection when the reflected angle θ_1 is larger than the threshold angle θ_{cutoff}. In this section, the shadowing function with one reflection is built following the idea of these two references.

Recall that this function is expressed as

$$S_M^1 \approx S_M^0 \times P_1, \tag{3.12}$$

where S_M^0 is the statistical monostatic shadowing function without reflection. In fact, in references [27,31], a normalization factor is employed instead of the function S_M^0. Discussion of this factor is outside of the scope of this chapter. Besides, to have a fair comparison between published models, S_M^0 is assumed to be the same function and it is similar to the Smith statistical monostatic shadowing function [5] averaged over the surface heights [32]. In this way, different models of the shadowing function with one reflection S_M^1 distinguish from each other mainly by the part P_1, which is the essence of the problem to solve.

In Chapter 1, we showed that the Smith statistical monostatic shadowing function averaged over the surface heights is

$$S_M^0(v) = \frac{1}{1+\Lambda(|v|)} \qquad \text{with} \qquad |\chi_0| \leq 90^\circ, \tag{3.13}$$

where $\tan \chi_0 = \gamma_0$ and

$$\begin{cases} v &= \dfrac{\cot\theta}{\sigma_\gamma\sqrt{2}} \\[2ex] \Lambda(v) &= \dfrac{\exp(-v^2) - v\sqrt{\pi}\,\text{erfc}(v)}{2v\sqrt{\pi}}, \\[2ex] \text{erfc}(v) &= \dfrac{2}{\sqrt{\pi}}\displaystyle\int_v^{+\infty} \exp(-x^2)\text{d}x \end{cases} \tag{3.14}$$

for a Gaussian process, and σ_γ is the surface RMS slope. Since the process is even, $\Lambda_-(v) = \Lambda(|v|)$ (case for which $v < 0$).

The shadowing function with one reflection following References [27,31] can be expressed as

$$S_M^1(\theta) = \int_{-\infty}^{+\infty} \frac{P_1(\theta_1)}{1+\Lambda(|v|)} p_\gamma(\gamma_0)\text{d}\gamma_0 \qquad \text{with} \qquad |\chi_0| \leq 90^\circ, \tag{3.15}$$

where P_1 is defined from (3.10) and (3.11). In addition, p_γ is the surface slope PDF assumed to be Gaussian.

P_1 depends on $\theta_1(\theta, \gamma_0)$ expressed from (3.6) as

$$\begin{cases} \sin\theta_1 &= -\dfrac{2\gamma_0\cos\chi_0}{\sqrt{1+\gamma_0^2}} - \sin\theta \\[2ex] \cos\theta_1 &= +\dfrac{2\cos\chi_0}{\sqrt{1+\gamma_0^2}} - \cos\theta \end{cases}, \tag{3.16}$$

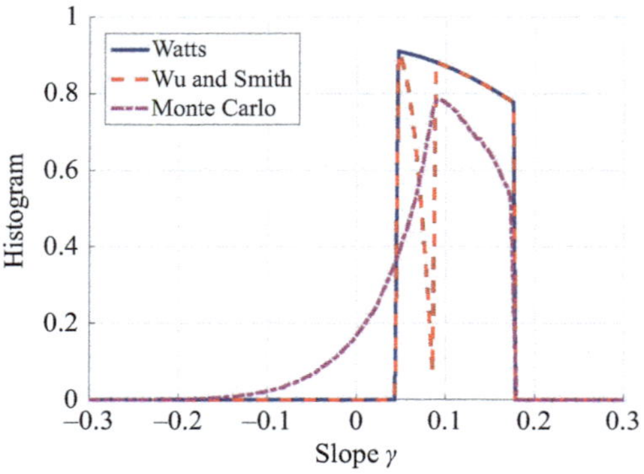

Figure 3.8 Marginal slope histogram of Watts et al. [31] and of Wu and Smith [27] versus γ_0 compared with that obtained from a Monte Carlo ray-tracing algorithm. The incidence angle $\theta = 80°$ and the RMS slope $\sigma_\gamma = 0.3$

where cos χ_0 is expressed by (3.4). Thus, inside (3.15), the kernel depends on θ and γ_0, and the integration over γ_0 can be performed without difficulty.

Marginal slope histogram

Figure 3.8 plots the marginal slope histogram versus the slope γ_0 and obtained from the Watts *et al.* [31] and Wu and Smith [27] models. It is defined by the kernel inside (3.15). The incidence angle $\theta = 80°$ and the surface RMS slope $\sigma_\gamma = 0.3$. The results are compared with those obtained by the Monte Carlo ray-tracing algorithm.

The marginal slope histogram is a relevant parameter to inspect which slopes are more likely to give rise to one surface reflection. It is especially important to calculate the sea surface infrared emissivity.

The marginal slope histogram given by Watts *et al.* [31] has a trapezoidal shape. For the slope $\gamma_0 = \cot(80°) \approx 0.1763$, which corresponds to the right edge of the trapezoidal window, the local incidence angle χ is 90°. Applying the reverse path way shown in Figure 3.2, the reflected ray $\hat{s}'$ propagates in the same direction as the incident ray $\hat{s}$. Then, $\theta_1 = \pi - \theta$ has the largest value. As the slope decreases, θ_1 decreases. When $\gamma_0 = \cot(85°) \approx 0.0875$, which corresponds to the peak of the Monte Carlo results, $\theta_1 = 90°$. In other words, the reflected ray propagates to the left horizontally.

As the slope continues to decrease, the reflected ray propagates upwards. At $\gamma_0 = \cot(87.5) \approx 0.0437$, $\theta_1 = 85°$, which corresponds to the left edge of the trapezoidal window. As the slope continues to decrease, θ_1 is smaller than the cut-off angle, which implies that the reflected ray does not intersect the surface again. Thus, the results of Watts *et al.* and of Wu and Smith decrease to zero.

The model of Watts *et al.* [31] predicts a similar curve to that obtained by the Monte Carlo ray-tracing algorithm. The model of Wu and Smith [27] successfully captures the drop at $\theta_1 = 90°$. However, they both show significant differences

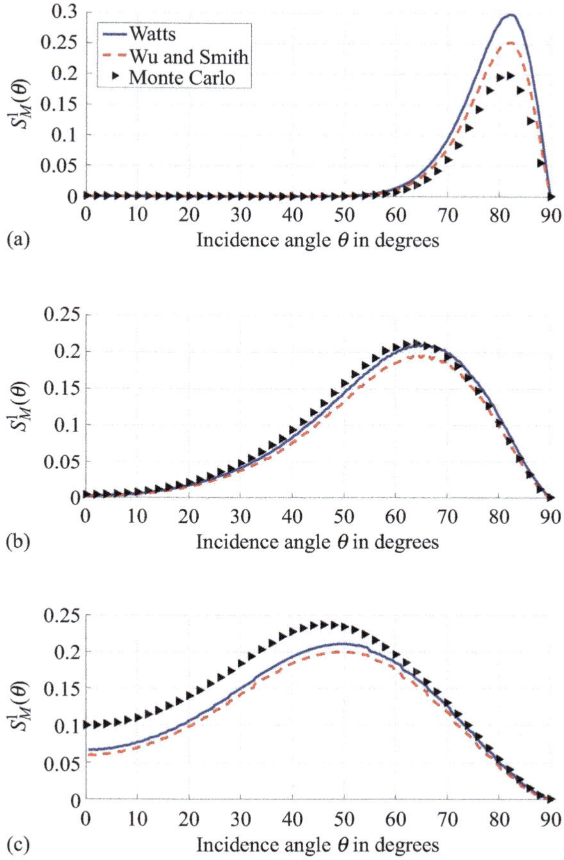

Figure 3.9 Comparison between the models of Watts et al. [31], Wu and Smith [27] and the Monte Carlo ray-tracing algorithm. (a) $\sigma_\gamma = 0.1$. (b) $\sigma_\gamma = 0.3$. (c) $\sigma_\gamma = 0.5$

for $\theta_1 < 90°$. For the model of Wu and Smith, P_1 is even larger at $\theta_1 = 85°$ than at $\theta_1 = 90°$, which is not physical. The Monte Carlo curve indicates that even for θ_1 larger than 85°, or for slope γ_0 smaller than 0.0437, it is still possible that the reflected ray intersects the surface again.

Average shadowing function with one reflection

Figure 3.9 plots the average shadowing function with one reflection computed from the models of Watts *et al.* [31] and Wu and Smith [27]. It gives the percentage of the surface area where one reflection occurs. It is calculated by performing the integration over γ_0, as shown in (3.15). The results are compared with those obtained from a Monte Carlo ray-tracing algorithm. Results for RMS slopes $\sigma_\gamma = 0.1$, $\sigma_\gamma = 0.3$ and $\sigma_\gamma = 0.5$ are shown.

These empirical models show a similar trend as that obtained from the Monte Carlo ray-tracing algorithm. An underestimation is found for $\sigma_\gamma = 0.5$, while an overestimation occurs for $\sigma_\gamma = 0.1$. In fact, one can adjust the value of the cut-off angle θ_{cutoff} to better fit with the Monte Carlo ray-tracing results. However, the value depends on σ_γ, which is a drawback and limits its application.

In the literature, the average shadowing function is directly derived instead of deriving the marginal height or/and slope PDF, which carries more information on the shadowing process but is less easier to calculate.

For instance, for $\sigma_\gamma = 0.3$ as shown in Figure 3.9(b), both the models of Watts *et al.* [31] and of Wu and Smith [27] show a quite well agreement with the Monte Carlo ray-tracing results. However, as shown in Figure 3.8, it is clear that these models do not well evaluate the slope distribution for one surface reflection coming from the right. The trough in the slope histogram of the model of Wu and Smith [27] has no significant impact because the integration over the surface slope hides this effect. This can lead to spurious error during the calculation of some slope-related parameters, such as the sea surface infrared emissivity and reflectivity.

3.1.4 Statistical model

3.1.4.1 Model of Masuda

Instead of defining an exact cut-off angle, Masuda [29] introduced a weighting function to weight the probability that a reflected ray $M_0(\hat{s}')$ with a zenith angle $\theta_1 < 90°$ intersects the surface again. This approach is applied to calculate the sea surface infrared emissivity with one reflection. It is clear that this weighting function is exactly P_1 defined from (3.1) and

$$P_1 = \begin{cases} 1 & \text{if} \quad \theta_1 \geq 90° \\ 1 - s(\theta_1) & \text{if} \quad \theta_1 < 90° \end{cases}. \tag{3.17}$$

The notation adopted in [29] is modified to agree with that used in this chapter. The term $s(\theta_1)$ is the normalization factor to determine the sea surface infrared emissivity strongly related to the shadowing function without reflection.

Masuda [29] interpreted $s(\theta)$ as the probability that an emitted ray originating from the surface along the zenith angle θ does not intercept the surface at another facet. Then $1 - s(\theta)$ can be interpreted as the probability that it does intercept the surface again, or in other words, an incident ray along θ originates from the rough surface.

It is important to point out that the function $s(\theta)$ actually refers to the percentage of the rough surface illuminated directly by the receiver along θ rather than a probability.

For the same reasons as those discussed in Section 3.1.3.3, the model of Masuda is rebuilt from using the height-averaged Smith shadowing function $S_M^0(\theta)$ instead of the original normalization factor. In other words, $s(\theta)$ is replaced by $S_M^0(\theta)$. To sum up, from (3.1), the model of Masuda can be reformulated as

$$\begin{aligned} S_M^1 &\approx S_M^0 \times P_1 \\ &= \begin{cases} S_M^0(\theta) & \text{if} \quad \theta_1 \geq 90° \\ S_M^0(\theta)\left[1 - S_M^0(\theta_1)\right] & \text{if} \quad \theta_1 < 90° \end{cases}. \end{aligned} \tag{3.18}$$

In addition, from the Smith formulation for S_M^0, S_M^1 can be written as

$$S_M^{1'}(v,v') \approx \begin{cases} \dfrac{1}{1+\Lambda(|v|)} & \text{if } |\chi_0| \le 90°, \theta_1 \ge 90° \\ \dfrac{1}{1+\Lambda(|v|)}\left[1 - \dfrac{1}{1+\Lambda(|v'|)}\right] & \text{if } |\chi_0| \le 90°, \theta_1 < 90° \end{cases}$$
$$= \begin{cases} \dfrac{1}{1+\Lambda(|v|)}, & \text{if } |\chi_0| \le 90°, \theta_1 \ge 90° \\ \dfrac{\Lambda(|v'|)}{[1+\Lambda(|v|)]\,[1+\Lambda(|v'|)]} & \text{if } |\chi_0| \le 90°, \theta_1 < 90° \end{cases}, \quad (3.19)$$

where Λ is defined by (3.14) and

$$v = \frac{\cot\theta}{\sigma_\gamma\sqrt{2}} \qquad v' = \frac{\cot\theta_1(\theta,\gamma_0)}{\sigma_\gamma\sqrt{2}}. \quad (3.20)$$

From the model of Masuda, the shadowing function with one reflection is then

$$S_M^1(\theta) = \int_{-\infty}^{+\infty} S_M^{1'}(v,v')p_\gamma(\gamma_0)\mathrm{d}\gamma_0 \qquad \text{with} \qquad |\chi_0| \le 90°, \quad (3.21)$$

where $p_\gamma(\gamma_0)$ is the surface slope PDF assumed to be Gaussian. The integrand (kernel) in (3.21) is function of the incident angle $\theta_1(\theta,\gamma_0)$ expressed by (3.16) and the slope of the first intersection point γ_0. The integration over γ_0 can be easily performed numerically from the trapezoidal rule.

3.1.4.2 Numerical results

Marginal slope histogram

Figure 3.10 compares the marginal slope histogram versus γ_0 computed from the Masuda [29] (integrand of (3.21)) and that computed from a Monte Carlo ray-tracing algorithm. The incident angle $\theta = 80°$ and the surface RMS slope $\sigma_\gamma = 0.3$.

The model of Masuda agrees well with that of the Monte Carlo ray-tracing algorithm in comparison to the results plotted in Figure 3.8. The slope $\gamma_0 = \cot(80°) \approx 0.1763$, above which the histogram vanishes, corresponds to case when the local incidence angle χ_0 is 90° and the reflected ray $\hat{s}'$ propagates downwards in the same direction as the incident ray $\hat{s}$. The slope $\gamma_0 = \cot(85°) \approx 0.0875$, at which the peak occurs on the both curves, corresponds to the case when $\theta_1 = 90°$ and the reflected ray propagates to the left horizontally. In this region $0.0875 < \gamma_0 < 0.1763$, the zenith angle of the reflected ray θ_1 is larger than 90°. Comparing (3.10), (3.11) and (3.17), the three models predict same results on P_1, which implies that the marginal slope histograms are also identical.

However, in the region for which $\gamma < 0.0875$ (or $\theta_1 < 90°$), the model of Masuda [29] shows a significant improvement. The model predicts the trend that

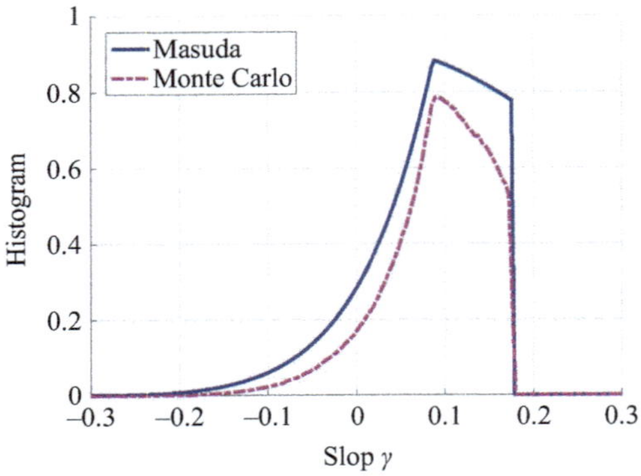

Figure 3.10 Marginal slope histogram versus γ_0 obtained from the Masuda [29] model and from a Monte Carlo ray-tracing algorithm. The incidence angle $\theta = 80°$ and the RMS slope $\sigma_\gamma = 0.3$

the probability of having a reflected ray intersecting the surface again grows thinner as θ_1, or the slope γ, grows smaller.

The drawback of the introduction of the cut-off angle is overcame but the Masuda model overestimates slightly the histogram.

Average shadowing function with one reflection

Figure 3.11 plots the average shadowing function with one reflection computed from the model of Masuda [29]. It corresponds to the percentage of the surface area where one reflection occurs. It is calculated by performing the integration over γ_0 in (3.21). The results are compared with those obtained from the Monte Carlo ray-tracing algorithm. The RMS slopes $\sigma_\gamma = 0.1$, $\sigma_\gamma = 0.3$ and $\sigma_\gamma = 0.5$.

The results of the Masuda model are weakly overestimated in comparison to the Monte Carlo ray-tracing ones. It is consistent with the marginal slope histogram shown in Figure 3.10 since the average shadowing function is obtained by averaging the marginal slope histogram over the surface slopes. The model of Masuda [29] is still not satisfactory enough if high accuracy is required. It is the case for the calculation of the sea surface infrared emissivity.

The main drawback is the overestimation of P_1 obtained by assuming that the function $s(\theta)$ is related to the percentage of surface area where one reflection occurs instead of considering as a probability that the reflected ray intersects the surface again. From a physical point of view, the percentage of points to be illuminated directly by the receiver should depend on the location, the height and the slope of this point. The average shadowing function loses this information because it is obtained by integrating over the heights and the slopes. This implies that all points regardless of their feature are treated the same. In the next section, an improved statistical model is introduced to address this issue.

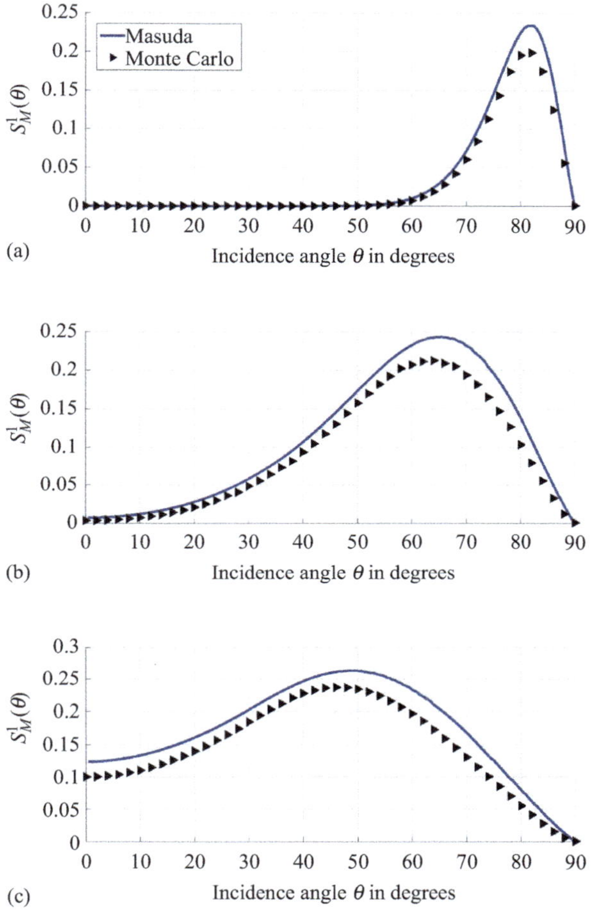

Figure 3.11 Comparison between the model of Masuda [29] and the Monte Carlo ray-tracing algorithm: (a) $\sigma_\gamma = 0.1$, (b) $\sigma_\gamma = 0.3$ and (c) $\sigma_\gamma = 0.5$

3.1.5 Improved statistical model

3.1.5.1 Model of Li *et al.*

Derivation

Li *et al.* [33] developed a model of shadowing function with one reflection from the statistical shadowing function of Smith [5] rather than that averaged over the surface heights. This model has a form similar to the one of Masuda [29], but its derivation is more rigorous and leads to better results.

As shown in Figure 3.12, the mixed path propagation principle is more convenient for the discussion. The incident and the reflected rays are treated as if they both

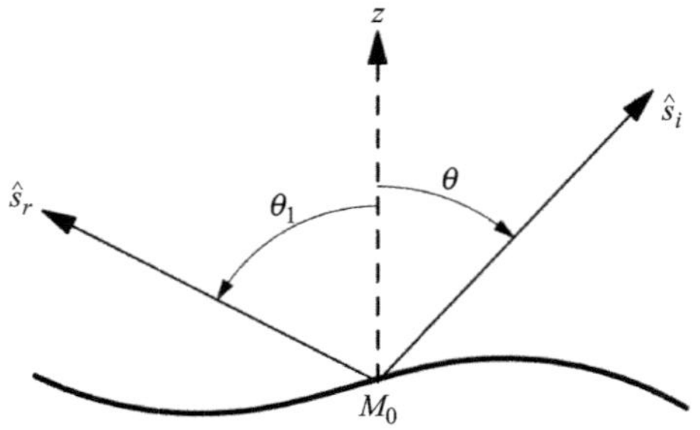

Figure 3.12 A mixed path reverse propagation: the incident and the reflected rays are treated as if they both originate from the intersection point M_0 and have incidence angles that satisfy the reflection law

originate from the intersection point M_0 and have incident or reflected angles that satisfy the reflection law. To be concise in formulation, Li *et al.* [33] defined that a zenith angle is positive if it is measured clockwise from zenith, and negative otherwise. For example, shown in Figure 3.12, θ is positive and θ_1 is negative.

From Figure 3.12, the statistical shadowing function with one reflection corresponds to the probability that the ray $\hat{s}_i(\theta)$ does not intersect the surface while the ray $\hat{s}_r(\theta_1)$ does. Mathematically, it can be defined as

$$\begin{aligned} S_M^1 &= p(\hat{s}_i(\theta) \text{ not intersect} \cap \hat{s}_r(\theta_1) \text{ intersects}) \\ &= p(\hat{s}_i(\theta) \text{ not intersect}) \\ &\quad \times p(\hat{s}_r(\theta_1)\text{intersects})|(\hat{s}_i(\theta) \text{ does not intersect}) \\ &= p(\hat{s}_i(\theta) \text{ not intersect}) \\ &\quad \times \left[1 - p(\hat{s}_r(\theta_1) \text{ not intersect} \mid \hat{s}_i(\theta) \text{ not intersect})\right] \\ &= p(a)[1 - p(b|a)] \\ &= S_M^0 \times P_1, \end{aligned} \tag{3.22}$$

where the event '$M_0(\theta)$ *does not intersect the surface*' is denoted symbolically as 'a' and the event '$M_0(\theta_1)$ *does not intersect the surface*' as 'b'. As expected, (3.22) has a similar form as (3.1). A more rigorous form of $p(a)$ and $p(b)$ is derived in the model of Li *et al.* [33].

Probability p(a)

There is no doubt that $p(a)$ corresponds to the statistical monostatic shadowing function without reflection which is addressed previously in Chapter 1. For sake of clarity, some equations of Chapter 1 are recalled here. From the Smith formulation [5] and for a surface of infinite length, the statistical monostatic shadowing function is given by

$$p(a) = A \times \exp\left[-\int_0^{+\infty} g(\mu|\gamma_0, \zeta_0; \tau)\mathrm{d}\tau\right], \tag{3.23}$$

where θ is the incidence angle and $\mu = \cot\theta$ the slope of the incident ray. In addition, the integration range is $\tau \in [0; +\infty[$ because the ray of initial point M_0 propagates to the right from $\tau = 0$ to $\tau = +\infty$.

For a given point M_0 of height ζ_0 and of slope γ_0, the conditional probability $g(\mu|\gamma_0, \zeta_0; \tau)$ gives the probability that the ray $\hat{s}(\theta)$ coming from M_0 intersects the surface again at a point of abscissa τ. From Chapter 1, it is expressed as

$$g(\mu|\gamma_0, \zeta_0; \tau) = \frac{\displaystyle\int_{\mu}^{+\infty} (\gamma - \mu) p(\zeta = \zeta_0 + \mu\tau, \gamma|\zeta_0, \gamma_0; \tau) \mathrm{d}\gamma}{\displaystyle\int_{-\infty}^{+\infty} \int_{-\infty}^{\zeta_0 + \mu\tau} p(\zeta, \gamma|\zeta_0, \gamma_0; \tau) \mathrm{d}\gamma \mathrm{d}\zeta}. \tag{3.24}$$

The calculation of $g(\mu|\gamma_0, \zeta_0; \tau)$ involves the conditional PDF $p(\zeta, \gamma|\zeta_0, \gamma_0; \tau)$ of the height ζ and the slope γ of the point at abscissa τ away from point M_0 of height ζ_0 and of slope γ_0.

This probability is assumed to be Gaussian and the correlation between the points $M_0(\zeta_0, \gamma_0)$ and $M(\zeta, \gamma)$ depends on the surface feature. In reference [28], the correlation is ignored to generate the surface, whereas in this chapter, the surface height autocorrelation function is assumed to be Gaussian for the Monte Carlo ray-tracing algorithm.

When the correlation is accounted for, as shown in Chapter 1, the integrations in (3.23) have to be performed numerically. In this chapter, a Gaussian autocorrelation is assumed to be consistent to the Monte Carlo ray-tracing algorithm.

If the correlation is ignored, the integrations in (3.23) can be performed analytically, leading to

$$p(a) = A \times F(\zeta_0)^{\Lambda(\mu)}, \tag{3.25}$$

where F stands for the height cumulative density function, and

$$\Lambda(\mu) = \frac{1}{\mu} \int_{\mu}^{+\infty} (\gamma - \mu) p_\gamma(\gamma) \mathrm{d}\gamma, \tag{3.26}$$

where p_γ is the surface slope PDF. For a Gaussian PDF, Λ is given by (3.14).

Mathematically, the factor A can be an arbitrary function. However, in a physical problem, A should be chosen according to the physical meaning of the shadowing function. The first feature to be ensured is that $p(a) = 1$ when $\theta = 0°$ because no shadow is produced by vertically incident rays as long as the surface is single valued. The second feature to be ensured is that $p(a) = 0$ when the local incident angle $\chi_0 > 90°$. Thus, A is chosen as the unit step function defined as

$$A = \Upsilon(\mu - \gamma_0) = \begin{cases} 0 & \text{if } \gamma_0 > \mu \\ 1 & \text{if } \gamma_0 \leq \mu \end{cases}. \tag{3.27}$$

Finally, the uncorrelated probability $p(a)$ is given by

$$p(a) = S_M^{0,\text{unco}} = \Upsilon(\mu - \gamma_0) F(\zeta_0)^{\Lambda(\mu)}. \tag{3.28}$$

Conditional probability $p(b|a)$

The conditional probability $p(b|a)$ is the probability that the ray of direction $\hat{s}_r(\theta_1)$ does not intersect the surface again given that the ray $\hat{s}_i(\theta)$ neither does. When the ray $\hat{s}_r(\theta_1)$ propagates downwards, or $\theta_1 > 90°$, it will certainly intersect the surface again somewhere; thus, $p(b|a) = 0$. For $\theta_1 < 90°$, it is expected that $p(b|a)$ is similar to the shadowing function without reflection with some modification due to the correlation between the events a and b. Mathematically, $p(b|a)$ has a similar form as $p(a)$ as shown in (3.23).

$$p(b|a) = \begin{cases} 0 & \text{if} \quad |\theta_1| > 90° \\ A' \exp\left[-\int_0^{+\infty} g(\mu_1|\gamma_0, \zeta_0; \tau)\mathrm{d}\tau\right] & \text{if} \quad 0 < \theta_1 < 90° \\ A' \exp\left[-\int_{-\infty}^{0} g_-(\mu_1|\gamma_0, \zeta_0; \tau)\mathrm{d}\tau\right] & \text{if} \quad -90° < \theta_1 < 0 \end{cases}. \quad (3.29)$$

In (3.29), the second line is the same as (3.23). This is because the ray $\hat{s}_r(\theta_1)$ propagates to the right, which is the same configuration as the ray $\hat{s}_i(\theta)$. In the third line, the integration range becomes $\tau \in [-\infty; 0]$ because the ray $\hat{s}_r(\theta_1)$ now propagates to the left from M_0 where the distance ranges from $\tau = 0$ to $\tau = -\infty$. The subscript in the term g_- indicates the ray propagating towards the negative side of the axis. It can be determined following the same way as Smith [5] for g, which is given from (3.24) as

$$g_-(\mu_1|\gamma_0, \zeta_0; \tau) = \frac{\int_{-\infty}^{\mu_1} (\mu_1 - \gamma)p(\zeta = \zeta_0 + \mu_1\tau, \gamma|\zeta_0, \gamma_0; \tau)\mathrm{d}\gamma}{\int_{-\infty}^{\infty}\int_{-\infty}^{\zeta_0+\mu_1\tau} p(\zeta, \gamma|\zeta_0, \gamma_0; \tau)\mathrm{d}\gamma\,\mathrm{d}\zeta}, \quad (3.30)$$

where $\mu_1 = \cot\theta_1$ is the slope of the ray $\hat{s}_r(\theta_1)$. As shown in Chapter 1, the function g_- can be performed from g by inverting the direction of the axis. In fact, if the slope PDF is even (case of Gaussian process with a zero mean value), $g_-(\mu_1|\gamma_0, \zeta_0; \tau)$ equal $g(-\mu_1| - \gamma_0, \zeta_0; \tau)$.

If the correlation is ignored, in (3.29), the integrations can be performed analytically (see Chapter 1) leading to

$$p(b|a) = \begin{cases} 0 & \text{if} \quad |\theta_1| > 90° \\ A'F(\zeta_0)^{\Lambda(\mu_1)} & \text{if} \quad 0 < \theta_1 < 90° \\ A'F(\zeta_0)^{\Lambda_-(\mu_1)} & \text{if} \quad -90° < \theta_1 < 0 \end{cases}, \quad (3.31)$$

where F stands for the height cumulative density function, Λ obtained from (3.26) and

$$\Lambda_-(\mu_1) = \frac{1}{\mu_1}\int_{-\infty}^{\mu_1} (\gamma - \mu_1)p_\gamma(\gamma_0)\mathrm{d}\gamma_0 = \frac{1}{|\mu_1|}\int_{|\mu_1|}^{\infty} (\gamma - |\mu_1|)p_\gamma(-\gamma_0)\mathrm{d}\gamma_0, \quad (3.32)$$

where p_γ is the surface slope PDF and $\mu_1 = -|\mu_1|$.

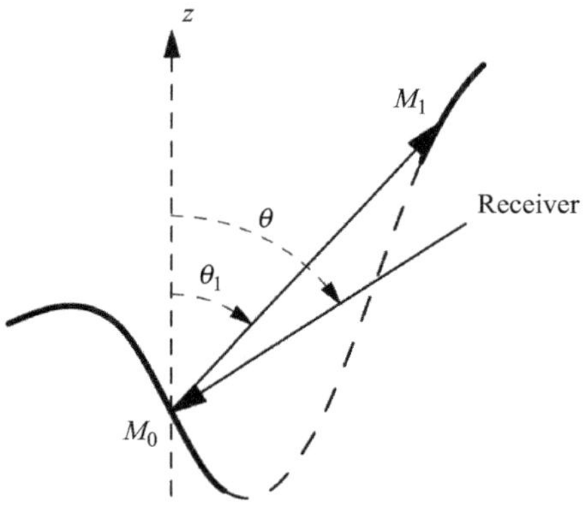

Figure 3.13 Special case when both the incident and reflected rays propagate to the right with $0 < \theta_1 < \theta$

For a Gaussian surface slope, Λ_- has the form

$$\begin{cases} \Lambda_-(\mu_1) = \Lambda(|\mu_1|) = \dfrac{\exp(-v_1^2) - v_1\sqrt{\pi}\,\mathrm{erfc}(v_1)}{2v_1\sqrt{\pi}} \\ v_1 = \dfrac{|\cot\theta_1|}{\sigma_\gamma\sqrt{2}} \end{cases}. \quad (3.33)$$

The determination of the factor A' depends heavily on the correlation between the events a and b. Recall that the ray $\hat{s}_r(\theta_1)$ is the reflected ray of $\hat{s}_i(\theta)$ at the point M_0. The angle θ can be an arbitrary angle ranging from 0° to 90°, thus there is a chance that θ goes larger than the slope of M_0 when the local incident angle is larger than 90°. Then the unit step function $\Upsilon(\mu - \gamma_0)$ is employed to prevent this from happening. However, θ_1 is the reflected angle which is determined by the incident direction θ and the slope of M_0 following the law of reflection. In this case, the constraint that the local reflected angle has not to be larger than 90° is automatically fulfilled. As a result, the same constraint has not to be applied to $p(b|a)$. This leads to

$$A' = 1. \quad (3.34)$$

Finally, $p(b|a)$ without correlation is

$$p(b|a) = \begin{cases} 0 & \text{for} \quad |\theta_1| > 90^\circ \\ F(\zeta_0)^{\Lambda(\mu_1)} & \text{for} \quad 0 <^\circ \theta_1 < 90^\circ \\ F(\zeta_0)^{\Lambda_-(\mu_1)} & \text{for} \quad -90^\circ < \theta_1 < 0^\circ \end{cases}. \quad (3.35)$$

Special case of 1D surfaces

The case shown in Figure 3.13 is investigated. The reflected ray $\hat{s}_r(\theta_1)$ propagates to the right as well as the incident ray $\hat{s}_i(\theta)$, but with $\theta_1 < \theta$.

For the conditional probability $p(b|a)$, it is already known that the event a occurs. In other words, it is known that the ray $\hat{s}_i(\theta)$ does not intersect the surface. Let us assume that the reflected ray $\hat{s}_r(\theta_1)$ intersects the surface again, then there must exist a facet M_1 which is 'higher' than the receiver as shown in Figure 3.13. Since the surface is single-valued and continuous, there must be a continuous surface lies between M_0 and M_1 shown as the dashed lines in Figure 3.13. It blocks the ray $\hat{s}_i(\theta)$ from the receiver, which conflicts with the knowledge that the ray $\hat{s}_i(\theta)$ does not intersect the surface. Thus, for $0 < \theta_1 < \theta$, the ray $\hat{s}_i(\theta)$ does not intersect the surface again and $p(b|a) = 1$.

In an extreme case when $\theta_1 = \theta$, the ray $\hat{s}_r(\theta_1)$ propagates in the reverse path of the ray $\hat{s}_i(\theta)$. This case corresponds to the backscattering direction. It is clear that if the ray $\hat{s}_i(\theta)$ does not intersect the surface elsewhere, neither does the ray $\hat{s}_r(\theta_1)$. Thus, $p(b|a) = 1$.

To sum up, the uncorrelated shadowing function with one reflection can be written as

$$S_M^{1,\text{unco}}(\theta, \gamma_0, \zeta_0) = \Upsilon(\mu - \gamma_0) F(\zeta_0)^{\Lambda(\mu)} \times \begin{cases} 1 & \text{if } |\theta_1| > 90^\circ \\ 0 & \text{if } 0^\circ < \theta_1 \leq \theta \\ 1 - F(\zeta_0)^{\Lambda(\mu_1)} & \text{if } \theta < \theta_1 \leq 90^\circ \\ 1 - F(\zeta_0)^{\Lambda_-(\mu_1)} & \text{if } -90^\circ \leq \theta_1 < 0^\circ \end{cases}, \tag{3.36}$$

where the angle θ_1 depends on γ_0 and θ.

The shadowing function with one reflection depends on the incident zenith angle θ, the surface slope γ_0 and the surface height ζ_0 of the point M_0. Comparing with the models of Masuda [29] (equation (3.18)), of Watts *et al.* [31] and of Wu and Smith [27] (equation (3.15)), the major difference is that the (3.36) depends on the height ζ_0.

The height ζ_0 is accounted for to calculate the probability $p(a)$ that the ray $\hat{s}_i(\theta)$ does not intersect the surface and, the probability $p(b|a)$ that $\hat{s}_r(\theta_1)$ does not neither, which agrees with the fact that the rays $\hat{s}_i(\theta)$ and $\hat{s}_r(\theta_1)$ originate from the same point. For the other models, the information on the height is unavailable.

Another advantage of the model of Li *et al.* [33] is that the surface correlation can be accounted for in the calculation of $p(a)$ and $p(b|a)$. When high accuracy is required, the correlation is needed. In the next subsection, simulations are presented, which show clearly an improvement of the model.

3.1.5.2 Numerical results

Marginal height histogram

The marginal height histogram reveals which heights are more likely to give rise to one surface reflection. The models of Masuda [29], Watts *et al.* [31] and Wu and Smith [27] cannot provide this information since they are not function of the height. The marginal height histogram of the model of Li *et al.* [33] is obtained by integrating (3.22) over the surface slopes γ_0. This leads to

$$\widetilde{p}_\zeta^1(\mu, \zeta_0) = p_\zeta(\zeta_0) \int_{-\infty}^{+\infty} S_M^1(\mu, \gamma_0, \zeta_0) p_\gamma(\gamma_0) \mathrm{d}\gamma_0, \tag{3.37}$$

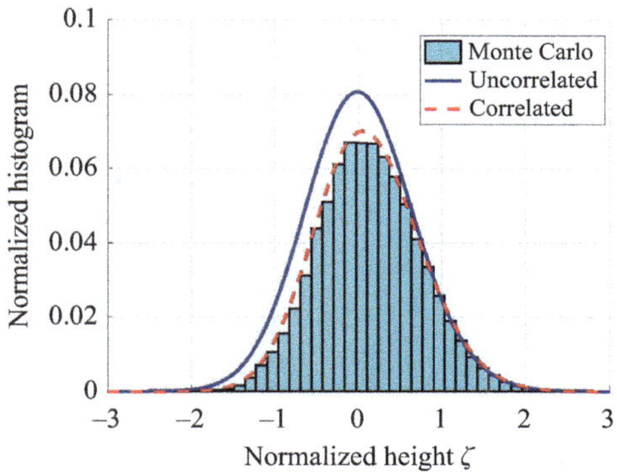

Figure 3.14 Marginal height histogram versus the normalized height $\zeta_0/(\sigma_\zeta\sqrt{2})$ calculated from the model of Li. The RMS slope $\sigma_\gamma = 0.2$ and the incidence angle $\theta = 80°$

where p_ζ and p_γ are the surface height and slope PDFs, respectively, assumed to be Gaussian for the simulations. Since the slope of the reflected ray $\mu_1 = \cot\theta_1$ is also function of the slope γ_0 (equation (3.16)), in (3.36), the integration has to be performed numerically. In Figure 3.14, the RMS slope $\sigma_\gamma = 0.2$ and the incidence angle $\theta = 80°$.

The correlation between the heights and slopes can also be taken into account by computing $p(a)$ and $p(b|a)$ with correlation via the functions g and g_- (details of the derivation reported in Section 1.4). The correlated results are also shown in Figure 3.14, and they are compared with those obtained from the Monte Carlo ray-tracing algorithm (reference solution). The surface height autocorrelation function is assumed to be Gaussian.

The uncorrelated model agrees well with that of the Monte Carlo ray-tracing algorithm with a slight overestimation. The correlation improves significantly the agreement.

Marginal slope histogram

The marginal slope histogram reveals which slopes are more likely to give rise to one surface reflection. All the models introduced previously [27,29,31] are also able to compute this function. In (3.22), from the model of Li *et al.* [33], it is obtained by integrating over the surface heights ζ_0. This leads to

$$\widetilde{p}^1_\gamma(\mu, \gamma_0) = p_\gamma(\gamma_0)\int_{-\infty}^{+\infty} S^1_M(\mu, \gamma_0, \zeta_0)p_\zeta(\zeta_0)\mathrm{d}\zeta_0, \tag{3.38}$$

where p_ζ and p_γ are the surface height and slope PDFs, respectively, which are assumed to be Gaussian for the simulations.

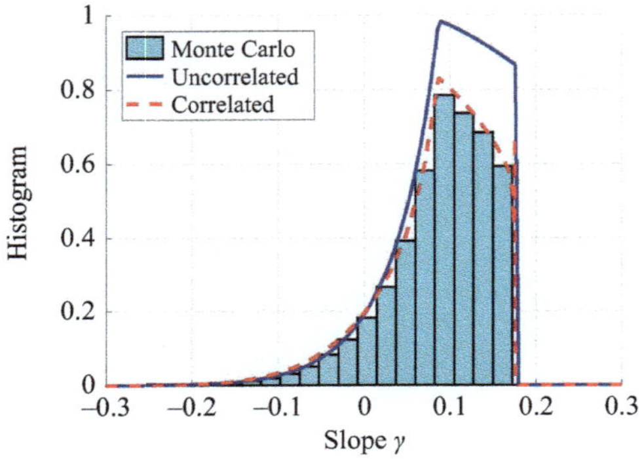

Figure 3.15 Marginal slope histogram by Li et al. *[33] compared with that obtained from a Monte Carlo ray-tracing algorithm versus the surface slopes γ_0. The incidence angle $\theta = 80°$ and the surface RMS slope $\sigma_\gamma = 0.3$*

The correlation between the heights and slopes can also be taken into account by computing $p(a)$ and $p(b|a)$ with correlation via the functions g and g_- functions (Section 1.4).

If this correlation is neglected, the integration can be performed analytically, leading to

$$\tilde{p}_\gamma^{\text{unco},\,1} = p_\gamma(\gamma_0) \times \begin{cases} \dfrac{\Upsilon(\mu - \gamma_0)}{\Lambda(v) + 1} & |\theta_1| > 90° \\ \dfrac{\Lambda_-(v_1)}{[\Lambda(v) + 1][\Lambda_-(v_1) + \Lambda(v) + 1]} & -90° \leq \theta_1 < 0° \\ \dfrac{\Lambda(v_1)}{[\Lambda(v) + 1][\Lambda(v_1) + \Lambda(v) + 1]} & \theta < \theta_1 \leq 90° \\ 0 & \text{otherwise} \end{cases}, \quad (3.39)$$

where $v = \cot\theta/(\sqrt{2}\sigma_\gamma)$ and $v_1 = \cot\theta_1(\theta, \gamma_0)/(\sqrt{2}\sigma_\gamma)$.

The comparison of (3.39) with (3.19) shows that the model of Li *et al.* [33] is similar to that of Masuda [29] with some differences. When $\theta_1 \geq 90°$, these two models match. When $\theta_1 < 90°$, the only difference is that the model of Li *et al.* [33] has an additional term $\Lambda(v)$ inside the denominator.

Simulation results of the marginal slope histogram by the model of Li *et al.* [33] is shown in Figure 3.15. The incident angle $\theta = 80°$ and the surface RMS slope $\sigma_\gamma = 0.3$ (same as in Figures 3.8–3.10 for easy comparison). The results are also compared with those obtained from the Monte Carlo ray-tracing algorithm.

As discussed in Figure 3.8, above the slope $\gamma_0 = \cot(80°) \approx 0.1763$, the strengths vanish, corresponding to the case when the local incidence angle χ_0 is 90° and the reflected ray $\hat{s}'$ propagates downwards in the same direction as the incident ray $\hat{s}$.

The slope $\gamma_0 = \cot(85^\circ) \approx 0.0875$, at which the maximum occurs for the both curves, corresponds to the case when $\theta_1 = -90^\circ$ and the reflected ray propagates to the left horizontally. In the region defined as $0.0875 < \gamma_0 < 0.1763$, the zenith angle of the reflected ray $|\theta_1|$ is larger than 90°. Comparing (3.10), (3.11) and (3.17) with (3.39), we can see that P_1 or $1 - p(b|a)$ is the same for all these models in this range. Thus, the marginal slope histograms are also the same if the surface correlation is ignored.

Figure 3.15 shows that an overestimation is found in this region by all these models. However, the introduction of the correlation into the model of Li *et al.* improves significantly the results.

In the region where $\gamma_0 < 0.0875$ ($\theta_1 < 90^\circ$), the model of Li *et al.* provides the best agreement among the four models, even when the surface correlation is ignored. The overestimation by Masuda [29] shown in Figure 3.10 is largely overcome here because of the extra term $\Lambda(v)$ in the denominator of (3.39).

Average shadowing function with one reflection

The average shadowing function with one reflection gives the percentage of the surface area where one reflection occurs. From the model of Li *et al.*, it is calculated by integrating the statistical shadowing function with one reflection $S_M^1(\mu, \gamma_0, \zeta_0)$ over the surface heights and slopes. This leads to

$$S_M^1(\mu) = \int_{-\infty}^{+\infty} \int_{-\infty}^{+\infty} S_M^1(\mu, \gamma_0, \zeta_0) p_\varsigma(\zeta_0) p_\gamma(\gamma_0) \mathrm{d}\zeta_0 \mathrm{d}\gamma_0. \tag{3.40}$$

Equivalently, the function $S_M^1(\theta)$ can also be obtained by integrating the marginal height histogram over the surface heights, or by integrating the marginal slope histogram over the surface slopes.

Figure 3.16 plots $S_M^1(\mu)$ versus θ with and without correlation. The results are also compared with those obtained from the Monte Carlo ray-tracing algorithm. The RMS slopes are $\sigma_\gamma = 0.1$, $\sigma_\gamma = 0.3$ and $\sigma_\gamma = 0.5$.

As we can see, the model of Li *et al.* shows a good agreement with that of the Monte Carlo ray-tracing algorithm even without considering the surface correlation. An overestimation is observed, but it is less significant than that of the model of Masuda [29] shown in Figure 3.11. It comes from the fact that the model of Li *et al.* is more precise in the region where the reflected ray $\hat{s}_r(\theta_1)$ propagates upwards as discussed in the previous subsection. The agreement is improved when the surface correlation is included.

To sum up, the model of Li *et al.* [33] is the most accurate among the four models discussed in this section. The main reason is that, it is derived from the statistical shadowing function of Smith [5] rather than a normalization factor not related to a direct probability.

By doing so, the model of Li *et al.* [33] is able to take into account the surface correlation. More importantly, it includes the fact that both the incident ray and the reflected ray intersect the surface at the same point M_0, which implies that both the probabilities $p(a)$ and $p(b|a)$ depend on the height of M_0. All the other models ignore this dependence.

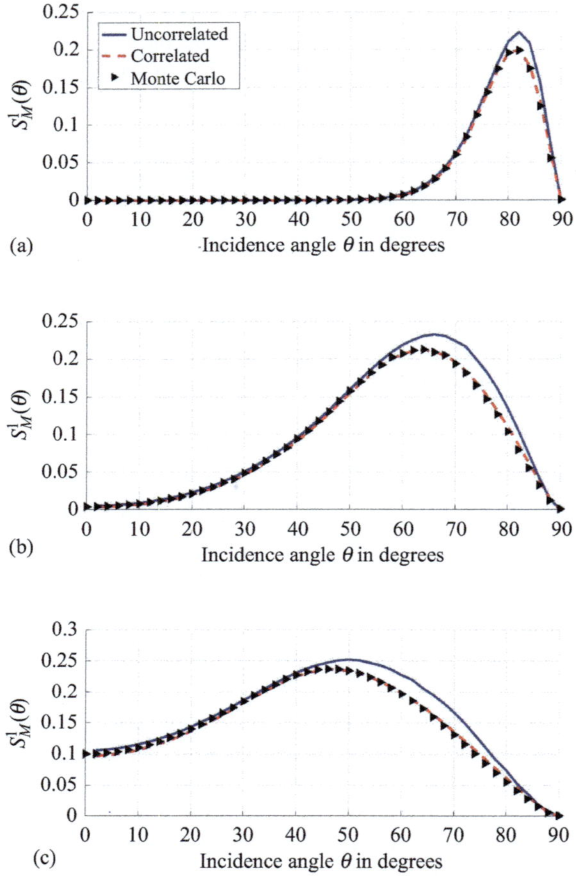

Figure 3.16 Comparison between the model of Li et al. [33] with the Monte Carlo ray-tracing algorithm. The RMS slopes are $\sigma_\gamma = 0.1$ in (a), $\sigma_\gamma = 0.3$ in (b) and $\sigma_\gamma = 0.5$ in (c)

3.1.6 Extension to include multiple reflections

3.1.6.1 Monostatic shadowing function with two reflections

Derivation

In this section, the shadowing function with two reflections is derived. The number of publications devoted to this topic is small [34,35] and in still under investigation.

In this section, we try to derive a monostatic shadowing function with two reflections, in the hope to progress in this issue and to give some inspiration to the readers for future investigations.

In the previous section, it is shown that the model of Li *et al.* [33] is a good candidate. It is the only model that predicts well the marginal slope histogram which gives rise to one surface reflection. This function is the key to some problems met

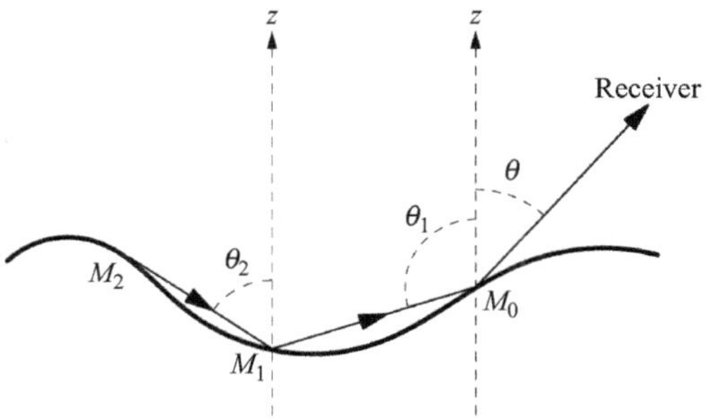

Figure 3.17 Case of a double surface reflections

in physics like the sea surface infrared emissivity. As a result, the proposed model is derived by generalizing the model of Li *et al.*

Figure 3.17 illustrates a case when two surface reflections occur. A ray is emitted by the point M_2. It intersects the surface at the point M_1 first, then the reflected ray is reflected again by the surface at the point M_0 before it reaches the receiver.

Applying the same way as that in the one reflection case, we define the following events:

- 'Ray $\hat{s}(\theta)$ coming from the point M_0 does not intersect the surface' is denoted symbolically as 'a'.
- 'Ray $\hat{s}(\theta_1)$ intersects the surface at the point M_1' is denoted symbolically as 'b'.
- 'Ray $\hat{s}(\theta_2)$ coming from the point M_2 does not intersect the surface' is denoted symbolically as 'c'.

Similar to (3.22), the shadowing function with two reflections can be defined as

$$\begin{aligned} S_M^2 &= p(\hat{s}(\theta) \text{ does not intersect} \cap \hat{s}(\theta_1) \text{ intersects at } M_1 \\ &\quad \cap \hat{s}(\theta_2) \text{ intersects}) \\ &= p(ab) \times p(\hat{s}(\theta_2) \text{ intersects}|ab) \\ &= p(ab) \times \left[1 - p(\hat{s}(\theta_2) \text{ does not intersect}|ab)\right] \\ &= p(ab)[1 - p(c|ab)]. \end{aligned} \tag{3.41}$$

The probability $p(ab)$ is exactly the statistical shadowing function with one reflection. The conditional probability $p(c|ab)$ can be derived in a similar way as $p(b|a)$ (one reflection case) in (3.22). It has the same form as $p(b|a)$ and expressed as

$$p(c|ab) = \begin{cases} 0 & \text{for} \quad |\theta_2| > 90^\circ \\ \exp\left[-\displaystyle\int_0^{+\infty} g(\mu_2|\gamma_1, \zeta_1; \tau)\mathrm{d}\tau\right] & \text{for} \quad 0^\circ < \theta_2 < 90^\circ \\ \exp\left[-\displaystyle\int_{-\infty}^{0} g_-(\mu_2|\gamma_1, \zeta_1; \tau)\mathrm{d}\tau\right] & \text{for} \quad -90^\circ < \theta_2 < 0 \end{cases}, \tag{3.42}$$

in which the factor A' equal one.

If the surface correlation is ignored, then $p(c|ab)$ is simplified as

$$p(c|ab) = \begin{cases} 0 & \text{for} \quad |\theta_2| > 90^\circ \\ F(\zeta_1)^{\Lambda(\mu_2)} & \text{for} \quad 0 < \theta_2 < 90^\circ \\ F(\zeta_1)^{\Lambda_{-}(\mu_2)} & \text{for} \quad -90 <^\circ \theta_2 < 0 \end{cases} . \tag{3.43}$$

To make the discussion easy to follow, the uncorrelated form shown in (3.43) is considered hereafter. In (3.43), the height and slope of the point M_1 is involved. As discussed previously, the shadowing function with one reflection involves the slope of the point M_0. As a result, the monostatic shadowing function with two reflections S_M^2 involves the heights and slopes of the points M_0 and M_1.

One critical question then arises. What is the correlation between the point M_0 and M_1? Although we have already explicitly stated that the statistical correlation of the heights and slopes between different surface points is ignored in an uncorrelated model, knowing that the event *ab* gives an extra information on the statistics of the point M_1. M_1 is not an arbitrary point of the surface and corresponds to the second intersection of an incident ray of zenith angle θ. Then, its height and slope PDFs differ from that of the whole surface.

Height PDF of M_1

For example, if the ray $\hat{s}(\theta_1)$ propagates downwards, it is obvious that the height of M_1 must be smaller than that of M_0, or $\zeta_1 < \zeta_0$. However, as the horizontal distance between M_0 and M_1 is unknown, it is impossible to express the height of M_1, ζ_1, as a function of that of M_0, ζ_0. In the literature, there is still no published works that address this issue.

As a result, assumptions have to be made. For example, when $|\theta_1| > 90^\circ$, or when the ray $\hat{s}(\theta_1)$ propagates downwards, one easy attempt is to assume the height PDF of M_1 is

$$p(\zeta_1) = \frac{p_\zeta(\zeta_1)}{F_\zeta(\zeta_0)} \Upsilon(\zeta_0 - \zeta_1) \qquad \text{for} \qquad |\theta_1| > 90^\circ, \tag{3.44}$$

where p_ζ is the surface height PDF and F_ζ is the surface height cumulative function.

On the contrary, when $|\theta_1| < 90^\circ$, the height PDF of M_1 can be

$$p(\zeta_1) = \frac{p_\zeta(\zeta_1)}{1 - F_\zeta(\zeta_0)} \Upsilon(\zeta_1 - \zeta_0) \qquad \text{for} \qquad |\theta_1| < 90^\circ. \tag{3.45}$$

If $|\theta_1|$ approaches 90° from the left and right, it is physical to expect that $p(\zeta_1)$ converges to the same unique value. However, (3.44) and (3.45) satisfy this conditions if $1 - F_\zeta(\zeta_0) = F_\zeta(\zeta_0) \Rightarrow F_\zeta(\zeta_0) = 1/2$, which occurs only for $\zeta_0 = 0$ (even height PDF). It is expected that a discontinuity can occur when calculating some height-related values.

In addition, it is obvious that ζ_1 is not an independent random variable when $|\theta_1| = 90^\circ$, since $\zeta_1 = \zeta_0$, which means that ζ_1 is related to ζ_0.

To sum up, the key to the problem is to describe the correlation between M_0 and M_1 since the derivation of the height PDF is crucial. The assumptions introduced in

(3.44) and (3.45) can be a good starting point but there it is not a rigorous mathematical way. This issue is open for the readers.

Slope PDF of M_1

Similarly, the slope PDF of M_1 is different from that of the whole surface. For example, if the ray $\hat{s}(\theta_1)$ propagates leftwards, the slope of M_1 must be smaller than the slope of $\hat{s}(\theta_1)$, $\gamma_1 \leq \cot\theta_1 = \mu_1$. This leads to

$$p(\gamma_1) = \frac{p_\gamma(\gamma_1)}{F_\gamma(\mu_1)}\Upsilon(\mu_1 - \gamma_1) \qquad \text{for} \qquad \theta_1 < 0, \tag{3.46}$$

where p_γ is the surface slope PDF and F_γ is the surface slope cumulative density function. On the contrary, if the ray $\hat{s}(\theta_1)$ propagates rightwards, then the slope PDF of M_1 is

$$p(\gamma_1) = \frac{p_\gamma(\gamma_1)}{1 - F_\gamma(\mu_1)}\Upsilon(\gamma_1 - \mu_1) \qquad \text{for} \qquad \theta_1 > 0^\circ, \tag{3.47}$$

Equations (3.46) and (3.47) are derived from two basic assumptions. To our knowledge, a more rigorous derivation is not available in the literature, and this issue is also open for the readers.

Height-averaged model

It is noticeable that the integration of the uncorrelated S_M^2 shadowing function over the heights of M_1, ζ_1, does not require the knowledge of its PDF. Then, the 'partial' height-averaged shadowing function is defined as

$$\begin{aligned}
\bar{S}_M^{2,\text{unco}} &= \int_{-\infty}^{+\infty} S_M^2 p_\zeta(\zeta_1)\mathrm{d}\zeta_1 \\
&= \int_{-\infty}^{+\infty} S_M^2 \mathrm{d}F(\zeta_1) \\
&= S_M^1 \times \begin{cases} 1 & \text{for} \quad |\theta_2| > 90^\circ \\ 1 - \dfrac{1}{\Lambda(v_2) + 1} & \text{for} \quad 0 < \theta_2 < 90^\circ \\ 1 - \dfrac{1}{\Lambda_-(v_2) + 1} & \text{for} \quad -90 <^\circ \theta_2 < 0 \end{cases},
\end{aligned} \tag{3.48}$$

in which S_M^1 does not depend on ζ_1. Equation (3.48) is called 'partially' height-averaged because it is only averaged over ζ_1 and the height of M_0, ζ_0, remains.

It is found that (3.48) is very similar to that of Masuda [29] shown in (3.18) and (3.19), except that S_M^0 is replaced by S_M^1 here. The ray $\hat{s}(\theta_2)$ is treated as an independent ray. After integration over the height ζ_1, the ray $\hat{s}(\theta_2)$ is considered as if it originates from a set of average surface points because the height information on M_1 is lost. Recall that the model of Masuda [29] overestimates the probability of observing one surface reflection. Similarly, the model shown by (3.48) should have the same feature.

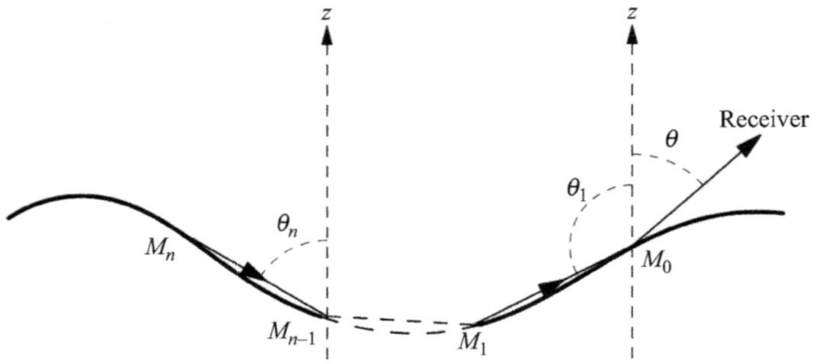

Figure 3.18 Multiple surface reflections under a monostatic configuration

3.1.6.2 Monostatic shadowing function with multiple reflections

As shown in the previous section, the statistical shadowing function with two reflections is not straightforward due to the difficulty to derive the height correlation between the first and the second intersection points. Then, the same difficulty occurs to extend the model to more reflections. To our knowledge, this issue has not be solved.

On the other hand, if high precision is not required and the heights of the successively intersection points are not required, the partially height-averaged shadowing function can be easily generalize from (3.48). This leads to

$$\overline{S}_M^{n,\text{unco}} = S_M^{n-1,\text{unco}} \times \begin{cases} 1 & \text{for} \quad |\theta_n| > 90^\circ \\ 1 - \dfrac{1}{\Lambda(v_n)+1} & \text{for} \quad 0 <^\circ \theta_n < 90^\circ \\ 1 - \dfrac{1}{\Lambda_-(v_n)+1} & \text{for} \quad -90 <^\circ \theta_n < 0^\circ \end{cases}, \tag{3.49}$$

where θ_n is the zenith reflected angle at the nth intersection point M_{n-1} shown in Figure 3.18. Equation (3.48) can be applied to calculate the sea surface infrared emissivity and reflectivity since they do not depend on the surface heights.

In fact, (3.48) is somewhat similar to the bistatic shadowing function with multiple reflections derived in [34], which is addressed in the next section.

3.2 Bistatic case from a 1D surface

3.2.1 Introduction of the problem

In this section, the bistatic shadowing function with multiple surface reflections is addressed. For a bistatic configuration, a ray is emitted by a transmitter, which encounters a series of surface reflections and then is picked up by a receiver. A receiver as well as a transmitter are involved, which are independent. The Bistatic shadowing function with multiple reflections is the key parameter to derive the sea surface infrared

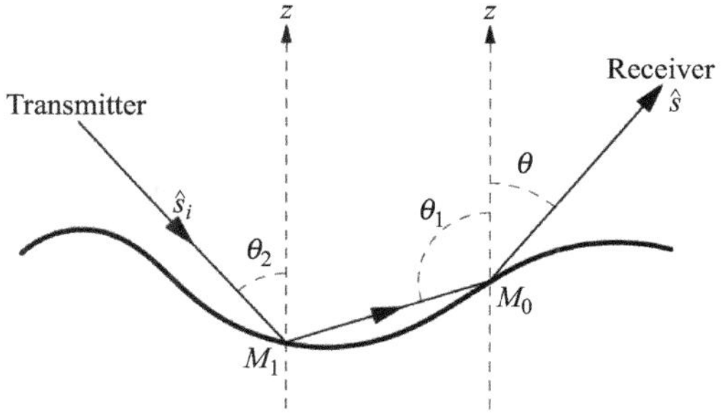

Figure 3.19 Illustration of two surface reflections with a bistatic configuration: the emitted ray form the transmitter and intersects the surface at the point M_1 and next at the point M_0. Then, the reflected ray propagates to the receiver

reflectivity and gives the probability that a ray emitted by a transmitter located above the rough surface is reflected N times before it is picked up by a receiver. Figure 3.19 illustrates the case when two surface reflections occur. A ray emitted by the transmitter propagates along the direction $\hat{s}_i$ with a zenith angle θ_2. It intersects the rough surface at the point M_1 where it is reflected and intersects the surface again at an another point M_0. The reflected ray at the point M_0 propagates along the direction $\hat{s}$ and is picked up by the receiver located in its path of propagation.

The configuration shown in Figure 3.19 is similar to that of the monostatic shadowing function with two reflections shown in Figure 3.17. The main difference is that the ray is emitted from the transmitter above the surface rather that originating from a facet on the surface.

Another significant difference is that the zenith angles θ_2 and θ are independent variables. From the inverse path principle, for a given rough surface and a given incident direction $-\hat{s}$, the direction of the reflected ray $-\hat{s}_i$ is also determined. Since θ_2 is an independent variable, it is obvious that only when the transmitter points towards the reflected ray, whose direction is obtained from the reflection laws, the received energy differs from zero. Conversely, for a given pair of θ_2 and θ, the slopes of M_1 and M_0 must satisfy the reflection laws so that the specular directions match with the angles θ_2 and θ.

For any number of surface reflections, it is obvious that no obstacle must occur between any two successive reflected rays and, between the last one from M_0 to the receiver. The probability that any surface point M_0 is viewed along the angle θ is described by the monostatic statistical shadowing function without reflection $S_M(\theta)$ derived in Chapter 1. Then, the bistatic statistical shadowing function with N reflections is derived from this probability.

For sake of clarity, the case of two reflections is first presented. The bistatic shadowing function with two reflections gives the probability that the incident ray along θ_2 intersects the rough surface twice and then the reflected ray is captured by the receiver in the direction θ. To be consistent with the derivation of the monostatic shadowing function, the reverse path principle is applied in which the ray is emitted by the receiver and captured by the transmitter.

The following events are then defined:

- 'Ray $\hat{s}(\theta)$ does not intersect the surface' is denoted symbolically as 'a'.
- 'Ray $\hat{s}(\theta_1)$ intersects the surface at point M_1' is denoted symbolically as 'b'.
- 'Reflected ray at M_1 propagates along θ_2' is denoted symbolically as 'c'.
- 'Ray $\hat{s}_i(\theta_2)$ does not intersect the surface' is denoted symbolically as 'd'.

Mathematically, the bistatic shadowing function with two reflections is expressed as

$$\begin{aligned} S_B^2 &= p(\hat{s}(\theta) \text{ does not intersect} \cap \hat{s}(\theta_1) \text{ intersects at } M_1 \\ &\quad \cap \text{ Reflected ray along } \theta_2 \cap \hat{s}_i(\theta_2) \text{ does not intersect}) \\ &= p(ab) \times p(\hat{s}_i(\theta_2) \text{ does not intersect}|ab)\delta(\theta_2 - \theta_r) \\ &= p(ab)[p(d|ab)]\delta(\theta_2 - \theta_r) \end{aligned} . \quad (3.50)$$

In (3.50), the probability $p(ab)$ is the monostatic statistical shadowing function with one reflection. The angle θ_r corresponds to the reflected direction at the point M_1 of the ray from M_0 to M_1. The Dirac delta function $\delta(\theta_2 - \theta_r)$ ensures that the reflected ray at M_1 propagates towards the transmitter.

Equation (3.50) shows that the bistatic shadowing function with two reflections S_M^2 is very similar to that of a monostatic case shown by (3.41). The main difference is that, the monostatic case requires that the ray $\hat{s}_i(\theta_2)$ intersects the surface while the bistatic case requires that it does not. As a result, S_B^2 can be derived following the same way as that for S_M^2. If so, similar difficulties as discussed in the last section are also expected.

There are not many publications in the literature addressing this topic. In Section 3.2.2, a numerical ray-tracing algorithm is developed to serve as a reference. In Section 3.2.3, the bistatic shadowing function of Lynch and Wagner [34] is reviewed. In Section 3.2.4, the bistatic shadowing function of Li *et al.* [30] is detailed.

3.2.2 Numerical tray-tracing algorithm

The derivation of the bistatic shadowing function with multiple reflections from a randomly rough surface is a difficult task. A Monte Carlo ray-tracing algorithm is a relevant tool to validate the closed expressions. All the models are based on the geometric optics approximation, but no additional assumptions are required for the ray-tracing algorithm. In addition, such an algorithm can help us to understand the shadowing mechanisms. In this section, a 1D surface is assumed, which means that the rough surface depends only on the abscissa x.

This algorithm is an extension of the one introduced previously in Section 3.1.2 where the monostatic shadowing function with one surface reflection is investigated. First, the rough surface is generated from a given surface height autocorrelation function (or its associated height spectrum) and by assuming a Gaussian process (of the surface heights and slopes). For the next simulations, the autocorrelation function is assumed to be Gaussian. Readers are free to use other autocorrelation functions or to assume that the surface points are uncorrelated [28].

The Monte Carlo process means that a large number N of randomly rough surfaces of same statistical features (only the seed of the Gaussian process changes) are generated, on which the ray-tracing algorithm is performed. Height and slope histograms of the points M_0 and M_1 and the associated mean values are computed by averaging over the results obtained from each surface. In this section, N equals 2,000 to reach the convergence.

3.2.2.1 Definition of the geometry

To be consistent with the ray-tracing algorithm introduced previously, the reverse propagation path is applied. In fact, in a bistatic configuration where the transmitter and receiver can swap, using the forward path or the reverse path, makes no difference thanks to the reciprocity principle. Figure 3.20 shows an example of two surface reflections when the reverse path is applied. A ray starting from the receiver propagates along the direction $\hat{s}_i$ and intersects the surface at the point M_0 where it is reflected in the specular direction $\hat{s}_{r1}$. The reflected ray intersects again the surface at the point M_1. At M_1, it is reflected again in the direction $\hat{s}_r$ and propagates to the transmitter. The subscript 'i' stands for incidence, and 'r' stands for reflection.

In Figure 3.20, if the incident zenith angle θ and the reflected zenith angle θ_2 are specified, the incident direction $\hat{s}_i$ and the final reflection direction $\hat{s}_r$ can be defined as

$$\hat{s}_i = -\begin{bmatrix}\sin\theta\\ \cos\theta\end{bmatrix}, \qquad \hat{s}_r = \begin{bmatrix}\sin\theta_2\\ \cos\theta_2\end{bmatrix}. \tag{3.51}$$

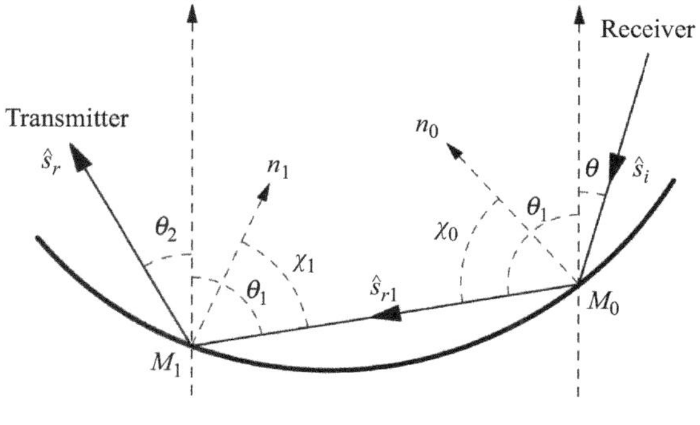

Figure 3.20 Reflection using the reverse path propagation: a ray starting from the receiver intersects the surface at the point M_0 and it is reflected in the specular direction at the point M_1 where it is reflected again towards the transmitter

Note that θ and θ_2 are positive if they are measured clockwise from the zenith and are negative otherwise. The unit normal vectors to the points M_0 and M_1 can be defined by their slope γ_0 and γ_1, respectively. This leads to

$$\hat{n}_0 = \frac{1}{\sqrt{1+\gamma_0^2}}\begin{bmatrix}-\gamma_0\\1\end{bmatrix}, \quad \hat{n}_1 = \frac{1}{\sqrt{1+\gamma_1^2}}\begin{bmatrix}-\gamma_1\\1\end{bmatrix}. \tag{3.52}$$

The local angles of incidence χ_0 and χ_1 are then given by

$$\begin{cases}\cos\chi_0 = -\hat{s}_i\cdot\hat{n}_0 = \dfrac{\cos\theta - \gamma_0\sin\theta}{\sqrt{1+\gamma_0^2}}\\[2ex] \cos\chi_1 = \hat{s}_r\cdot\hat{n}_1 = \dfrac{\cos\theta_2 - \gamma_1\sin\theta_2}{\sqrt{1+\gamma_1^2}}\end{cases}. \tag{3.53}$$

Since physically the local angle of incidence must not exceed 90° (otherwise the incident ray comes from the back of the surface), it is always required that $\cos\chi_0 \geq 0$, or equivalently from (3.53)

$$\cos\theta - \gamma_0\sin\theta \geq 0 \Rightarrow \gamma_0 \leq \cot\theta, \tag{3.54}$$

and similarly for the point M_1

$$\gamma_1 \leq \cot\theta_2. \tag{3.55}$$

Equations (3.54) and (3.55) both reveal that the slope of the surface facet has to never exceed the slope of the incident ray to intersect the facet, and hence a reflection occurs.

With the knowledge of the incident direction $\hat{s}_i$ and the surface normal vector $\hat{n}_0$, the specular reflected direction can then be calculated as

$$-2\hat{n}_0(\hat{n}_0\cdot\hat{s}_i) = -\hat{s}_i + \hat{s}_{r1} \Rightarrow \hat{s}_{r1} = 2\cos\chi_0\hat{n}_0 + \hat{s}_i, \tag{3.56}$$

and similarly for the point M_1

$$-2\hat{n}_1(\hat{n}_1\cdot\hat{s}_{r1}) = -\hat{s}_{r1} + \hat{s}_r \Rightarrow \hat{s}_r = -2\cos\chi_1\hat{n}_1 + \hat{s}_{r1}. \tag{3.57}$$

It is clear that the incident angle θ and the slopes γ_0 and γ_1 must satisfy the above equations such that the transmitter angle checks $\theta_2 = \theta_r$.

3.2.2.2 Implementation

In this subsection, the implementation of the ray-tracing algorithm with multiple surface reflections under a bistatic configuration is detailed. The developed algorithm is a continuation of that presented in Section 3.1.2.2.

As shown in Figure 3.20, we consider that the receiver and the transmitter are located on the right- and left-hand sides of the rough surface, respectively. It is easy to swap the locations of the receiver and the transmitter by reversing the direction of the horizontal axis.

First a large number of rough surfaces is generated and the same ray-tracing algorithm is applied over all the surfaces in order to reach the convergence. This allows us to locate all the surface points which are illuminated directly by the receiver and on which the reflected rays intersect the surface again. This process is fully described in Section 3.1.2.2 and the readers are invited to refer to Section 3.1.2.2 for the implementation of this step. Referring to Figure 3.20, the aim of this step is to find all the groups of points M_0 and M_1.

The second step is to determine the actual reflected angle θ_r at each point M_1 obtained in the previous step and determine if it equals the specified angle θ_2. In fact, numerically θ_r does not exactly equal θ_2. Then, a small region $\Delta\theta$ around the given reflected angle θ_2 is considered. In other words, if the following criterion is fulfilled

$$|\theta_r - \theta_2| < \Delta\theta, \tag{3.58}$$

the reflected ray $\hat{s}_r(\theta_r)$ is captured by the transmitter of angle θ_2. At the end of step two, all groups of M_0 and M_1 are removed if their reflected angle θ_r does not satisfy (3.58).

The third step is to determine if the reflected ray at each point M_1 obtained in the previous step intersects the surface again. The specular reflection direction $\hat{s}_r$ is derived from (3.57). The reflected ray at M_1 can then be drawn and its equation is given by

$$\begin{cases} z = (x - x_1)\cot\theta_2 + z_1 \\ x < x_1 \text{ if } \hat{s}_r \text{ propagates to the left} \\ x > x_1 \text{ if } \hat{s}_r \text{ propagates to the right} \end{cases}, \tag{3.59}$$

where (x_1, z_1) are the Cartesian coordinates of the point M_1.

Then, this ray is plotted to determine whether it intersects the surface at a point other than M_1. If so, more than two reflections occur, which is not considered in this step. The corresponding group of points M_0 and M_1 is then removed. If the reflected ray does not intersect the surface again, then this group of points M_0 and M_1 is marked. A detailed illustration of the algorithm is shown in Figure 3.21.

In fact, the third step of the ray-tracing algorithm is very similar to that described in Section 3.1.2.2, except that, the ray-tracing starts at the second intersection point M_1, while in Section 3.1.2.2, it starts at the first point M_0. In addition, we are looking for the case when the reflected ray does not intersect the surface again.

The statistical properties of the point M_0 can be studied, such as distributions of its heights and slopes, which point out what slopes or heights are more likely to lead to two surface reflections. This corresponds to the marginal height or slope histograms of the bistatic shadowing function with two reflections. Looking at the height or slope histograms of M_1 help us to develop an analytical model.

In addition, the ratio of the number of groups N_1 of points M_0 and M_1, or equally the number of remaining points M_0, over the total number of the surface points N_{total} corresponds to the average bistatic shadowing function with two reflections

$$\bar{S}_B^2(\theta, \theta_2) = \frac{N_1}{N_{\text{total}}}. \tag{3.60}$$

The average bistatic shadowing function with two reflections gives the percentage of the surface where two successive surface reflections occur.

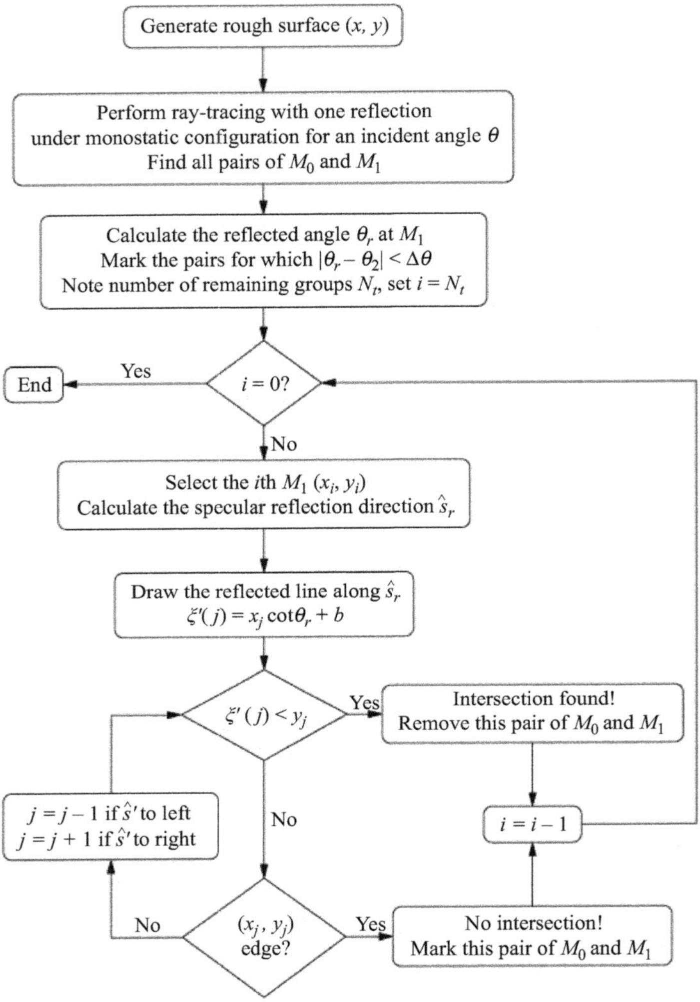

Figure 3.21 Illustration of the ray-tracing algorithm with two surface reflections under a bistatic configuration

If more reflections occur, the corresponding bistatic ray-tracing algorithm with N reflections can be developed by adopting the monostatic one with $N - 1$ reflections and then performing step two, as described above. For N reflections, all intersection points, from M_0 to M_{N-1}, have to be grouped. Similarly, the associated marginal height or slope histograms as well as the corresponding average shadowing function can be performed.

As claimed before, from an analytical model, when two or more reflections are involved, the abscissa of the second intersection point M_1 is unknown. This can cause

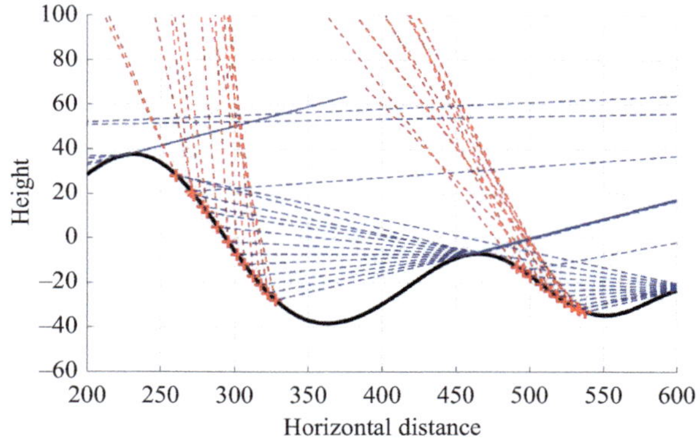

Figure 3.22 One example of the ray-tracing algorithm under a bistatic configuration with two surface reflections. The incident rays originating from the receiver and the reflected rays at the point M_0 are marked in blue, whereas the reflected rays at the point M_1 is marked in red. The second intersection point is denoted by the red symbol '+'. $\theta = 80°$

some problems for the calculations of the ensemble averages. From a ray-tracing algorithm, the statistics of any intersection point is available. Besides, since a lot of groups of points are removed after every intersection, the computational consumption of the ray-tracing process does not increase significantly as the number of reflections increases. Thus, it is rather easier to determine a shadowing function, monostatic or bistatic alike, when the number of reflections is large, than to derive an analytical model.

3.2.2.3 Numerical results

Illustration

Figure 3.22 shows a piece of a generated rough surface, on which the ray-tracing algorithm with two reflections is performed. The blue lines are the incident rays originated from the receiver and the corresponding reflected rays emanating from the first intersection point. Then, these rays intersect the surface again at those points marked by the red symbol '+' and the reflected rays at the second intersection points are marked in red.

In Figure 3.22, the direction of the reflected ray $\hat{s}_r$ at the second intersection point M_1 is not forced to a specific value θ_2. In other words, the second step in the ray-tracing algorithm, which requires that the reflected ray $\hat{s}_r$ satisfies $\theta_r = \theta_2$ is ignored. As a result, in Figure 3.22, the angles of the final reflected rays vary. If θ_2 is specified, the red reflected rays in Figure 3.22 has to be filtered again to find out the ones with an angle θ_r satisfying $|\theta_2 - \theta_r| < \Delta\theta$.

In Figure 3.22, all the reflected rays at the first intersection propagate to the left. In fact, left-propagated reflected rays are dominant when the observation angle θ is large. To have a reflected ray propagating to the right, the slope of the first intersection

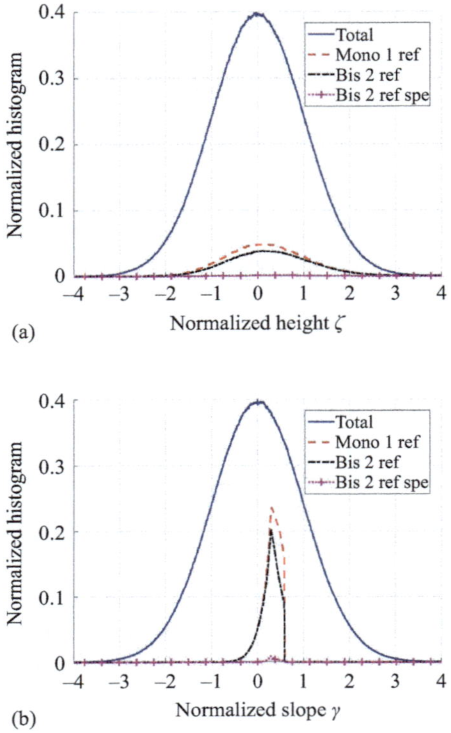

Figure 3.23 Height (a) and slope (b) histograms of the first intersection point by the Monte Carlo ray-tracing algorithm. $\theta = 80°$, $\sigma_\gamma = 0.3$ *and* $N = 5{,}000$

has to be very large, which is not likely to occur. If θ decreases, then a reflected ray which propagates to the right start to show up, as the required slope becomes smaller.

Figure 3.22 also shows that although the range of the reflected angle θ_r is not specified, the directions of the final reflected rays do not vary much. Instead, θ_r is distributed in a small range not far from the global specular reflection direction. More discussion will be given in Section 3.2.2.3.

Histograms of the first intersection point

Figure 3.23 shows the surface height and slope histograms of the first intersection point obtained by the Monte Carlo ray-tracing algorithm with two reflections under a bistatic configuration. The incident angle $\theta = 80°$, the RMS slope $\sigma_\gamma = 0.3$ and the number of realizations $N = 5,000$. The height and slope are normalized by their RMS value σ_ξ and σ_γ, respectively. The histograms are also normalized so that their integration over the normalized heights or slopes of the whole surface equals one.

In Figure 3.23, to show clearly the impact of an additional reflection, the histograms obtained for a monostatic configuration (Section 3.1.2.2) is also plotted.

For the bistatic results, the ones without specifying the angle θ_2 as well as the ones for $\theta_2 = 70^\circ \pm 1^\circ$ are also shown for comparison. In addition, the histogram of the whole surface is plotted, which follows a Gaussian profile

The height histogram of the first intersection point of the monostatic configuration with one reflection has a bell-shape dashed line, but its magnitude is largely reduced, which implies that the number of points leading to one surface reflection is only a small portion of the whole surface. At the second intersection, the second reflected rays are traced and only those which do not intersect the surface are kept. As a result, the magnitude of the histogram is reduced again (dash-dot line). Applying the additional restriction that $\theta_r \in [69^\circ; 71^\circ]$, the magnitude of the histogram (dot-plus line) decreases significantly.

In general, the addition of the event that the second reflected ray does not intersect the surface again does not change the shape of the height histogram associated to the first intersection point.

The slope histogram is shown in Figure 3.23(b). As expected, the one of the total surface follows a Gaussian distribution. The one for the first intersection point (dashed line) is non-zero only in a small range. As discussed in the previous section, the normalized slopes smaller than the one that occurs at the peak correspond to reflected rays propagate downwards, while the other slopes correspond to the case when they propagates upwards.

After adding the requirement that the second reflected ray does not intersect the surface, the slope histogram decreases mainly when the first reflected ray propagates downwards. Indeed, the second intersection point M_1 has a lower height, which reduces the probability that the ray does not intersect the surface. Applying the additional restriction $\theta_r \in [69^\circ; 71^\circ]$, the magnitude of the histogram (dot-plus line) decreases significantly with a change of its shape.

Histograms of the second intersection point

Figure 3.24 plots the surface height and slope histograms of the second intersection point obtained from a Monte Carlo ray-tracing algorithm with two reflections under a bistatic configuration. The simulations parameters are the same as in Figure 3.23. The results are also compared with the ones obtained for a monostatic configuration with one reflection.

Note that the height and slope histograms of the second intersection points shown here do not correspond to the height and slope probability density function of the second intersection point discussed in Section 3.1.6.1. The main difference is that in Section 3.1.6.1, the height of the first intersection point as well as the propagation direction of the first reflected ray are known, and the second intersection point fulfilling the above criteria is considered. In this section, all of the second intersection points are accounted for.

The aim is to study the statistics of the second intersection points, expecting to help the readers to solve similar problems in the future.

The surface height histograms are shown in Figure 3.24(a). The one (dashed line) of the second intersection point M_1 in Figure 3.20 has a bell shape, which is very similar to that of the first intersection. After the second intersection, some points are

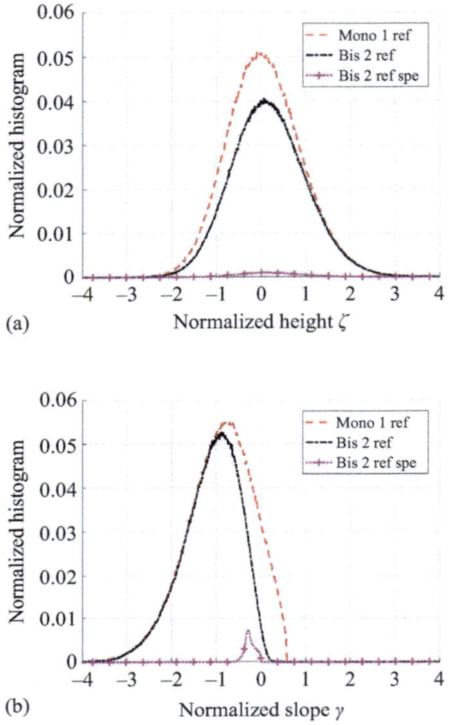

Figure 3.24 Surface height (a) and slope (b) histograms of the second intersection point obtained from a Monte Carlo ray-tracing algorithm. The simulations parameters are the same as in Figure 3.23

filtered out because of the requirement that the second reflected rays do not intersect the surface again. As a result, the magnitude of the histogram is reduced (dash-dot line). Applying the restriction $\theta_r \in [69°; 71°]$, the magnitude of the histogram (dot-plus line) decreases significantly. Similar to the first intersection point, the shape of the height histogram of the second intersection point is always similar to a bell shape. This may also reveal that there is no strong preference in height for the second intersection regarding to the occurrence of two successive surface reflections.

The surface slope histograms are shown in Figure 3.24(b). The slope histogram of the second reflection point looks quite different to that of the first intersection. Applying the ray-tracing algorithm for the second reflected ray reduces the magnitude of the slope histogram but do not change the shape much. The shape of the slope histogram looks somewhat similar to a log-normal distribution but obviously a more precise physical-based model is required.

Histogram of the second reflected angle θ_2

For a given incidence angle θ, the probability that two successive reflections occur with the second reflected ray leaving the surface at a zenith angle θ_2 is of interest.

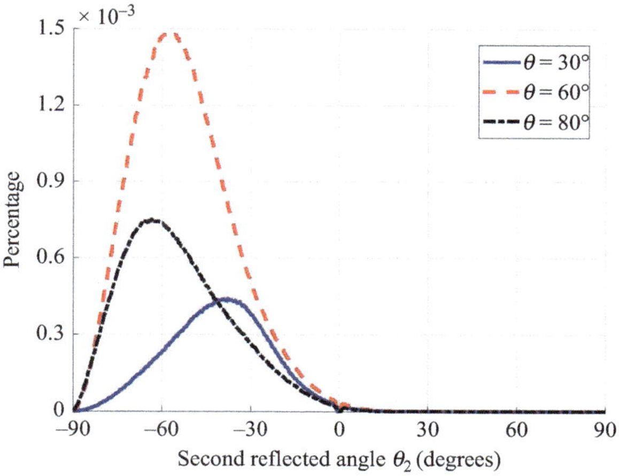

Figure 3.25 Histogram of the second reflected angle θ_2. The RMS slope $\sigma_\gamma = 0.3$, the number of surface realizations $N = 3,000$ and the incidence angles $\theta = 30°$, $60°$ and $80°$

As mentioned previously, it is impossible to fix both θ and θ_2 in a numerical computation. As a result, the probability that the reflected angle θ_r falls between $\theta_2 - \Delta\theta$ and $\theta_2 + \Delta\theta$ is approximated as the one that θ_r equals θ_2. In this section, $\Delta\theta = 0.2°$ is considered. This probability is related to the histogram of θ_2 with a sampling step of 0.4°.

The histogram of θ_2 is plotted in Figure 3.25 and it is obtained from the ray-tracing algorithm. The RMS slope $\sigma_\gamma = 0.3$, the number of surface realizations $N = 3,000$ and the incidence angles $\theta = 30°$, $60°$ and $80°$. The histograms are normalized by the total number of surface points. In other words, for each θ_2, the strength gives the percentage of the surface points over the whole surface for which the second reflected angle ranges from $\theta_2 - 0.2°$ to $\theta_2 + 0.2°$.

The shapes of the histogram depend on the incidence angle θ and do not seem to fit to a common distribution function. In Figure 3.25, θ_2 is distributed mostly in the negative region, meaning that most of the time the second reflected rays propagate to the left-hand side of the surface. The peaks of the curves are difficult to predict. In general, it occurs near the global specular reflection direction $\theta_2 = -\theta$.

In addition, the magnitudes of the curves depend on the incident angle θ. For instance, for $\theta = 60°$, the magnitude is the most significant, meaning that two successive reflections are most likely to occur. In fact, this particular value of θ depends also on the RMS slope. It is the purpose of the following subsection.

Percentage of area leading to two reflections

For a given incident angle θ, the probability that two successive reflections occur with the second reflected ray leaving the surface regardless of its direction is also a relevant parameter. It reveals which incident angles θ are more likely to give rise to two surface reflections. In a ray-tracing algorithm, it is obtained by dividing the

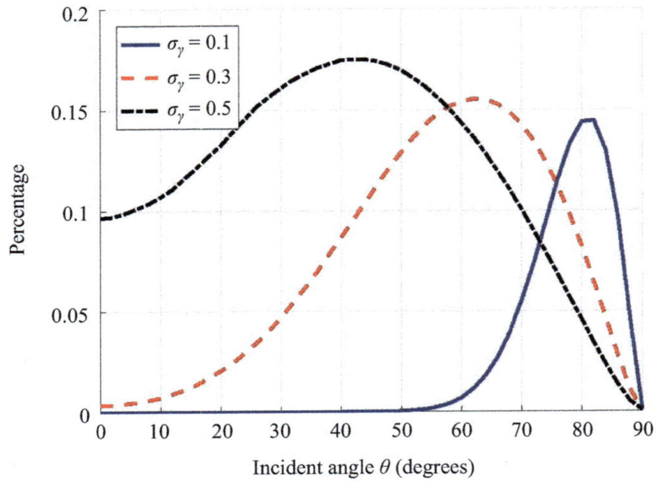

Figure 3.26 Percentage of surface leading to two successive reflections versus the incidence angle θ. The RMS slope $\sigma_\gamma = 0.1$, 0.3 and 0.5

number of the remaining groups N_1 of M_0 and M_1 by the number of points of the whole surface. This leads to

$$\bar{S}_B^2(\theta) = \frac{N_1}{N_{\text{total}}}. \tag{3.61}$$

This value corresponds to the hemispherical average of the bistatic shadowing function with two reflections.

This function is plotted in Figure 3.26 versus the incidence angle θ. The RMS slopes $\sigma_\gamma = 0.1, 0.3$ and 0.5. In fact, this result is similar to the percentage of surface leading to one reflection under a monostatic configuration plotted in Figure 3.6.

As we can see in Figure 3.26, the strengths are small when θ is small and increase as θ increases. Double surface reflections tend to occur when the incident angle θ is large. For $\sigma_\gamma = 0.1$, double surface reflection is only found when θ is larger than 60°. The range of θ becomes larger when σ_γ increases. Only for large RMS slopes, the strengths are significant when θ is small.

Besides, the maximum slightly increases as the RMS slope σ_γ increases. For $\sigma_\gamma = 0.5$, the maximum is about 0.2, which means that two successive surface reflections occur on about 20 per cent of the surface points.

3.2.3 Model of Lynch and Wagner

3.2.3.1 Derivation

Lynch and Wagner [34] developed a bistatic shadowing function with two surface reflections from the height-averaged Smith monostatic shadowing function [5]. In this model, 1D surfaces are considered. The distributions of the surface heights and slopes are assumed to be Gaussian. In this section, this model is summarized for the case where the transmitter and the receiver are located on the left- and right-hand sides of the surface, respectively.

In the model of Lynch and Wagner [34], the reverse path is also applied (see Figure 3.20). To express the bistatic shadowing function with two reflections S_B^2, the model is based on the same derivation presented in Section 3.2.1. In the following, the notation in [34] is modified to agree with that in this book. Four events, very similar to those introduced in Section 3.2.1, are defined as

- 'Ray $\hat{s}_i(\theta)$ does not intersect the surface' is denoted symbolically as 'a'.
- 'Ray $\hat{s}_{r1}(\theta_1)$ intersects the surface at the point M_1' is denoted symbolically as 'b'.
- 'The slope of M_1 reflects the incident ray in the direction of θ_2' is denoted symbolically as 'c'.
- 'Ray $\hat{s}_r(\theta_2)$ does not intersect the surface' is denoted symbolically as 'd'.

Then

$$S_B^2 = p(abcd) = p(a)p(b|a)p(c|ab)p(d|abc). \tag{3.62}$$

Lynch and Wagner [34] defined the probability $p(a)$ as the uncorrelated height-averaged Smith shadowing function given by

$$p(a) = \Upsilon(\mu - \gamma_0)\frac{1}{1 + \Lambda(|v|)}, \tag{3.63}$$

where

$$\left\{\begin{aligned} v &= \frac{\cot\theta}{\sigma_\gamma\sqrt{2}} \\ \Lambda(v) &= \frac{\exp(-v^2) - v\sqrt{\pi}\,\mathrm{erfc}(v)}{2v\sqrt{\pi}} \\ \mathrm{erfc}(v) &= \frac{2}{\sqrt{\pi}}\int_v^{+\infty}\exp(-t^2)\mathrm{d}t \end{aligned}\right., \tag{3.64}$$

where σ_γ is the surface RMS slope. In addition, since the process is assumed to be Gaussian (thus even), $\Lambda_-(v) = \Lambda(|v|)$ (case for which $v < 0$).

The conditional probability $p(b|a)$ is given by [34]

$$\begin{aligned} p(b|a) &= 1 - p(\bar{b}|a) = 1 - \frac{p(\bar{b}a)}{p(a)} \\ &= \left\{\begin{array}{ll} \dfrac{\Lambda(|v_1|)}{1 + \Lambda(|v|) + \Lambda(|v_1|)} & \text{for } -90^\circ < \theta_1 \le 0 \\ 0 & \text{for } 0^\circ < \theta_1 < \theta \\ 1 - \dfrac{1 + \Lambda(|v|)}{1 + \Lambda(|v_1|)} & \text{for } \theta < \theta_1 \le 90^\circ \\ 1 & \text{for } |\theta_1| > 90^\circ \end{array}\right., \end{aligned} \tag{3.65}$$

where $v_1 = \cot\theta_1/(\sigma_\gamma\sqrt{2})$. Note that for 1D surfaces, the zenith angles are oriented with the clockwise direction being the positive direction. The angle θ_1 can be positive or negative, depending on the incident angle θ and on the slope γ_0 of the point M_0.

The conditional probability $p(c|ab)$ requires that the ray $\hat{s}_{r1}$ is reflected into the specular direction $\hat{s}_r(\theta_2)$. In other words, the surface slope at the point M_1 must satisfy

$$\hat{n}_1^{\text{spe}} = \frac{1}{\sqrt{1+\left(\gamma_1^{\text{spe}}\right)^2}}\begin{bmatrix}-\gamma_1^{\text{spe}}\\ 1\end{bmatrix} = \frac{\hat{s}_r - \hat{s}_{r1}}{\left\|\hat{s}_r - \hat{s}_{r1}\right\|}, \tag{3.66}$$

where γ_1^{spe} stands for the slope of M_1 which reflects the ray $\hat{s}_{r1}(\theta_1)$ in the specular direction $\hat{s}_r(\theta_2)$. The conditional probability $p(c|ab)$ is then given by a Dirac delta function as

$$p(c|ab) \approx \delta\left(\gamma_1 - \gamma_1^{\text{spe}}\right). \tag{3.67}$$

Considering the dependence on the events a and b, Lynch and Wagner [34] stated that the slope of M_1 has to check the condition that the local angle of incidence at M_1 satisfies $|\chi_1| < 90°$. Inversely, it is assumed that any slope checking this condition equals the slope of M_1. Thus, the slope PDF of M_1 is defined as

$$p(\gamma_1) = \frac{\Upsilon(\mu_1 - \gamma_1)}{\displaystyle\int_{-\infty}^{\mu_1} p_\gamma(t)\mathrm{d}t} p_\gamma(\gamma_1), \tag{3.68}$$

where $\mu_1 = \cot\theta_1$ and p_γ is the surface slope PDF.

The function $p(d|abc)$, which is the conditional probability that M_1 is viewed from the transmitter along the $\hat{s}_r$ direction, given that the ray of direction $\hat{s}$ does not intersect the surface and $\hat{s}_{r1}$ intersects the surface at M_1, and it is reflected into the specular direction $\hat{s}_r$. Lynch and Wagner [34] pointed out that it is not straightforward to derive the correlation of a, b and c with d. Thus, their dependence is ignored, and $p(d|abc)$ is approximated by

$$p(d|abc) \approx p(d) = \begin{cases} \dfrac{1}{1+\Lambda(|v_2|)} & \text{for } \; -\pi/2 < \theta_1 < 0, |\theta_2| < \pi/2 \\ \dfrac{1}{1+\Lambda(|v_2|)} & \text{for } \; \theta_1 < -\pi/2, -\pi/2 < \theta_2 < 0 \\ 1, & \text{for } \; \theta_1 < -\pi/2, 0 < \theta_2 < \theta \\ \dfrac{1}{1+\Lambda(|v_2|)} & \text{for } \; \theta_1 < -\pi/2, \theta < \theta_2 < \theta_1 \\ \dfrac{1}{1+\Lambda(|v_2|)} & \text{for } \; 0 < \theta_1 < \pi/2, |\theta_2| < \pi/2 \\ \dfrac{1}{1+\Lambda(|v_2|)} & \text{for } \; \theta_1 > \pi/2, \theta_1' < \theta_2 < \pi/2 \\ 0 & \quad\text{otherwise} \end{cases}, \tag{3.69}$$

where $v_2 = \cot\theta_2/(\sqrt{2}\sigma_\gamma)$.

The bistatic illumination function with two reflections S_B^2 is then obtained by substituting (3.63), (3.65), (3.67) and (3.69) into (3.62). Lynch and Wagner [34] showed that after taking into account the second reflection of the surface, the energy conservation condition was better met. Since this model was developed from the height-averaged illumination function, it is not possible to take into account the correlation between the surface heights and slopes.

In the following section, the model is compared with the Monte Carlo ray-tracing algorithm.

3.2.3.2 Numerical results

Slope histogram of M_1 for a given θ_2

From the model of Lynch and Wagner [34], (3.62)–(3.69), the bistatic illumination function with two reflections S_B^2 depends on the incident angle θ, the reflected angle θ_2 and the slope γ_0 of first intersection point M_0. In order to compare with the results obtained from the Monte Carlo ray-tracing algorithm, the reflected angle $\theta_r \in [\theta_2 - 1°; \theta_2 + 1°]$ is considered. The slope histogram of M_0 is then given by

$$\widetilde{p}_{\gamma_0}^{2,\text{spe}}(\theta, \gamma_0) = p_\gamma(\gamma_0) \int_{\gamma_1 \to \theta_r = \theta_2 \pm 1°} S_B^2(\theta, \gamma_0, \gamma_1) p_\gamma(\gamma_1) \mathrm{d}\gamma_1. \tag{3.70}$$

It corresponds to the probability that the slope γ_0 of the first intersection point M_0 gives rise to a pair of two successive reflections.

Figure 3.27 plots the slope histogram versus the normalized slope γ/σ_γ. The incidence angle $\theta = 80°$, the reflected angle $\theta_2 = -80°$ and the surface RMS slope $\sigma_\gamma = 0.3$.

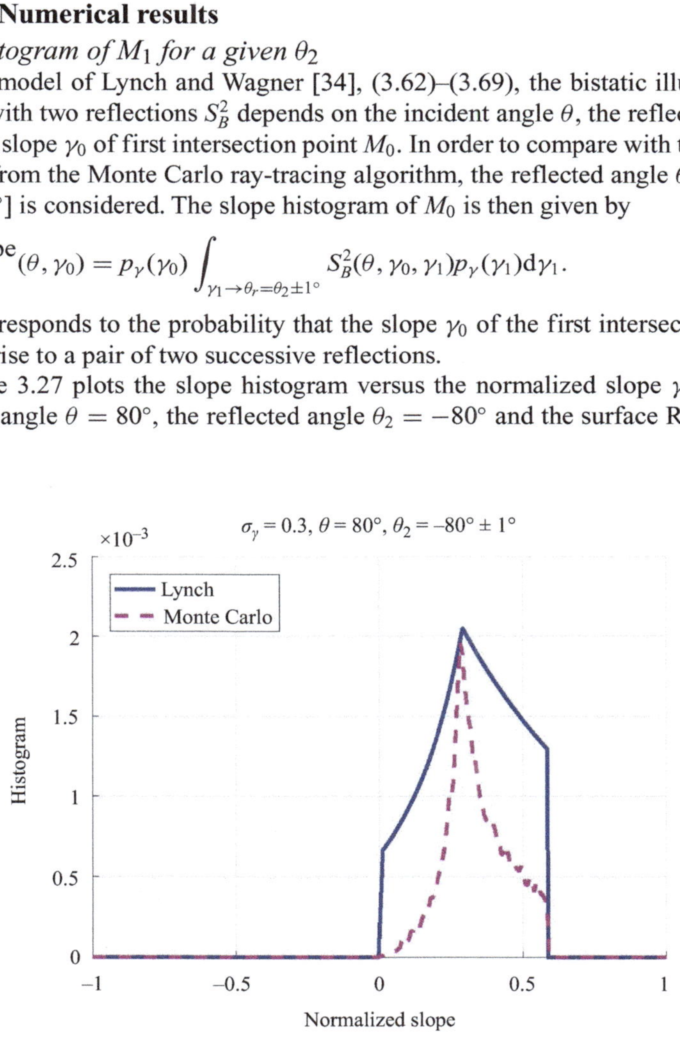

Figure 3.27 Slope histogram of the bistatic illumination function with two reflections S_B^2 from Lynch and Wagner [34] model and versus the normalized slope γ/σ_γ. The incidence angle $\theta = 80°$, the reflected angle $\theta_2 = -80°$ and the surface RMS slope $\sigma_\gamma = 0.3$

As we can see, the results are overestimated in comparison to the Monte Carlo ones, and the range of the slope γ_0 is well predicted. In fact, according to the configuration, the model of Lynch and Wagner [34] may also underestimate the slope histogram. Readers can refer to Section 3.2.3.2 for more discussion.

The peak located around the normalized slope $\gamma/\sigma_\gamma \approx 0.3$ corresponds to the case when the first reflected angle is $\theta_1 = -90°$. In other words, on the right of the peak, the first reflected ray propagates downwards, while on the left, it propagates upwards. The location of the peak is well predicted.

Slope histogram of M_1 for any θ_2

If the value of θ_2 is not specified, the slope of the second intersection point can vary so that $-90° < \theta_2 < 90°$. In this case, the slope histogram gives the probability that the slope γ_0 of the first intersection M_0 leads to a pair of two successive reflections without a restriction on the angle range θ_r. It can be obtained as

$$\widetilde{p}^{2,\text{all}}_{\gamma_0}(\theta,\gamma_0) = p_\gamma(\gamma_0)\int_{\gamma_1\to\theta_r\in[-90°;90°]} S_B^2(\theta,\gamma_0,\gamma_1)p_\gamma(\gamma_1)\mathrm{d}\gamma_1. \tag{3.71}$$

Figure 3.28 plots the slope histogram of M_0 versus the normalized slope γ/σ_γ. The incidence angle $\theta = 80°$ and the surface RMS slope $\sigma_\gamma = 0.3$ (same as in Figure 3.27 where the integration range over γ_1 is determined such that $\theta_r \in [-81; -79]$, instead of $\theta_r \in [-90°; 90°]$ in Figure 3.28).

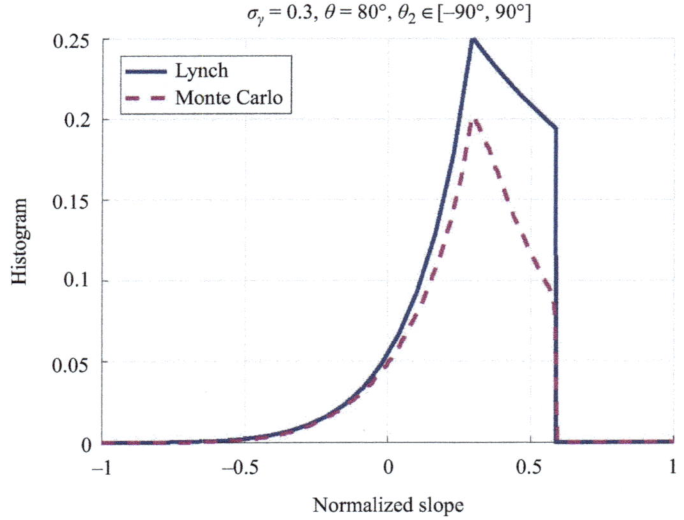

Figure 3.28 Slope histogram of the bistatic illumination function with two reflections S_B^2 from Lynch and Wagner [34] model and versus the normalized slope γ/σ_γ. The incidence angle $\theta = 80°$, the reflected angles $\theta_2 \in [-90°; 90°]$ and the RMS slope $\sigma_\gamma = 0.3$

As we can see, the model slightly overestimates the results in comparison to those obtained from the Monte Carlo ray-tracing algorithm, which is consistent with Figure 3.27. The peak located around the normalized slope $\gamma/\sigma_\gamma \approx 0.3$ is also well predicted, which corresponds to the case when the first reflected angle is $\theta_1 = -90^\circ$. The shapes of the curves are quite similar. On the right of the peak (downwards propagation), the overestimation is more significant than that at on the left (upwards propagation).

Histogram of the second reflected angle θ_2

The histogram of the second reflected angle can be obtained by integrating the bistatic shadowing function S_B^2 defined in (3.62) over the slope γ_0 of the first intersection point M_0 and over the slope γ_1 of the second intersection point M_1, for which the reflected angle θ_2 equals the desired value θ_2'. For the same reason as discussed in Section 3.2.2.3, for a given θ, the reflected angle θ_r is defined as $\theta_r \in [\theta_2 - 0.2^\circ, \theta_2 + 0.2^\circ]$ so that the histogram can be calculated numerically. Then

$$\widetilde{S}^2_{B,\theta_2\pm0.2^\circ} = \int_{-\infty}^{+\infty}\int_{\gamma_1\rightarrow\theta_r\in\theta_2\pm0.2^\circ} S_B^2(\theta,\gamma_0,\gamma_1)p_\gamma(\gamma_0)p_\gamma(\gamma_1)\mathrm{d}\gamma_1\mathrm{d}\gamma_0. \tag{3.72}$$

Equivalently, it can also be obtained by integrating (3.70) over γ_0 after modifying the range of θ_2 accordingly.

Figure 3.29 plots the histogram of the reflected angle θ_2 computed from the model of Lynch and Wagner [34] and from the Monte Carlo ray-tracing algorithm. The surface RMS slope $\sigma_\gamma = 0.3$. The incidence angle $\theta = 30^\circ$ in (a), $\theta = 60^\circ$ in (b) and $\theta = 80^\circ$ in (c).

As we can see, the strengths are of the same order and the agreement is the best for $\theta = 60^\circ$. Similar results are found for other surface RMS slopes.

In Figure 3.27 an overestimation is found for $\theta = 80^\circ$, which is consistent with Figure 3.29(c) at $\theta_2 = -80^\circ$. In fact, in Figure 3.29(c), the strengths are overestimated for all values of θ_2 when $\theta = 80^\circ$. However, for $\theta = 30^\circ$ and $\theta = 60^\circ$, an underestimation can be found depending of the value of θ_2.

For $\theta = 30^\circ$, the shapes of the histograms do not agree well, whereas for $\theta = 60^\circ$ and $\theta = 80^\circ$, the agreement is better. Overall, the model of Lynch and Wagner [34] gives satisfactory results.

Hemispherical averaged bistatic shadowing function

The hemispherical averaged bistatic shadowing function with two reflections gives the percentage of the surface area leading to two reflections. For a given incidence angle θ, it can be obtained by integrating the bistatic shadowing function S_B^2, defined in (3.62), over the slopes γ_0 and γ_1 of the first M_0 and second M_1 intersection points, respectively. Then

$$\bar{S}_B^2(\theta) = \int_{-\infty}^{+\infty}\int_{-\infty}^{+\infty} S_B^2(\theta,\gamma_0,\gamma_1)p_\gamma(\gamma_0)p_\gamma(\gamma_1)\mathrm{d}\gamma_1\mathrm{d}\gamma_0. \tag{3.73}$$

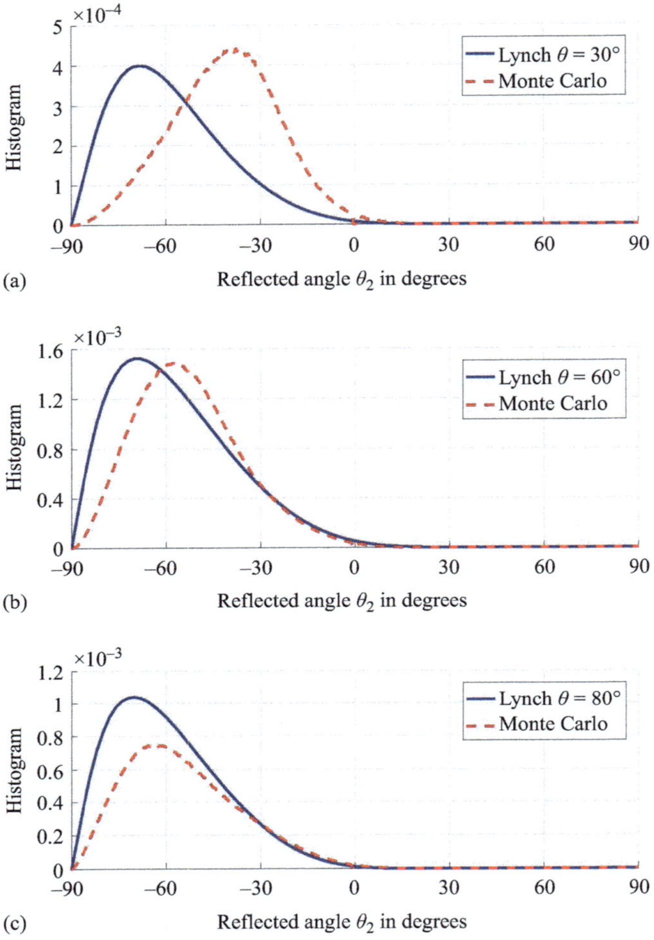

Figure 3.29 Histogram of the reflected angle θ_2 of the bistatic illumination function with two reflections S_B^2. The surface RMS slope $\sigma_\gamma = 0.3$. The incidence angle $\theta = 30°$ in (a), $\theta = 60°$ in (b), and $\theta = 80°$ in (c)

Equivalently, it can also be obtained by integrating (3.71) over the surface slope γ_0.

Figure 3.30 plots the hemispherical averaged bistatic shadowing function with two reflections $S_B^2(\theta)$ computed from the model of Lynch and Wagner [34] and the Monte Carlo ray-tracing algorithm. The surface RMS slope $\sigma_\gamma = 0.1$ in (a), $\sigma_\gamma = 0.3$ in (b), and $\sigma_\gamma = 0.5$ in (c).

A satisfactory agreement is found between the two approaches, especially when the RMS slope is small. For small incidence angles θ, the model of Lynch and Wagner

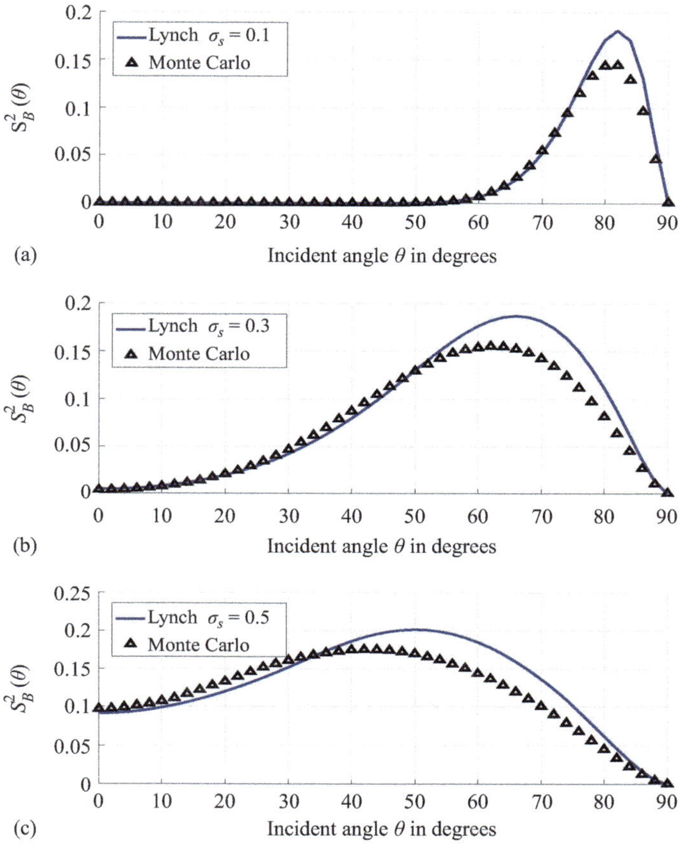

Figure 3.30 Hemispherical averaged bistatic shadowing function with two reflections S_B^2. The surface RMS slope is $\sigma_\gamma = 0.1$ in (a), $\sigma_\gamma = 0.3$ in (b) and $\sigma_\gamma = 0.5$ in (c)

slightly underestimates the results, while an overestimation is found for intermediate and large incidence angles.

To sum up, the model of Lynch and Wagner [34] predicts the slope histograms of the first intersection reasonably well. Although an overestimation or an underestimation can be found according to the chosen configuration, the range of the slope is well predicted. The histogram of the reflected angle θ_2 is also well predicted except for the case when the incident angle is small. The hemispherical averaged results also agree quite well with the Monte Carlo method. The relations between the four events *a*, *b*, *c* and *d* defined in Section 3.2.3.1 are not well considered when deriving the conditional probabilities.

For example, the link between the heights of the first M_0 and second M_1 intersection points is ignored. When the ray at M_0 with an incidence angle θ_1 propagates upwards, the height of M_1 should be higher than that of M_0. However, in the model of Lynch and Wagner [34], this statement cannot be considered because it is based on the height-averaged Smith shadowing function, which does not depend on the heights of the intersection points. In the next section, another analytical model is developed, which accounts for this point.

3.2.4 *Model of Li* et al.

3.2.4.1 Derivation for 1D surfaces

The model of Li *et al.* [30] is similar to that of Lynch and Wagner [34] presented in the previous section. The main difference is that, this model is based on the statistical monostatic shadowing function of Smith [5], while the model of Lynch and Wagner [34] is also based on this function but averaged over the surface heights.

First of all, the four events a, b, c and d defined in Section 3.2.3.1 are introduced. The bistatic shadowing function with two reflections is then expressed as

$$S_B^2 = p(abcd) = p(ab)p(c|ab)p(d|abc). \tag{3.74}$$

Determination of $p(ab)$

Note that the product $p(a)p(b|a)$ equals the monostatic illumination function with one surface reflection S_M^1 derived in Section 3.1.5.1. It is not repeated here and this model is able to consider the correlation between surface slopes and heights and requires numerical integrations. The uncorrelated model can be expressed as

$$\begin{aligned} S_M^{1,\text{unco}}(\theta, \gamma_0, \zeta_0) &= \Upsilon(\mu - \gamma_0) F(\zeta_0)^{\Lambda(\mu)} \\ &\times \begin{cases} 1 & \text{if} \quad |\theta_1| > 90^\circ \\ 0 & \text{if} \quad 0^\circ < \theta_1 \leq \theta \\ 1 - F(\zeta_0)^{\Lambda(\mu_1)} & \text{if} \quad \theta < \theta_1 \leq 90^\circ \\ 1 - F(\zeta_0)^{\Lambda_-(\mu_1)} & \text{if} \quad -90^\circ \leq \theta_1 < 0^\circ \end{cases}, \end{aligned} \tag{3.75}$$

where γ_0 and ζ_0 are the surface slopes and heights of the first intersection point M_0. The term $\mu = \cot\theta$ is the slope of the incident ray and $\mu_1 = \cot\theta_1$ is the slope of the first reflected ray (see Figure 3.20).

Determination of $p(c|ab)$

The conditional probability $p(c|ab)$ is related to the slope PDF of the point M_1. Considering the dependence on the events a and b, Li *et al.* [30] assumed that any slope checking this condition can equally be the slope of M_1. This hypothesis is the same as introduced by Lynch and Wagner [34]. Thus, $p(c|ab)$ is defined as

$$p(c|ab) \approx \delta\left(\gamma_1 - \gamma_1^{\text{spe}}\right), \tag{3.76}$$

and the slope PDF of M_1 is

$$p(\gamma_1) = \frac{\Upsilon(\mu_1 - \gamma_1)}{\displaystyle\int_{-\infty}^{\mu_1} p_\gamma(t)\mathrm{d}t} p_\gamma(\gamma_1), \tag{3.77}$$

where $\mu_1 = \cot\theta_1$ and p_γ is the surface slope PDF.

Determination of $p(d|abc)$

The conditional probability $p(d|abc)$ expresses the probability that $\hat{s}_r(\theta_2)$ does not intersect the surface again given that the ray $\hat{s}$ does not intersect the surface and that the ray $\hat{s}_{r1}$ intersects the surface at M_1, and it is reflected into the specular direction $\hat{s}_r$.

Assuming that the events abc and d are independent, $p(d|abc)$ is approximated as

$$p(d|abc) \approx p(d) = S_M^0(\gamma_1, \zeta_1, \theta_2), \tag{3.78}$$

where S_M^0 is the Smith shadowing function [5], and γ_1 and ζ_1 are the surface slope and height of the second intersection point M_1. The Smith shadowing function is also able to consider the correlation between surface slopes and heights and requires then numerical integrations. The uncorrelated model is expressed as

$$S_M^0(\gamma_1, \zeta_1, \theta_2) = F(\zeta_1)^{\Lambda(\mu_2)}, \tag{3.79}$$

where $\mu_2 = \cot\theta_2$ is the slope of the ray $\hat{s}_r(\theta_2)$.

3.2.4.2 Height probability PDF of M_1

Equation (3.79) depends on the height of M_1, which is discussed in Section 3.1.6. Here the same problem is encountered. The fact that the point M_1 is higher or lower than M_0 affects heavily the conditional probability $p(d|abc)$. But this information cannot be accounted for since the horizontal distance between M_0 and M_1 is unknown.

For the case where the height PDF of M_1 is required, for example, to calculate the average height of M_1, readers may refer to (3.46) and (3.47). For the case where this information is not needed, for example, to calculate the sea surface infrared reflectivity, the model of Li *et al.* [30] can be averaged over the heights of M_1. This is discussed in the following section.

3.2.4.3 Uncorrelated height-averaged model

It is noticeable that the integration of the uncorrelated S_B^2 function of Li *et al.* [30] over the heights of M_1 does not require the knowledge of its PDF. This leads to the 'partially' height-averaged model.

Ignoring the correlation between the surface heights and slopes of two different surface points, the height-averaged uncorrelated conditional probability $p(ab)$ is expressed as

$$p(ab) = \begin{cases} 0 & \text{if} \quad 0^\circ < \theta < \theta \\ \dfrac{1}{1+\Lambda(v)} & \text{if} \quad |\theta_1| > 90^\circ \\ \dfrac{\Lambda(v_1)}{[1+\Lambda(v)]\,[1+\Lambda(v)+\Lambda(v_1)]} & \text{if} \quad \theta < \theta_1 \le 90^\circ \\ \dfrac{\Lambda_-(v_1)}{[1+\Lambda(v)]\,[1+\Lambda(v)+\Lambda_-(v_1)]} & \text{if} \quad -90^\circ \le \theta_1 < 0^\circ \end{cases}, \tag{3.80}$$

where $v = \mu/(\sqrt{2}\sigma_\gamma) = \cot\theta/(\sqrt{2}\sigma_\gamma)$ and $v_1 = \mu_1/(\sqrt{2}\sigma_\gamma) = \cot\theta_1/(\sqrt{2}\sigma_\gamma)$.

The conditional probability $p(c|ab)$ does not depend on the surface heights and $p(d|abc)$ can be expressed in the same form as in (3.69).

For a surface with an even slope probability density function, $\Lambda_-(v_1) = \Lambda(|v_1|)$. Then, the conditional probability $p(ab)$ of the model of Lynch and Wagner [34], expressed by (3.63) and (3.65), is similar to (3.80), except for the case $\theta < \theta_1 \leq 90°$. For moderate-to-large incidence angles θ, this case occurs for a very large surface slope γ_0 and the corresponding strength is small. Since the other two functions $p(c|ab)$ and $p(d|abc)$ are the same, it is expected that the model of Lynch and Wagner [34] produces very similar values to those of the uncorrelated model of Li *et al.* [30].

3.2.4.4 Numerical results

Slope histogram of M_1 for a given θ_2

From the model of Li *et al.* [30], the bistatic shadowing function with two reflections S_B^2 depends on the incidence angle θ, the surface height ζ_0 and slope γ_0 of the first intersection point, and the surface height ζ_1 and slope γ_1 (related to θ_2) of the second intersection point. For specified incidence θ and reflected θ_2 angles, the slope histogram of the first intersection gives the probability that which slope γ_0 of the first intersection point M_0 gives rise to a pair of two successive reflections. It is given by

$$\widetilde{p}_{\gamma_0}^{2,\text{spe}}(\theta, \gamma_0) = \int\int\int_{\gamma_1 \to \theta_r = \theta_2 \pm 1°} S_B^2 p_4(\zeta_0, \gamma_0, \zeta_1, \gamma_1) \mathrm{d}\zeta_0 \mathrm{d}\zeta_1 \mathrm{d}\gamma_1, \tag{3.81}$$

where p_4 is the joint PDF of the random variables $\zeta_0, \gamma_0, \zeta_1$ and γ_1. If the correlation is neglected, then $p_4 = p_\zeta(\zeta_0)p_\gamma(\gamma_0)p_\zeta(\zeta_1)p_\gamma(\gamma_1)$, where p_ζ and p_γ are the surface height and slope PDFs, respectively.

Figure 3.31 plots the slope histogram of S_B^2 versus the normalized slope γ_0/σ_γ. The incidence angle $\theta = 80°$, the reflected angle $\theta_2 = -80°$ and the surface RMS slope $\sigma_\gamma = 0.3$.

The same calculation is performed following the model of Lynch and Wagner [34], and the results are shown in Figure 3.27 (Section 3.2.3.2). For this configuration, the histogram does not vanish for $\theta_1 < 0°$ by both two models. This explained why the model of Lynch and Wagner and that of the uncorrelated one of Li *et al.* give same results (see discussion following (3.80)).

Figure 3.31 also shows that the correlation improves the agreement with the results obtained from a Monte Carlo ray-tracing algorithm. But the strengths around the peak (for which $\theta_1 = -90°$) are underestimated. As a result, the inclusion of the surface correlation between the surface heights and slopes does not solve fully the issue. In fact, the correlations between the four events a, b, c and d defined at the beginning of Section 3.2.3.1 play an important role in the derivation of the bistatic shadowing function with two reflections and can explain this deviation.

Slope histogram of M_1 for any θ_2

Without specifying a value of θ_2, the slope of the second intersection point changes so that $-90° < \theta_2 < 90°$. The slope histogram is then

$$\widetilde{p}_{\gamma_0}^{2,\text{all}}(\theta, \gamma_0) = \int\int\int_{\gamma_1 \to |\theta_r| < 90°} S_B^2 p_4(\zeta_0, \gamma_0, \zeta_1, \gamma_1) \mathrm{d}\zeta_0 \mathrm{d}\zeta_1 \mathrm{d}\gamma_1. \tag{3.82}$$

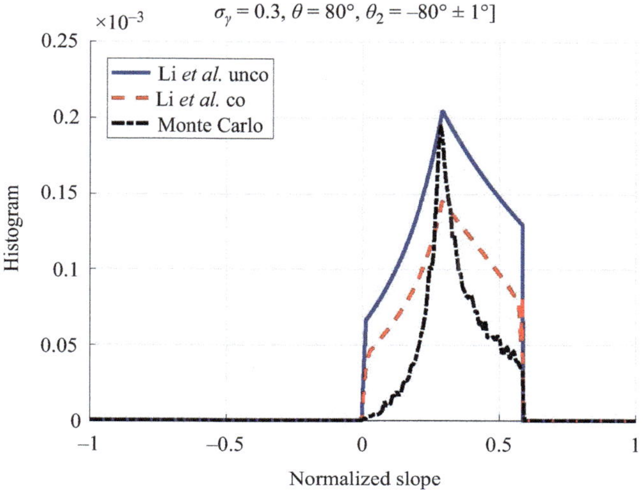

Figure 3.31 Slope histogram of the bistatic illumination versus the normalized slope γ_0/σ_γ. The incidence angle $\theta = 80°$, the reflected angle $\theta_2 = -80°$ and the surface RMS slope $\sigma_\gamma = 0.3$

From the model of Li *et al.* [30], Figure 3.32 plots the slope histogram of the first intersection point M_0 without a specified reflected angle θ_2 and versus the normalized slope γ_0/σ_γ. The incident angle $\theta = 80°$ and the surface RMS slope $\sigma_\gamma = 0.3$.

As discussed previously, the results from the uncorrelated model of Li *et al.* [30] are the same as those of the model of Lynch and Wagner [34]. Figure 3.32 shows that the uncorrelated results are overestimated in comparison to the Monte Carlo ones and the inclusion of the surface correlation between the surface heights and slopes improves the results.

For $\theta_2 \in [-90°; 90°]$, the model of Li *et al.* [30] with surface correlation predicts quite well the slopes leading to two surface reflections. However, it is still possible that a difference occurs for each individual θ_2 as shown in Figure 3.31. The integration over the surface slopes hides this information.

Histogram of the second reflected angle θ_2

From the model of Li *et al.* [30], the histogram of the second reflected angle is obtained by integrating the bistatic shadowing function S_B^2, defined by (3.74), over the surface heights (ζ_0, ζ_1) and slopes (γ_0, γ_1) of the first and second intersection points (M_0, M_1), respectively, and that the reflected angle θ_r equals the desired value θ_2.

For the same reason as discussed in Section 3.2.2.3, for a given θ, the reflected angle θ_r can range $\theta_r \in [\theta_2 - 0.2°; \theta_2 + 0.2°]$ so that the histogram can be calculated numerically. It is given by

$$\widetilde{S}^2_{B,\theta_2 \pm 0.2°} = \int\int\int\int_{\gamma_1 \to \theta_r \in \theta_2 \pm 0.2°} S_B^2 p_4(\zeta_0, \gamma_0, \zeta_1, \gamma_1) \mathrm{d}\zeta_0 \mathrm{d}\zeta_1 \mathrm{d}\gamma_1 \mathrm{d}\gamma_0. \tag{3.83}$$

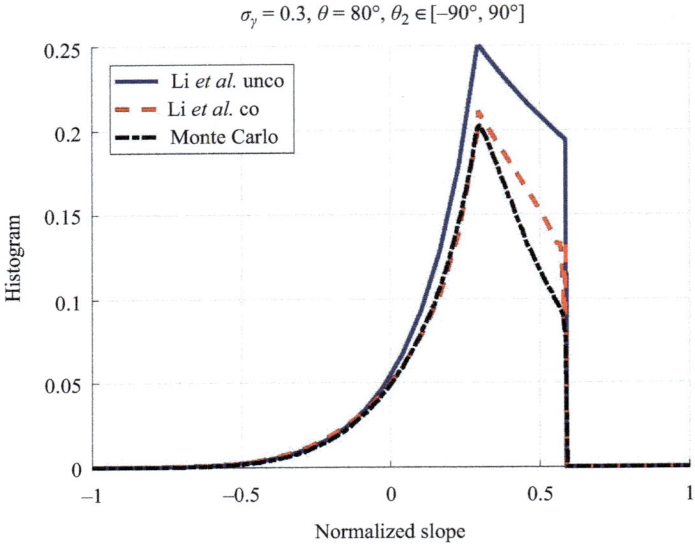

Figure 3.32 Slope histogram of the bistatic illumination function with two reflections S_B^2 versus the normalized slope γ_0/σ_γ. The incidence angle $\theta = 80°$, the reflected angle $\theta_2 \in [-90°; 90°]$ and the RMS slope $\sigma_\gamma = 0.3$

Figure 3.33 plots the histogram of the reflected angle θ_2. The surface RMS slope is $\sigma_\gamma = 0.3$, the incident angle $\theta = 30°$ in (a), $\theta = 60°$ in (b) and $\theta = 80°$ in (c). The configuration is the same as that in Figure 3.29 (with the model of Lynch and Wagner [34]).

Although the mathematical expression of the uncorrelated model of Li *et al.* [30] and that of the model of Lynch and Wagner [34] is slightly different, their numerical results are extremely similar for this configuration. As discussed previously for $\theta = 30°$, the shapes of the uncorrelated curve do not agree well with that of the Monte Carlo ray-tracing results and, a significant overestimation is observed for $\theta = 80°$. Taking into account the surface correlation between the surface heights and slopes do not improve significantly the agreement, except for $\theta = 80°$.

Hemispherical averaged bistatic shadowing function

From the model of Li *et al.* [30], the hemispherical bistatic shadowing function with two reflections is obtained by integrating the bistatic shadowing function S_B^2, defined by (3.74), over the surface heights (ζ_0, ζ_1) and slopes (γ_0, γ_1) of the first and second intersection points (M_0, M_1), respectively. It is expressed as

$$\bar{S}_B^2(\theta) = \int\int\int\int S_{BP}^2 p_4(\zeta_0, \gamma_0, \zeta_1, \gamma_1) \mathrm{d}\zeta_0 \mathrm{d}\zeta_1 \mathrm{d}\gamma_1 \mathrm{d}\gamma_0. \tag{3.84}$$

It can also be obtained by integrating (3.82) over γ_0.

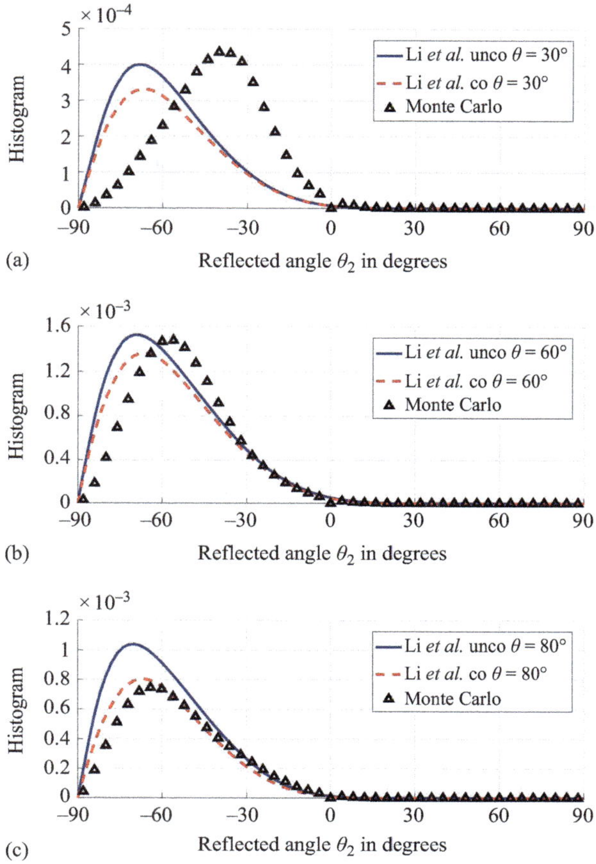

Figure 3.33 Histogram of the reflected angle θ_2 of the bistatic illumination function with two reflections S_B^2 from Li et al. [30] with an incident angle $\theta = 80°$. Surface RMS slope is $\sigma_\gamma = 0.3$. The incident angle is $\theta = 30°$ in (a), $\theta = 60°$ in (b) and $\theta = 80°$ in (c). Results are compared with the Monte Carlo ray-tracing method

Figure 3.34 plots the hemispherical averaged bistatic shadowing function with two reflections versus the angle θ_2. The surface RMS slope is $\sigma_\gamma = 0.1$ in (a), $\sigma_\gamma = 0.3$ in (b) and $\sigma_\gamma = 0.5$ in (c).

As discussed previously, the numerical results of the uncorrelated model of Li *et al.* [30] and those of the model of Lynch and Wagner [34] are similar for this configuration. A good agreement is found between the uncorrelated model and the Monte Carlo ray-tracing algorithm. Taking into account the surface correlation between the heights and the slopes, the agreement is better for large incidence angles θ. For small θ and large σ_γ, the improvement is not as significant with a slightly underestimation.

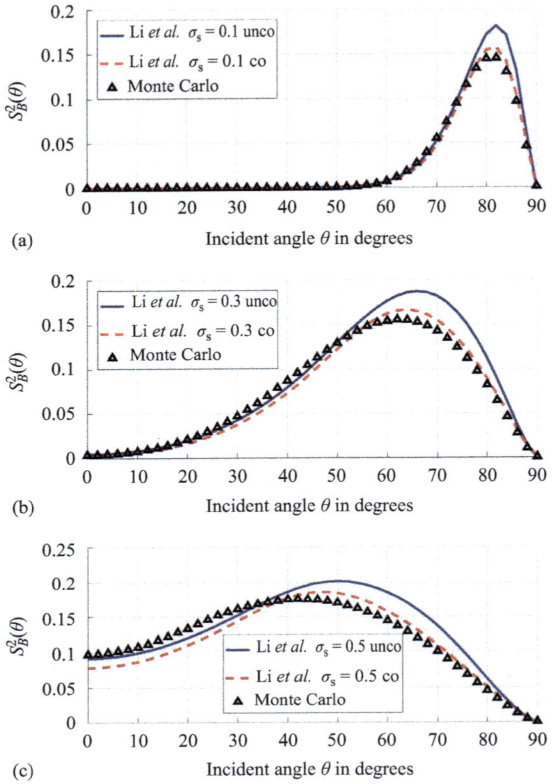

Figure 3.34 Hemispherical averaged bistatic shadowing function with two reflections S_B^2 versus the angle θ_2. (a) RMS surface slope $\sigma_\gamma = 0.1$, (b) $\sigma_\gamma = 0.3$. (c) $\sigma_\gamma = 0.5$. Results are compared with the Monte Carlo ray-tracing algorithm

To sum up, the uncorrelated model of Li *et al.* [30] is very similar to the model of Lynch and Wagner [34]. The main advantage of the model of Li *et al.* is that it can take into account the surface correlation and depends on the surface heights. In addition, in most of the case, the correlation improves comparisons with the Monte Carlo ray-tracing algorithm.

3.2.5 Extension to include multiple reflections

3.2.5.1 Derivation

In this section, the bistatic shadowing function with n reflections (shown in Figure 3.35) is addressed. From (3.74), the bistatic shadowing function with two reflections is based on the monostatic shadowing function with one reflection as

$$S_B^2 = p(ab)p(c|ab)p(d|abc) = S_M^1 p(c|ab)p(d|abc). \tag{3.85}$$

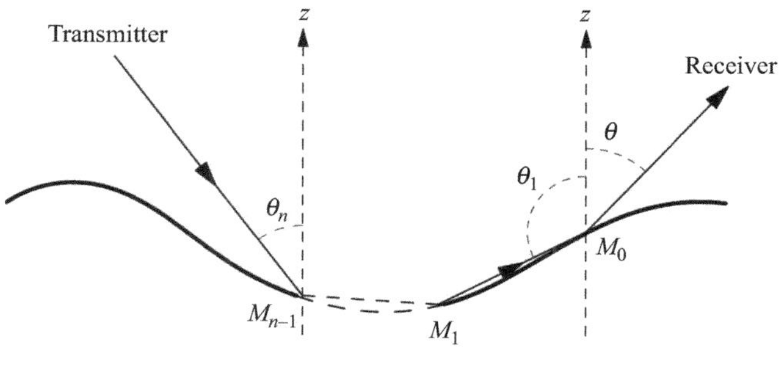

Figure 3.35 Multiple surface reflections under a bistatic configuration

The bistatic shadowing function with n reflections can be obtained in the same way. This leads to

$$S_B^n = S_M^{n-1} p(c|ES_M^{n-1}) p(d|ES_M^{n-1}c), \tag{3.86}$$

where S_M^{n-1} is the monostatic shadowing function with n reflections, which is derived in Section 3.1.6. ES_M^{n-1} stands for the event that ensures $n-1$ reflections occur under a monostatic configuration. The events c and d are redefined as

- 'The slope of M_{n-1} reflects the incident ray into the direction θ_n' is denoted symbolically as 'c'.
- 'Ray $\hat{s}(\theta_n)$ does not intersect the surface' is denoted symbolically as 'd'.

As discussed previously, the correlation between the events ES_M^{n-1} and c is difficult to model. Then, the conditional probability $p(c|ES_M^{n-1})$ is assumed to be

$$p(c|ES_M^{n-1}) = \frac{p_\gamma\left(\gamma_{n-1}^{\text{spe}}\right)}{1 - (1/2)\text{erfc}\left(|\mu_n|/\sqrt{2}\sigma_\gamma\right)}, \tag{3.87}$$

where $\mu_n = \cot\theta_n$ and p_γ is the surface slope PDF. The variable $\gamma_{n-1}^{\text{spe}}$ stands for the slope of M_{n-1} which reflects the ray $\hat{s}(\theta_{n-1})$ into the specular direction $\hat{s}(\theta_n)$.

Similarly, the conditional probability $p(d|ES_M^{n-1}c)$ is given by

$$p(d|ES_M^{n-1}c) \approx p(d) = S_M^0(\gamma_{n-1}, \zeta_{n-1}, \theta_n), \tag{3.88}$$

where S_M^0 is the Smith shadowing function [5] and γ_{n-1} and ζ_{n-1} are the surface slope and height of the last intersection point M_{n-1}.

3.2.5.2 Height of the intersection points

As discussed previously, the heights of the intersection points are unknown, except for M_0. Depending on the value of the reflected angle θ_j, it is known that the height of the jth intersection point M_j is larger or smaller than that of the previous intersection point M_{j-1}, but the exact relation depends on their horizontal distance which is unknown.

This problem is already addressed when deriving the monostatic shadowing function with multiple reflections, as well as when deriving the bistatic shadowing function with two reflections. The answer to this problem is open to the readers. As shown in the next chapter, for the calculation of the surface infrared emissivity, which does not depend on surface heights, an alternative is to use the uncorrelated height averaged bistatic shadowing function.

3.3 Extension to a 2D surface

For a two-dimensional (2D) surface, in this section, the shadowing function with multiple reflections is investigated. As shown in Chapter 2, the 2D monostatic shadowing function without reflection can be expressed from that of 1D surface by redefining the X-axis, which is obtained by making a rotation of the original abscissa x over the azimuthal angle ϕ.

3.3.1 Monostatic shadowing function with one reflection

3.3.1.1 Definition of the new system of coordinates

Figure 3.36 illustrates a reflection from a 2D surface. A ray originates from its source point M_1 propagates along the direction $\hat{s}'$ with a zenith angle $\theta_1 \in [0°; 180°]$ and then intersects the surface at another point M_0 where it is reflected towards the receiver along the direction $\hat{s}$.

Under the Snell–Descartes laws, the ray $\hat{s}$ propagates in the plane defined by the z-axis and the vector $\hat{s}$. The intersection of this plane with the 2D surface gives a 1D cross section. Then, the problem of a 2D monostatic shadowing reduces to a 1D cross section by redefining a new system of coordinates as

$$\begin{cases} \gamma_X = \gamma_x \cos\phi + \gamma_y \sin\phi \\ \gamma_Y = -\gamma_x \sin\phi + \gamma_y \cos\phi \end{cases}. \tag{3.89}$$

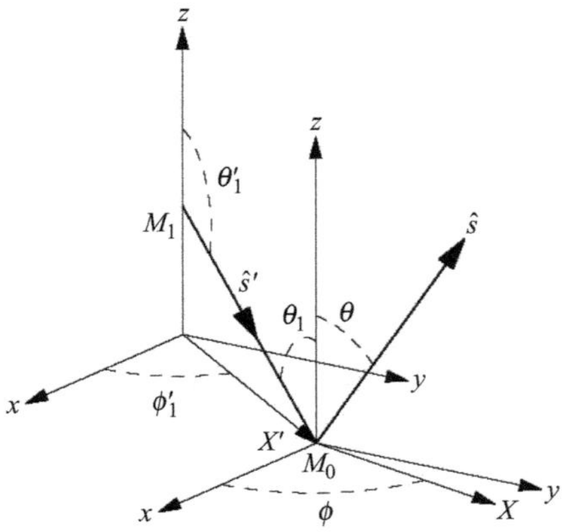

Figure 3.36 Illustration of one surface reflection from 2D surfaces: the emitted ray form the source point M_1 propagates along the $\hat{s}'$ and intersects the surface at M_0 where it is reflected towards the receiver along the direction $\hat{s}$

Since the surface slope distributions along the x and y directions are usually known, the slope distribution along the new X- and Y-axis can be derived following the way addressed in Chapter 2.

The same strategy is also applied to the ray $\hat{s}'$. Given the direction $\hat{s}$ and that of the normal to the surface $\hat{n_0}$ at the point M_0, we have

$$\hat{s} = \begin{bmatrix} \sin\theta\cos\phi \\ \sin\theta\sin\phi \\ \cos\theta \end{bmatrix} \qquad \hat{n}_0 = \frac{1}{\sqrt{1+\gamma_{x0}^2+\gamma_{y0}^2}} \begin{bmatrix} -\gamma_{x0} \\ -\gamma_{y0} \\ 1 \end{bmatrix}, \tag{3.90}$$

and the direction $\hat{s}'$ can be expressed as

$$\hat{s}' = -2\hat{n}_0(\hat{n}_0 \cdot \hat{s})\hat{n}_0 + \hat{s} = \begin{bmatrix} \sin\theta_1'\cos\phi_1' \\ \sin\theta_1'\sin\phi_1' \\ \cos\theta_1' \end{bmatrix}. \tag{3.91}$$

The system of coordinates for the ray $\hat{s}'$ is then defined as

$$\begin{cases} \gamma_{X'} = \gamma_x\cos\phi_1' + \gamma_y\sin\phi_1' \\ \gamma_{Y'} = -\gamma_x\sin\phi_1' + \gamma_y\cos\phi_1' \end{cases}. \tag{3.92}$$

The surface slope PDF along the new X'- and Y'-axis can also be derived. The monostatic shadowing function with one reflection can then be expressed in these two system of coordinates [32], as discussed in the following section.

3.3.1.2 Derivation

The model of Li *et al.* [33] derived in Section 3.1.5.1 shows the best agreement with the Monte Carlo ray-tracing algorithm among the published models. It has been extended to 2D surfaces for the derivations of the sea surface infrared emissivity [32]. From Section 3.1.5.1, the monostatic shadowing function with one reflection can be expressed as

$$\begin{aligned} S_M^1 &\approx p(\hat{s}\text{ does not intersect}) \times p(\text{Incident ray from surface}) \\ &= S_M^0 \times P_1. \end{aligned} \tag{3.93}$$

The first term S_M^0 is the Smith shadowing function [5] expressed as

$$S_M^0(\gamma_{X_0}, \zeta_0, \theta) = \Upsilon(\mu - \gamma_{X_0})\exp\left[-\int_0^{+\infty} g(\mu|\gamma_{X_0}, \zeta_0; \tau)\mathrm{d}\tau\right], \tag{3.94}$$

where

$$g(\mu|\gamma_{X_0}, \zeta_0; \tau) = \frac{\displaystyle\int_\mu^{+\infty} (\gamma_X - \mu)p(\zeta = \zeta_0 + \mu\tau, \gamma_X|\zeta_0, \gamma_{X_0}; \tau)\mathrm{d}\gamma_X}{\displaystyle\int_{-\infty}^{+\infty}\int_{-\infty}^{\zeta_0+\mu\tau} p(\zeta, \gamma_X|\zeta_0, \gamma_{X_0}; \tau)\mathrm{d}\gamma_X\mathrm{d}\zeta}, \tag{3.95}$$

where γ_{X_0} is the slope of M_0 along the X direction and $\mu = \cot\theta$ is the slope of the ray $\hat{s}(\theta)$ along the X direction.

For 2D surfaces, ignoring the surface correlation between the surface heights and slopes, the uncorrelated shadowing function S_M^0 is given by

$$S_M^0(\gamma_{X_0}, \zeta_0, \theta) = \Upsilon(\mu - \gamma_{X_0})F(\zeta_0)^{\Lambda(\mu)}, \tag{3.96}$$

where

$$\Lambda(\mu) = \frac{1}{\mu}\int_{\mu}^{+\infty}(\gamma_X - \mu)p_{\gamma_X}(\gamma_X)\mathrm{d}\gamma_X, \tag{3.97}$$

with p_{γ_X} the marginal slope PDF along the X direction. For a Gaussian process, it is shown in Chapter 2 that

$$\begin{cases} v & = & \dfrac{\mu}{\sigma_{\gamma X}\sqrt{2}} \\ \Lambda & = & \dfrac{\exp(-v^2) - v\sqrt{\pi}\,\mathrm{erfc}(v)}{2\sqrt{\pi}} \\ \mathrm{erfc}(v) & = & \dfrac{2}{\sqrt{\pi}}\displaystyle\int_v^{+\infty}\exp(-t^2)\mathrm{d}t \end{cases}. \tag{3.98}$$

For 2D surfaces, in (3.93), the function P_1 can be obtained following the same way but with respect to the X'-axis. Then

$$P_1 = \begin{cases} 1 & \text{if} \quad \theta_1 < -90^\circ \\ 1 - \exp\left[-\displaystyle\int_{-\infty}^{0} g_-(\mu_1|\gamma_{X_0'}, \zeta_0; \tau)\mathrm{d}\tau\right] & \text{if} \quad -90^\circ < \theta_1 < 0^\circ \end{cases}, \tag{3.99}$$

with $\gamma_{X_0'}$ the slope of M_0 along the X' direction and $\mu_1 = \cot\theta_1$ the slope of the ray $\hat{s}(\theta_1)$ along the X' direction. In addition, g_- is given by

$$g_-(\mu_1|\gamma_{X_0'}\zeta_0; \tau) = \frac{\displaystyle\int_{-\infty}^{\mu_1}(\mu_1 - \gamma_{X'})p(\zeta = \zeta_0 + \mu_1\tau, \gamma_{X'}|\zeta_0, \gamma_{X_0'}; \tau)\mathrm{d}\gamma_{X'}}{\displaystyle\int_{-\infty}^{\infty}\int_{-\infty}^{\zeta_0+\mu_1\tau} p(\zeta, \gamma_{X'}|\zeta_0, \gamma_{X_0'}; \tau)\mathrm{d}\gamma_{X'}\mathrm{d}\zeta}. \tag{3.100}$$

Ignoring the surface correlation between surface heights and slopes, P_1 is expressed as

$$P_1 = \begin{cases} 1 & \text{for} \quad \theta_1 > 90^\circ \\ 1 - F(\zeta_0)^{\Lambda_-(\mu_1)} & \text{for} \quad -90^\circ < \theta_1 < 0^\circ \end{cases}, \tag{3.101}$$

where

$$\Lambda_-(\mu_1) = \frac{1}{\mu_1}\int_{-\infty}^{\mu_1}(\gamma_{X'} - \mu_1)p_{\gamma_{X'}}(\gamma_{X'})\mathrm{d}\gamma_{X'}, \tag{3.102}$$

and $p_{\gamma_{X'}}$ is the marginal PDF of the surface slope along the X'-axis. For a Gaussian PDF, Λ_- is

$$v_1 = \frac{\mu_1}{\sigma_{\gamma_{X'}}\sqrt{2}} \qquad \Lambda_-(v_1) = \Lambda(|v_1|), \tag{3.103}$$

with $\sigma_{\gamma_{X'}}$ being the surface RMS slope along the X'-axis.

Comparing (3.29) with (3.35) (Section 3.1.5.1), it is noticeable that for the 1D case, the reflected angle θ_1 can be positive and negative. For the 2D case, θ_1 is always negative thanks to the definition of the X'-axis, which simplifies the mathematical expression of P_1.

Another significant difference between the 2D and 1D cases is that, for the 2D case (equations (3.94) and (3.99)), the special case when X and X' are parallel and $|\theta_1| < \theta$ is not considered. Under this configuration, the 2D model is reduced to a 1D model and $P_1 = 0$ as discussed at the end of Section 3.1.5.1. The associated 2D expression then causes a discontinuity unless the surface correlation along the azimuth direction is considered. Readers are referred to Chapter 2 for discussion on the inclusion of the azimuthal correlation.

3.3.2 Bistatic shadowing function with two reflections

3.3.2.1 Definition of the new system of coordinates

Figure 3.37 illustrates two successive reflections from a 2D surface. A ray comes from its source point, propagates along the direction $\hat{s}''$ with a zenith angle $\theta_2 \in [0°; 180°]$, and then intersects the surface at the point M_1 where it is reflected into a new direction $\hat{s}'$. The reflected ray propagates and intersects the surface again at the point M_0 where it is reflected towards the receiver along the direction $\hat{s}$.

Applying the same way as in the previous section, the 2D bistatic shadowing function with two reflections can be derived from that of 1D surface. The definition of the systems of coordinates XOY and $X'OY'$ is addressed in Section 3.3.1.1. Denoting the surface normal at the point M_1 as

$$\hat{n}_1 = \frac{1}{\sqrt{1+\gamma_{x1}^2+\gamma_{y1}^2}}\begin{bmatrix} -\gamma_{x1} \\ -\gamma_{y1} \\ 1 \end{bmatrix}, \tag{3.104}$$

the direction $\hat{s}''$ can be expressed as

$$\hat{s}'' = -2\hat{n}_1(\hat{n}_1 \cdot \hat{s}')\hat{n}_1 + \hat{s}' = \begin{bmatrix} \sin\theta_2' \cos\phi_2' \\ \sin\theta_2' \sin\phi_2' \\ \cos\theta_2' \end{bmatrix}. \tag{3.105}$$

The system of coordinates for the ray $\hat{s}''$ is defined as

$$\begin{cases} \gamma_{X''} = \gamma_x \cos\phi_2' + \gamma_y \sin\phi_2' \\ \gamma_{Y''} = -\gamma_x \sin\phi_2' + \gamma_y \cos\phi_2' \end{cases}. \tag{3.106}$$

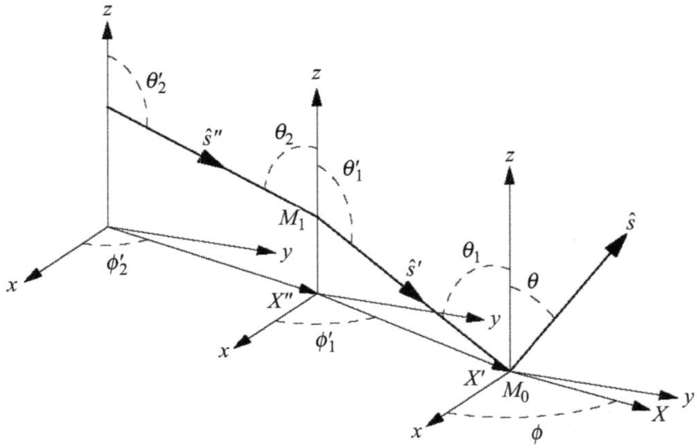

Figure 3.37 Illustration of two surface reflections from 2D surfaces: the emitted ray from the source point M_2 propagates along the $\hat{s}''$ direction and intersects the surface at M_1 where it is reflected into the direction $\hat{s}'$. The reflected ray intersects the surface again at M_0 where it is reflected towards the receiver along the direction $\hat{s}$

3.3.2.2 Derivation

Li *et al.* [36] investigated the bistatic shadowing function with two reflections from a 2D surfaces for the calculation of sea surface infrared reflectivity. This model is obtained by extending the model, valid for a 1D surface (Section 3.2.4). In this section, this model is presented with notations in agreement with this book.

From Section 3.2.4, the bistatic shadowing function with two reflections can be expressed as

$$\begin{aligned} S_B^2 &= p(abcd) \\ &= p(ab)p(c|ab)p(d|abc) \\ &= S_M^1 p(c|ab)p(d|abc), \end{aligned} \tag{3.107}$$

where the events are defined in Section 3.2.3.1.

The first term is the monostatic shadowing function with one reflection from 2D surfaces. It is addressed in Section 3.3.1, and its expression is given by (3.93). The third term can be expressed approximately by the monostatic shadowing function without reflection $S_M^0(\gamma_{X_1''}, \zeta_1, \theta_2)$, which is given by (3.96).

The second term $p(c|ab)$ requires that the ray $-\hat{s}'$ is reflected into the specular direction $-\hat{s}''$ (reverse path principle). In other words, the surface slope at the point M_1 must fulfil

$$\hat{n}_1^{\text{spe}} = \frac{1}{\sqrt{1 + \left(\gamma_{x_1}^{\text{spe}}\right)^2 + \left(\gamma_{y_1}^{\text{spe}}\right)^2}} \begin{bmatrix} -\gamma_{x_1}^{\text{spe}} \\ -\gamma_{y_1}^{\text{spe}} \\ 1 \end{bmatrix} = \frac{\hat{s}' - \hat{s}''}{\left\|\hat{s}' - \hat{s}''\right\|}. \tag{3.108}$$

The conditional probability $p(c|ab)$ is then given by the product of two Dirac delta functions as

$$p(c|ab) \approx \delta\left(\gamma_{x_1} - \gamma_{x1}^{\text{spe}}\right)\delta\left(\gamma_{y_1} - \gamma_{y1}^{\text{spe}}\right). \tag{3.109}$$

The fact that $-\hat{s}''$ is a reflected ray at the point M_1 gives an extra information on the probability density of the slope of the point M_1. The condition that $\gamma_{X_1'} \leq \mu_1$ must be fulfilled. The slope probability density function of M_1 is then given by

$$p(\gamma_{X_1'}) = \frac{\Upsilon(\mu_1 - \gamma_{X_1'})}{\displaystyle\int_{-\infty}^{\mu_1} p_{\gamma_{X'}}(t)\mathrm{d}t} p_{\gamma_{X'}}(\gamma_1), \tag{3.110}$$

where $p_{\gamma_{X'}}$ is the surface marginal slope PDF in the X' direction.

3.4 Conclusion

In this chapter, the monostatic and bistatic shadowing functions with multiple reflections are addressed. Several well-known models are reviewed. It is shown that the monostatic shadowing function with one reflection of Li *et al.* [33] shows the best agreement with the Monte Carlo ray-tracing algorithm. As for the bistatic shadowing function with two reflections, the model of Li *et al.* [30] shows the best agreement thanks to its ability to take into account the surface correlation between the surface heights and slopes.

However, since these two models both require knowledge of the heights of the intersection points, it is difficult to extend these models to include multiple reflections.

Instead, when the physical quantity to be determined does not depend on the surface heights, it is possible to include multiple reflections by ignoring surface correlation between the surface heights and slopes. Then, the uncorrelated height averaged shadowing function of the model of Li *et al.* [30] or, similarly that of Lynch and Wagner [34], can be applied.

In the following chapter, some physical applications of the shadowing function are investigated.

Chapter 4
Some applications

Christophe Bourlier[1] and Hongkun Li[1]

4.1 Introduction

When dealing the general problem of the wave scattering (emitted from an optics sensor or a Radar) by an object (in this chapter, it is a random rough surface), the calculation of the scattered field is not straightforward. For canonical geometries, like spheres, circular cylinders, etc., a closed-form expression of the scattered field can be found by introducing special functions [37].

Full-wave techniques, such as the method of moments, have been developed that rigorously incorporate all electromagnetic scattering mechanisms (edge diffraction, multiple reflection, shadowing effect, etc.) and can provide highly accurate results for rough surface scattering [15,16,18,38] (for instance). However, full-wave techniques are computationally intensive and, as a result, it is usually not feasible to employ a full-wave approach when we want to get results quickly. Consequently, approximate techniques are often the only practical alternative.

One class of such methods is the geometrical optics (GO) approximation, which can be obtained from the well-known Kirchhoff approximation [17] (for instance). This approach assumes that the frequency tends to infinity and the incident field can be modelled as a collection of rays. Here, the rays are assumed to be not complex, that is the contribution of the creeping waves is neglected. In other words, the field in the shadow zone vanishes [39].

For incidence angles near the horizon, GO can be corrected by accounting for the fact that a part of the rough surface is not illuminated from the transmitter or/and not observed from the receiver. Then, from the shadowing function derived in Chapters 1 and 2, three applications are addressed in this chapter:

- The microwave radar scattering from a rough sea surface at low grazing angles.
- The sea infrared radiation (emissivity and reflectivity).
- The connection between the sea infrared reflectivity and the BRDF (bidirectional reflectance distribution function) employed in the computer graphics domain.

[1]IETR Laboratory, CNRS, Nantes, France

4.2 Microwave radar scattering

In this section, a simple way is presented to incorporate the shadowing effect in the reflection coefficient to deal with the microwave wave scattering from a rough sea surface at low grazing angles.

Although the use of GO is questionable for this specific application [16], this model is used for its simplicity of implementation [38,40–45] and gives satisfactory results in comparison to the method of moments [38,43].

4.2.1 Unshaded Ament reflection coefficient

The Ament reflection coefficient modifies the strength field dynamics in the specular direction and for the zone close to the ground it is equivalent to the propagation over a flat surface.

The reflection coefficient defined by Ament [40] to take into account the surface roughness in the specular direction is

$$\mathscr{R}_A(\chi) = \mathscr{R}_0(\chi) \int_{-\infty}^{+\infty} e^{-jQ\zeta} p_\zeta(\zeta_0) \mathrm{d}\zeta_0, \tag{4.1}$$

where $\mathscr{R}_0$ is the Fresnel reflection coefficient of a flat surface defined as

$$\mathscr{R}_0(\theta) = \begin{cases} \dfrac{n_1 \cos\theta - n_2 \cos\theta_t}{n_1 \cos\theta + n_2 \cos\theta_t} & \text{TE } (H) \text{ polarization} \\ \\ \dfrac{n_2 \cos\theta - n_1 \cos\theta_t}{n_2 \cos\theta + n_1 \cos\theta_t} & \text{TM } (V) \text{ polarization} \end{cases}, \tag{4.2}$$

where $n_1 = 1$ (refractive index of the upper medium assumed to be vacuum), $n_2 = \sqrt{\varepsilon_r}$ (refractive index of the lower medium, which is the sea) and θ_t the refraction angle defined as $\theta_t = \arcsin(n_1 \sin(\theta)/n_2)$. Polarization TE (transverse electric) means that the magnetic incident field lies in the incidence plane $(\hat{x}, \hat{z})$ (and the electric field is transverse to this plane), whereas polarization TM (transverse magnetic) means that the electric incident field lies in the incidence plane $(\hat{x}, \hat{z})$ (and the magnetic field is orthogonal to this plane). For a plane wave, the electric field $\vec{E}$ is related to the magnetic field $\vec{H}$ by the relationship $\vec{H} = \hat{k} \wedge \vec{E}/Z_0$, where the unitary vector $\hat{k}$ stands for the direction of the incidence wave ($\hat{k} = \hat{x}\sin\theta - \hat{z}\cos\theta$) and $Z_0 \approx 120\pi$ is the wave impedance in free space. In addition, $Q = 2k_0 \cos\theta = 2k_0 \sin\chi$ (Rayleigh parameter), in which $\chi = \pi/2 - \theta$ and $Q\zeta$ is the phase correction in presence of roughness and in the specular direction (see Figure 4.1 with $\theta_{\mathrm{inc}} = \theta$ and $\zeta = 0\ \forall x$).

If the shadow is ignored and assuming a Gaussian surface height PDF, p_ζ, with zero mean value and σ_ζ standard deviation, the integration over ζ, corresponding to the derivation of the characteristic function, leads to

$$\mathscr{R}_A(\chi) = \mathscr{R}_0(\chi) \exp\left(-\frac{Q^2\sigma_\zeta^2}{2}\right) = \mathscr{R}_0(\chi) \exp\left(-2k_0^2\sigma_\zeta^2 \sin^2\chi\right). \tag{4.3}$$

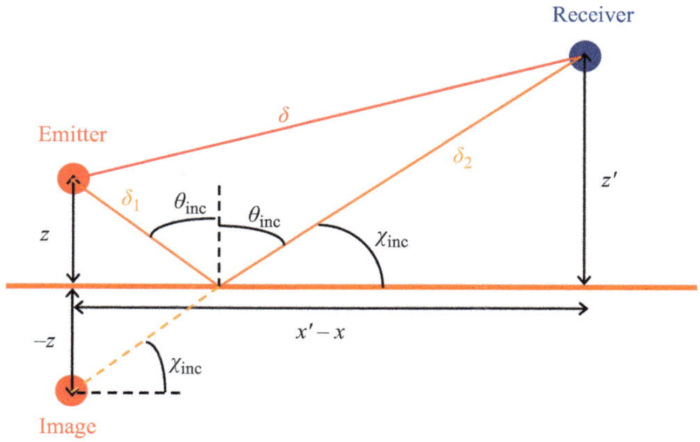

Figure 4.1 Illustration of the problem

For $Q\sigma_\zeta \ll 1$, the above equation equals the Fresnel reflection coefficient. In this case and ignoring the shadowing effect, the optical path correction due to the roughness can be neglected.

To improve the Ament model, the shadowing is accounted for.

4.2.2 Illuminated height PDF

4.2.2.1 Derivation

To improve the Ament model, the (unshadowed) height PDF p_ζ is substituted for the shadowed one, $\breve{p}(\chi;\zeta)$, defined as

$$\breve{p}(\chi;\zeta_0) = \frac{\displaystyle\int_{-\infty}^{+\infty} p_\gamma(\gamma_0)S(\chi;\zeta_0,\gamma_0)\mathrm{d}\gamma_0}{\displaystyle\int_{-\infty}^{+\infty}\int_{-\infty}^{+\infty} p_{\zeta,\gamma}(\zeta_0,\gamma_0)S(\chi;\zeta_0,\gamma_0)\mathrm{d}\zeta_0\mathrm{d}\gamma_0} p_\zeta(\zeta_0), \tag{4.4}$$

in which p_γ and p_ζ are the slope and the height PDFs and $p_{\zeta,\gamma}$ is the joint height PDF of the surface heights and slopes. $S(\chi;\zeta_0,\gamma_0)$ is the statistical illumination function in the specular direction of an arbitrary point of the surface of height ζ_0 and slope γ_0 illuminated from a grazing incidence angle χ. At the denominator, the double integral corresponds to the normalisation function such as $\int_{-\infty}^{+\infty}\breve{p}(\chi;\zeta_0)\mathrm{d}\zeta_0 = 1$. If the points of the surface are all illuminated, then $S = 1$ and the illuminated height PDF equals the height PDF, $\breve{p}(\chi;\zeta_0) = p_\zeta(\zeta_0)$.

As shown in Figure 4.2, for high incidence angles θ_{inc} (or low grazing angle χ_{inc}), a part of the surface is shadowed which implies that the height standard deviation $\breve{\sigma}_\zeta$ of illuminated surface decreases in comparison to σ_ζ and the height mean value $\breve{m}_\zeta$ differs from zero.

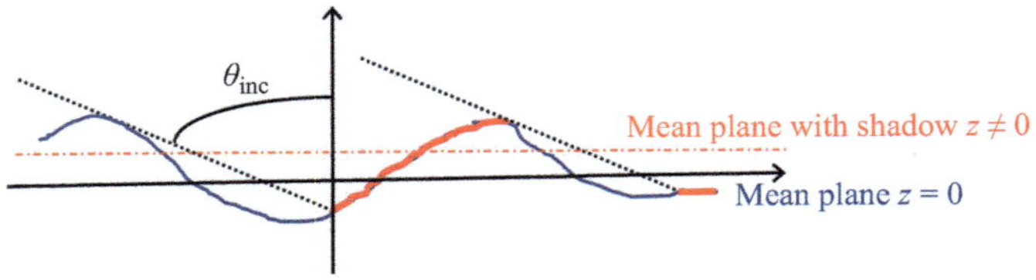

Figure 4.2 Illustration of the shadowing effect for the monostatic case

From Chapter 1, in the forward (or specular) direction and for *any uncorrelated* random process, the use of the Wagner, S_W, and the Smith, S_S, formulations lead to

$$\begin{cases} S_W(\chi;\zeta_0) = \left(\exp\left\{-\Lambda(\chi,\sigma_\gamma)\left[1-F_\zeta(\zeta_0)\right]\right\}\right)^2 \\ S_S(\chi;\zeta_0) = \left\{\left[F_\zeta(\zeta_0)\right]^{\Lambda(\chi,\sigma_\gamma)}\right\}^2 \end{cases}. \tag{4.5}$$

For a bistatic configuration, S_W and S_S are defined as the product of two monostatic statistical illumination functions. In the forward direction, since the incidence angles of the transmitter and the receiver are equal in absolute value, the monostatic statistical illumination function with respect to the transmitter and the receiver is equal (the surface is assumed to be even over its abscissa). This explains the power 2 in the above equation. We can note that S_W and S_S do not depend on the surface slope γ. F_ζ stands for the height cumulative function defined as

$$F_\zeta(\zeta') = \int_{-\infty}^{\zeta'} p_\zeta(\zeta)d\zeta, \tag{4.6}$$

where p_ζ is the height PDF. Λ (Chapter 1) is expressed as

$$\Lambda(\chi,\sigma_s) = \frac{1}{\tan\chi}\int_{+\tan\chi}^{+\infty}(\gamma-\tan\chi)p_\gamma(\gamma)d\gamma, \tag{4.7}$$

where p_γ is the slope PDF and the grazing angle $\chi \geq 0$. Substituting (4.5) into (4.4) and integrating over the heights ζ and the slopes γ, we show for any uncorrelated process that

$$\begin{cases} \breve{p}_W(\chi;\zeta_0) = p_\zeta(\zeta_0)\dfrac{2\Lambda}{1-\exp(-2\Lambda)}\exp\left\{-2\Lambda\left[1-F_\zeta(\zeta_0)\right]\right\} \\ \breve{p}_S(\chi;\zeta_0) = p_\zeta(\zeta_0)(1+2\Lambda)\left[F_\zeta(\zeta_0)\right]^{2\Lambda} \end{cases}. \tag{4.8}$$

For Gaussian statistics, we have

$$p_\zeta(\zeta) = \frac{1}{\sigma_\zeta\sqrt{2\pi}}\exp\left(-\frac{\zeta^2}{2\sigma_\zeta^2}\right) \quad p_\gamma(\gamma) = \frac{1}{\sigma_\gamma\sqrt{2\pi}}\exp\left(-\frac{\gamma^2}{2\sigma_\gamma^2}\right), \tag{4.9}$$

where σ_ζ and σ_γ stand for the standard deviations of the surface heights and slopes (the mean values are assumed to be equal to zero), respectively. From (4.7) and (4.6),

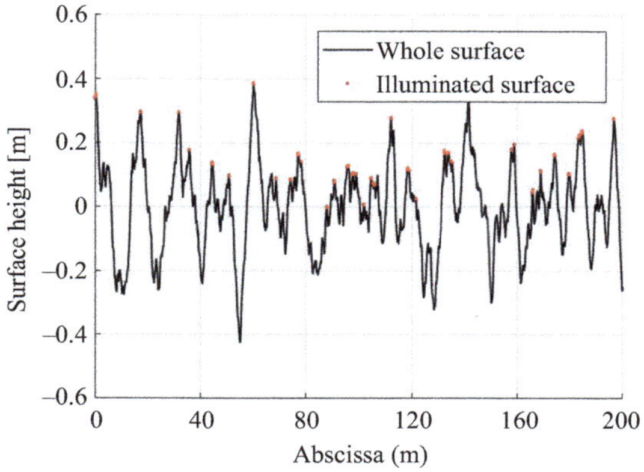

Figure 4.3 Piece of the generated sea surface heights ζ_0 versus its abscissa over $x \in [0;200]$ m. The illuminated surface is also plotted. $\chi = 2^o$ and $u_{10} = 5$ m/s

this leads to

$$\begin{cases} \Lambda(\chi, \sigma_s) = \Lambda(v) = \dfrac{\exp(-v^2) - v\sqrt{\pi}\,\mathrm{erfc}(v)}{2v\sqrt{\pi}} & v = \dfrac{\tan\chi}{\sqrt{2}\sigma_\gamma} \\ F_\zeta(\zeta_0) = 1 - \dfrac{1}{2}\mathrm{erfc}\left(\dfrac{\zeta_0}{\sqrt{2}\sigma_\zeta}\right) & \end{cases}, \tag{4.10}$$

where and the grazing angle $\chi \geq 0$.

In conclusion, the illuminated height PDF for any uncorrelated process, $\breve{p}_{W,S}(\chi;\zeta_0) \equiv \breve{p}_{W,S}(v;\zeta_0)$, depends on the surface height ζ_0 and on the parameter v.

4.2.2.2 Numerical results

Figure 4.3 plots a piece of the generated sea surface heights ζ_0 versus its abscissa over $x \in [0; 200]$ m. The illuminated surface is also plotted. $\chi = 2^o$ and $u_{10} = 5$ m/s. As we can see, only the top of the surface is both illuminated from the transmitter and observed by the receiver located in the forward direction.

The sea surface is generated from a spectral method (Chapter 1 or [16]) and the Elfouhaily *et al.* [46] spectrum is used. More precisely, since a one-dimensional (1D) sea surface is considered, only the isotropic part of the spectrum is used. This model is governed by the wind speed u_{10} defined at 10 m above the sea mean level and the sea is assumed to be fully developed (the inverse of the wave age equals 0.84).

Figure 4.4 plots the height PDF versus the surface heights ζ_0. $\chi = 2^o, u_{10} = 5$ m/s, the sea surface length is 10 km and the number of sample is $n = 2^{20} = 1,048,576$. In the legend, the label means the following:

- 'Un' is the surface height PDF expressed by (4.9).
- 'MC' is the surface height PDF of the illuminated points computed from a Monte Carlo process explained in Chapter 1.

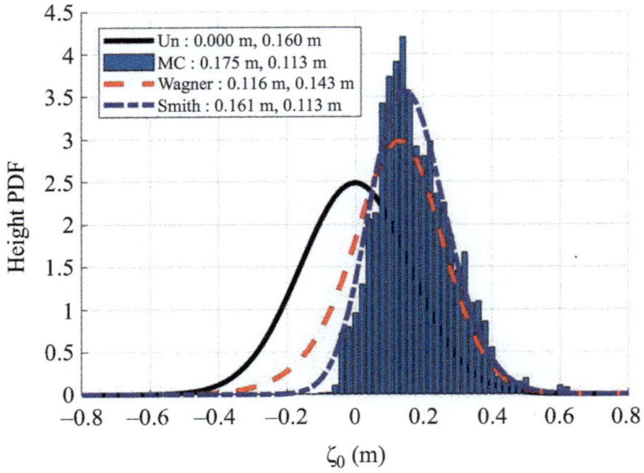

Figure 4.4 Height PDF versus the surface heights ζ_0. $\chi = 2^o$, $u_{10} = 5$ m/s, the sea surface length is 10 km and the number of samples is $n = 2^{20} = 1{,}048{,}576$

- 'Smith' is the surface height PDF of the illuminated points computed from the Smith formulation (Equation (4.8)).
- 'Wagner' is the surface height PDF of the illuminated points computed from the Wagner formulation (Equation (4.8)).

In addition, the two numbers are the mean value $\breve{m}_\zeta$ and the standard deviation $\breve{\sigma}_\zeta$ of the unshaded heights, respectively. As expected, the illuminated height PDF is shifted towards larger heights and its width decreases in comparison to the (unshaded) height PDF p_ζ. The mean value and the standard deviation show that the Smith approach is the more accurate, and the associated height PDF matches well with that obtained from the Monte Carlo procedure.

Figure 4.4 reveals also that the shape of the illumination height PDF is close to a Gaussian PDF. Then, $\breve{p}(\zeta_0; \chi)$ can be approximated as

$$\breve{p}_G(\zeta_0; \phi) = \frac{1}{\breve{\sigma}_\zeta\sqrt{2\pi}} \exp\left[-\frac{(\zeta_0 - \breve{m}_\zeta)^2}{2\breve{\sigma}_\zeta^2}\right], \tag{4.11}$$

where the mean value $\breve{m}_\zeta$ and variance $\breve{\sigma}_\zeta^2$ are expressed as

$$\begin{cases} \breve{m}_\zeta = \displaystyle\int_{-\infty}^{+\infty} \zeta_0 \breve{p}(\zeta_0; \chi) d\zeta_0 \\ \breve{\sigma}_\zeta^2 = \displaystyle\int_{-\infty}^{+\infty} (\zeta_0 - \breve{m}_\zeta)^2 \breve{p}(\zeta_0; \chi) d\zeta_0 \end{cases}. \tag{4.12}$$

At the top, Figure 4.5 plots the Smith height PDF versus the surface heights ζ_0, whereas at the bottom, the difference against the Smith Model (Equation (4.8))

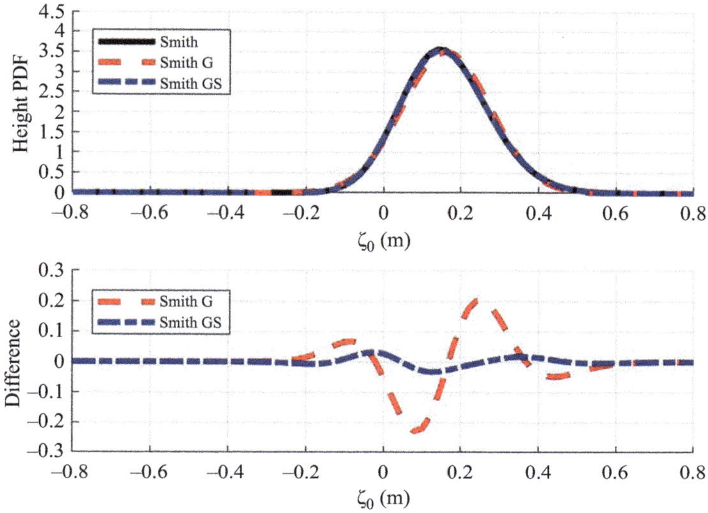

Figure 4.5 *Top: Smith height PDF versus the surface heights ζ_0. Bottom: Difference against the Smith Model (Equation* (4.8)*). $\chi = 2^o$, $u_{10} = 5$ m/s, the sea surface length is 10 km and the number of samples is $n = 2^{20} = 1{,}048{,}576$*

is plotted. $\chi = 2^o$, $u_{10} = 5$ m/s, the sea surface length is 10 km and the number of samples is $n = 2^{20} = 1,048,576$. In the legend, the label means

- 'Smith G' is the surface height PDF of the illuminated points computed from the Smith formulation by assuming a Gaussian profile (Equation (4.11)).
- 'Smith GS' is the surface height PDF of the illuminated points computed from the Smith formulation by assuming a Gram–Charlier PDF truncated at the third order (Equation (4.13)).

As we can see, the Smith height PDF is close to its Gaussian profile. To decrease the difference, higher order statistics are introduced. From Chapter 1, considering the Gram–Charlier distribution truncated at third order (introduction of the skewness), the illuminated height PDF can be modelled as

$$\breve{p}_{GS}(\zeta_0; \phi) = \frac{1}{\breve{\sigma}_\zeta \sqrt{2\pi}} \exp\left(-\frac{f^2}{2}\right)\left[1 + \frac{\lambda_\zeta}{6}\left(f^3 - 3f\right)\right], \tag{4.13}$$

where $f = (\zeta_0 - \breve{m}_\zeta)/\breve{\sigma}_\zeta$, $\lambda_\zeta = \mu_\zeta/\breve{\sigma}_\zeta^3$ and

$$\mu_\zeta = \int_{-\infty}^{+\infty} (\zeta_0 - \breve{m}_\zeta)^3 \breve{p}(\zeta_0; \chi) d\zeta_0. \tag{4.14}$$

Figure 4.5 shows that by introducing the skewness, the height PDF better match.

4.2.2.3 Modelling of the statistics moments

In Chapter 1, we showed that the average shadowing function depends only on the parameter $v = \tan\chi/(\sigma_\gamma\sqrt{2})$, which allows us to decrease the freedom degrees from 2 (χ, σ_γ) to one v.

Making the variable transformations $h_0 = \zeta_0/(\sigma_\zeta\sqrt{2})$ and $s_0 = \gamma_0/(\sigma_\gamma\sqrt{2})$, all the mathematical expressions derived in the previous subsections remain unchanged, in which p_ζ and p_γ become $p_h(h_0) = e^{-h_0^2}/\sqrt{\pi}$ and $p_s(s_0) = e^{-s_0^2}/\sqrt{\pi}$, respectively. In addition, the statistical moments up to the third order become

$$\begin{cases} \breve{m}_h = \int_{-\infty}^{+\infty} h_0 \breve{p}(h_0; v) dh_0 \\ \breve{\sigma}_h^2 = \int_{-\infty}^{+\infty} (h_0 - \breve{m}_h)^2 \breve{p}(h_0; v) dh_0 \\ \mu_h = \int_{-\infty}^{+\infty} (h_0 - \breve{m}_h)^3 \breve{p}(h_0; v) dh_0. \end{cases}, \tag{4.15}$$

which depend only on v and are dimensionless.

Figure 4.6 plots the mean value $\breve{m}_h$, the standard deviation $\breve{\sigma}_h$ and the skewness $\lambda_h = \mu_h/\breve{\sigma}_h^3$ of the illuminated surface (in the forward direction) versus the parameter v. $u_{10} = 5$ m/s, the sea surface length is 40 km and the number of samples is $n = 2^{22} = 4{,}194{,}304$.

The standard deviation predicted from the Smith model matches very well with that computed from the MC procedure, and a good agreement is obtained between the mean values. The skewness gives satisfactory results.

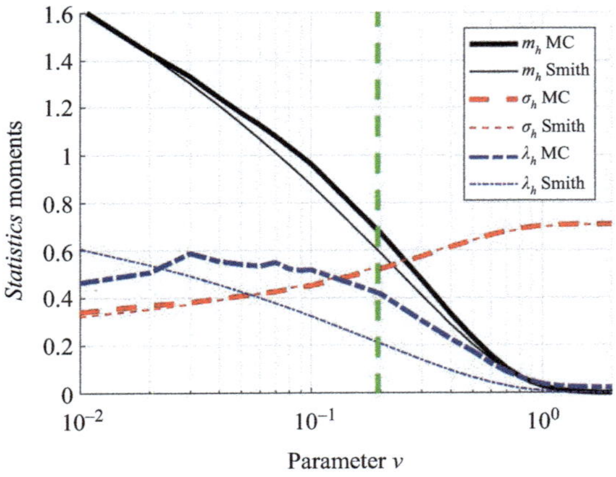

Figure 4.6 Mean value $\breve{m}_h$, standard deviation $\breve{\sigma}_h$ and skewness $\lambda_3 = \mu_3/\breve{\sigma}_h^3$ of the illuminated surface (in the forward direction) versus the parameter v. $u_{10} = 5$ m/s, the sea surface length is 40 km and the number of samples is $n = 2^{22} = 4{,}194{,}304$

For a real case, if $u_{10} = 5$ m/s, then $\sigma_\zeta = 0.161$ m, $\sigma_\gamma = 0.165$ and $v = \tan\chi/(\sigma_\gamma\sqrt{2}) \approx 4.291\tan\chi$. In addition, if $\chi = 2.6^{\circ}$, then $v \approx 0.2 = v_0$. This value is reported in Figure 4.6 by the vertical dashed-line. Then, the values $(\breve{m}_\zeta, \breve{\sigma}_\zeta)$ in meter are obtained from those of $(\breve{m}_h, \breve{\sigma}_h)$ multiplied by $\sqrt{2}\sigma_\zeta$. This rule does not hold for λ_ζ because it is defined from a ratio (dimensionless). From Figure 4.6, $\breve{m}_h = \{0.6716, 0.5879\}$, $\breve{\sigma}_h = \{0.5216, 0.5317\}$ and $\breve{\lambda}_h = \{0.4175, 0.2090\}$ from MC and Smith approaches, respectively, which implies that $\breve{m}_\zeta = \{0.1525, 0.1335\}$ m, $\breve{\sigma}_\zeta = \{0.1185, 0.1208\}$ m and $\breve{\lambda}_\zeta = \breve{\lambda}_h$.

4.2.3 Ament reflection coefficient with shadow

As shown previously, for a rough surface illuminated by a plane wave at grazing angles, the shadowing effect increases the height mean value $\breve{m}_\zeta(v(\chi,\sigma_\gamma),\sigma_\zeta)$ and decreases the RMS height $\breve{\sigma}_\zeta(v(\chi,\sigma_\gamma),\sigma_\zeta)$ of the illuminated points of the surface. They depend on the slope standard deviation, σ_γ, the grazing incident angle, χ, and the height standard deviation, σ_ζ. The Ament reflection coefficient does not account for this phenomenon. It only modifies the strength field dynamics in the specular direction, and for the zone close to the ground, it is equivalent to the propagation over a flat surface.

To derive a reflection coefficient accounting for the shadowing effect, the height distribution $p_\zeta(\zeta_0)$ is substituted in (4.1) by the illumination height PDF, $\breve{p}(\chi;\zeta_0)$, expressed from (4.4). Considering the Smith model given by (4.8), the reflection coefficient with shadow is

$$\begin{aligned}\breve{\mathscr{R}}_A(\chi,v) &= \mathscr{R}_0(\chi)\int_{-\infty}^{+\infty} e^{-jQ\zeta_0}\breve{p}(\chi;\zeta_0)\mathrm{d}\zeta_0 \\ &= \mathscr{R}_0(\chi)\frac{1+2\Lambda}{\sqrt{\pi}}\int_{-\infty}^{+\infty} e^{-jQ\sqrt{2}\sigma_\zeta h_0 - h_0^2}\left[1-\frac{\operatorname{erfc}(h_0)}{2}\right]^{2\Lambda}\mathrm{d}h_0, \quad (4.16)\end{aligned}$$

where $h_0 = \zeta_0/(\sigma_\zeta\sqrt{2})$ and Λ depends on v.

To avoid the numerical computation of the integral over h_0, a second approach is proposed, assuming that the normalized height PDF has a Gaussian profile expressed from (4.11). Then

$$\breve{\mathscr{R}}_{A,G}(\chi,v) = \mathscr{R}_0(\chi)\exp\left(-jQ\breve{m}_\zeta - \frac{Q^2\breve{\sigma}_\zeta^2}{2}\right). \quad (4.17)$$

Unlike (4.3), (4.16) and (4.17) have a non-zero phase term, due to the fact that the mean value of the illuminated height, $\breve{m}_\zeta$, does not vanish. In addition, since $\breve{\sigma}_\zeta \leq \sigma_\zeta$, $|\breve{\mathscr{R}}_A| \geq |\mathscr{R}_A|$ and $|\breve{\mathscr{R}}_{A,G}| \geq |\mathscr{R}_A|$.

Figure 4.5 reveals that by introducing the skewness, the height PDF better match. This leads from (4.13) to

$$\breve{\mathscr{R}}_{A,GS}(\chi,v) = \breve{\mathscr{R}}_{A,G}(\chi,v)\left(1+\frac{jQ\lambda_\zeta\breve{\sigma}_\zeta}{6}\right). \quad (4.18)$$

4.2.4 Propagation factor

4.2.4.1 Definition

The propagation factor, F, is generically defined as the magnitude of the field at the receiver in the presence of clutter divided by the magnitude of the field at the receiver if it were in vacuum. In our case, clutter will mean the field radiated by the ocean surface. Then

$$F = \frac{\psi_{\text{inc}} + \psi_{\text{sca}}}{\psi_{\text{inc}}}, \tag{4.19}$$

where ψ_{inc} is the incident field and ψ_{sca} the field scattered by the surface.

For a smooth surface of infinite length, the reflected field can be obtained from the image theory leading to $\psi_{\text{sca}} = \mathscr{R}_0(\theta_{\text{inc}})\psi_{\text{inc,image}}$, where $\psi_{\text{inc,image}}$ is the image of the incident field, θ_{inc} the angle defined on the surface with respect to the positive direction $\hat{z}$ (see Figure 4.1) and $\mathscr{R}_0$ the Fresnel reflection coefficient defined by (4.2).

The modulus of the propagation factor can be written as

$$F = \sqrt{1 + |\mathscr{R}_0|^2 + 2\,|\mathscr{R}_0| \cos\left(k_0\delta + \phi_{\mathscr{R}_0}\right)}, \tag{4.20}$$

where $\phi_{\mathscr{R}_0}$ is the phase of the reflection coefficient $\mathscr{R}_0$ evaluated at the angle θ_{inc}. In addition, $k_0 = 2\pi/\lambda_0$ is the wave number ($\lambda_0 = c/f$, the wavelength related to the radar frequency f by $\lambda_0 = c/f$, in which $c \approx 3 \times 10^8$ m/s) and δ is the path length difference between the reflected field and the incident fields defined (Figure 4.1) as

$$\delta = \delta_1 + \delta_2 - \sqrt{(x' - x)^2 + (z - z')^2}, \tag{4.21}$$

where the transmitter and the receiver have coordinates (x, z) and (x', z'), respectively. In addition

$$\delta_1 = \frac{z}{\sin\chi_{\text{inc}}}, \quad \delta_2 = \frac{z'}{\sin\chi_{\text{inc}}}, \tan\chi_{\text{inc}} = \frac{z' + z}{x' - x}. \tag{4.22}$$

From (4.20), the dynamic of $|F|^2$, ΔF_2, is expressed as

$$\begin{cases} |F|_{\min} = |1 - |\mathscr{R}_0|| \\ |F|_{\max} = |1 + |\mathscr{R}_0|| \\ \Delta F_2 = |F|^2_{\max} - |F|^2_{\min} = 4\,|\mathscr{R}_0| \end{cases}. \tag{4.23}$$

For a rough sea surface, the reflection coefficient can be computed numerically from the method of moments or analytically from the small perturbation method [38]. Here, the approximated model of Ament is used, in which the shadow in accounted for.

4.2.4.2 Numerical results for 1D sea surfaces

Figure 4.7 plots the propagation factor modulus in dB scale ($20\log_{10}|F|$) versus the receiver height z' for a given range $x' = x'_0$ given in the title of the sub-figures. The frequency is $f = 3$ GHz, the wind speed is $u_{10} = 5$ m/s and the transmitter height is $z = 10$ m of abscissa $x = 0$. For the computation of the MC solution, the sea surface length is 10 km and the number of samples is 1,048,576. The polarization is TE (or H)

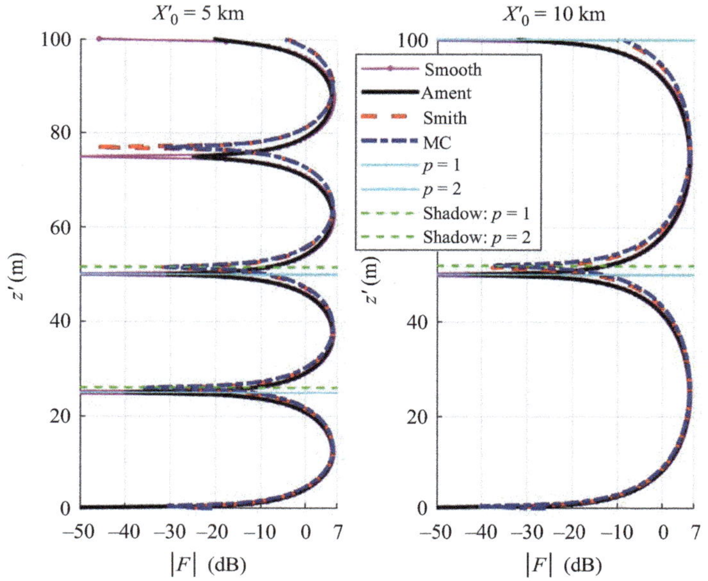

Figure 4.7 Propagation factor modulus in dB scale ($20\log_{10}|F|$) versus the receiver height z' for a given range $x' = x'_0$ given in the title of the sub-figures. The frequency is $f = 3$ GHz, the wind speed is $u_{10} = 5$ m/s and the transmitter height is $z = 10$ m of abscissa $x = 0$. For the computation of the MC solution, the sea surface length is 10 km and the number of samples is 1,048,576. The polarization is TE (or H) and the surface can be considered as perfectly conducting (metallic) for microwaves frequencies [38], which means that $\mathscr{R}_0 = -1$

and the surface can be considered as perfectly conducting (metallic) for microwaves frequencies [38], which means that $\mathscr{R}_0 = -1$.

As we can see, $|F|$ is maximum when the incident and scattered fields are in phase (constructive interferences), whereas it is minimum when the two fields are in opposite phase (destructive interferences). Since $\mathscr{R}_0 = -1$, from (4.23), $|F_{\max}| = 2 = 6$ dB and $|F_{\min}| = 0 = -\infty$ dB. In comparison to a smooth surface, the Ament model predicts smaller levels due to the term $\exp(-2k_0^2\sigma_\zeta^2 \sin^2\chi) < 1$ in (4.3) and does not modify the positions of the minima because this term is a real positive number.

When the shadowing effect is included (Equation (4.16)), the strengths of the minima slightly decreases (because $\breve{\sigma}_\zeta \leq \sigma_\zeta \Rightarrow \exp(-Q^2\breve{\sigma}_\zeta^2) \geq \exp(-Q^2\sigma_\zeta^2)$) and are in good agreement with those obtained from MC. In addition, the shadowing produces a shift of the minima positions towards the higher heights and match well with those computed from MC. As the receiver height grows, the deviation between MC and Smith results weakly increases.

Figure 4.8 plots the propagation factor modulus in dB scale ($20\log_{10}|F|$) versus the receiver height z' for a given range $x' = x'_0$ given in the title of the sub-figures. The simulation parameters are the same as in Figure 4.7.

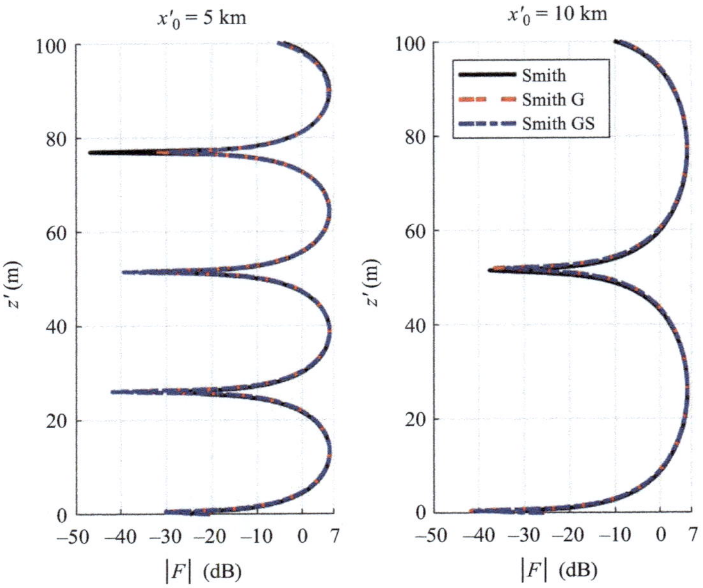

Figure 4.8 Propagation factor modulus in dB scale ($20\log_{10}|F|$) versus the receiver height z' for a given range $x' = x'_0$ given in the title of the sub-figures. The simulation parameters are the same as in Figure 4.7

In the legend, the label means

- 'Smith G' the reflection coefficient is computed from the Smith formulation by assuming a Gaussian profile (Equation (4.17)).
- 'Smith GS' the reflection coefficient is computed from the Smith formulation by assuming a non-Gaussian profile (Equation (4.18)), in which the skewness is included.

As we can see, the Gaussian approximation gives good results and the skewness has a minor impact on the strengths.

From (4.20), the positions of the minima satisfied $k_0\delta + \phi_{\mathscr{R}_0} = \pi + 2p\pi$, where p is an integer, which implies that the corresponding receiver heights $\{z'_{\text{min}}\}$ satisfied

$$z'_{\text{min}} = \frac{(x' - x)(\pi + 2p\pi - \phi_{\mathscr{R}_0})}{2k_0 z}, \tag{4.24}$$

where

$$\delta \approx \frac{2zz'}{x' - x} \text{ for } |x' - x| \gg h_{1,2}. \tag{4.25}$$

In Figure 4.8, the horizontal full lines report the values of z'_{min} ($\{25, 50\}$ m for $x_0 = 5$ km and $\{50, 100\}$ m for $x_0 = 10$ km) for $p = \{1, 2\}$.

From (4.17), including the shadow, (4.24) becomes

$$z'_{S,\min} = \frac{z}{z - \breve{m}_\zeta(z'_{\min})}\left[\breve{m}_\zeta(z'_{\min}) + z'_{\min}\right]. \tag{4.26}$$

In Figure 4.8, the horizontal dashed lines report the values of $z'_{S,\min}$ ({26.06, 51.56} m for $x_0 = 5$ km and {51.89, 102.95} m for $x_0 = 10$ km) for $p = \{1, 2\}$. Then, $z'_{S,\min} - z'_{\min}$ can exceed 1 m which is not negligible for practical applications.

4.2.4.3 Numerical results for 2D sea surfaces

The Ament reflection coefficient depends statistically only on the height standard deviation σ_ζ. Then, for a two-dimensional (2D) sea surface, it is independent of the wind azimuthal direction ϕ.

When the shadow in accounted for, the Ament reflection coefficient depends in addition to the slope standard deviation σ_γ, function of ϕ. More precisely, this dependence comes from Λ in (4.16).

In Chapter 2, this term was derived by assuming the Cox and Munk slope distribution [21] given by (2.17), leading to (2.32) for Λ. Also, Λ was derived for an exponential slope PDF expressed by (2.16), yielding (2.36) for Λ.

In polar coordinates, the general form of the sea height spectrum is given by [46]

$$\hat{S}(k, \psi) = (2\pi)\hat{S}_0(k)[1 + \hat{\Delta}(k)\cos(2\psi)], \tag{4.27}$$

where $\hat{S}_0$ stands for the isotropic part and $\hat{\Delta}$ the anisotropic part. In addition, k is the wave number and ψ the azimuthal direction with respect to that of the wind. To be consistent with (2.45), the spectrum $\hat{S}(k, \psi)$ is multiplied by $(2\pi)^2$. The substitution of (2.47) into (2.45) leads to

$$\sigma^2_{\gamma_x} = \frac{\alpha + \beta}{2} \quad \sigma^2_{\gamma_y} = \frac{\alpha - \beta}{2}, \tag{4.28}$$

where

$$\begin{cases} \alpha = \displaystyle\int_0^\infty k^2 \hat{S}(k, \psi) \mathrm{d}k \\ \beta = \displaystyle\frac{1}{2}\int_0^\infty k^2 \hat{\Delta}(k)\hat{S}(k, \psi)\mathrm{d}k \end{cases}. \tag{4.29}$$

Moreover, the parameters related to the skewness and the kurtosis are expressed by [47]

$$\begin{cases} c_{21} = (0.86u_{12} - 1 \pm 3)10^{-2} \\ c_{03} = (3.3u_{12} - 4 \pm 12)10^{-2} \end{cases} \quad \begin{cases} c_{04} = 0.23 \pm 0.41 \\ c_{40} = 0.40 \pm 0.23 \\ c_{22} = 0.12 \pm 0.06 \end{cases}, \tag{4.30}$$

where the wind speed $u_{12} \approx u_{10}$ is measured at 12.5 m above the sea.

For an exponential slope PDF given by (2.16), Λ is expressed from (2.36).

Figure 4.9 plots the propagation factor modulus in dB scale ($20\log_{10}|F|$) versus the receiver height z' for a given range $x' = x'_0$ given in the title of the sub-figures. The frequency is $f = 3$ GHz, the wind speed is $u_{10} = 5$ m/s and the transmitter height

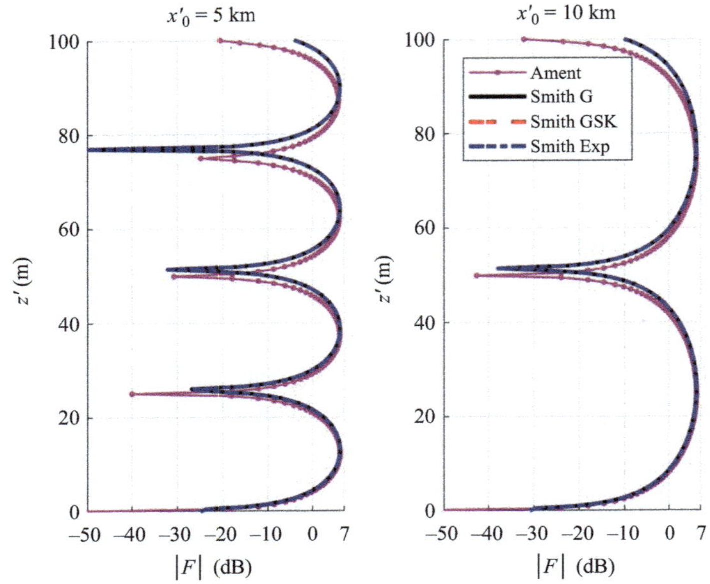

Figure 4.9 Propagation factor modulus in dB scale ($20\log_{10}|F|$) versus the receiver height z' for a given range $x' = x'_0$ given in the title of the sub-figures. The frequency is $f = 3$ GHz, the wind speed is $u_{10} = 5$ m/s and the transmitter height is $z = 10$ m of abscissa $x = 0$. The polarization is TE (or H) and the surface can be considered as perfectly conducting (metallic) for microwaves frequencies [38], which means that $\mathscr{R}_0 = -1$

is $z = 10$ m of abscissa $x = 0$. The polarization is TE (or H) and the surface can be considered as perfectly conducting (metallic) for microwaves frequencies [38], which means that $\mathscr{R}_0 = -1$. From (4.28), $\sigma_{\gamma_x} = 0.130$ and $\sigma_{\gamma_y} = 0.106$.

In the legend, the label means

- 'Ament', model of Ament without shadow.
- 'Smith G', model of Smith by assuming a Gaussian slope PDF.
- 'Smith GSK', model of Smith by assuming that the Cox and Munk (Gram-Charlier distribution) slope PDF.
- 'Smith Exp', model of Smith by assuming an exponential slope PDF.

Figure 4.9 shows that the results computed from the three slope PDFs coincide, which means that the reflection coefficient is few sensitive to the slope PDF. For this case, $z'_{\rm min} = \{26.00, 51.45\}$ m and $z'_{\rm min} = \{51.79, 102.74\}$ m for $x'_0 = 5$ km and $x'_0 = 10$ km, respectively.

Figure 4.10 plots the propagation factor modulus in dB scale ($20\log_{10}|F|$) versus the azimuthal direction ϕ and for a given receiver height $z' = z_0 = z'_{S,\rm min}$ given in the title of the sub-figures. The frequency is $f = 3$ GHz, the wind speed is $u_{10} = 5$ m/s,

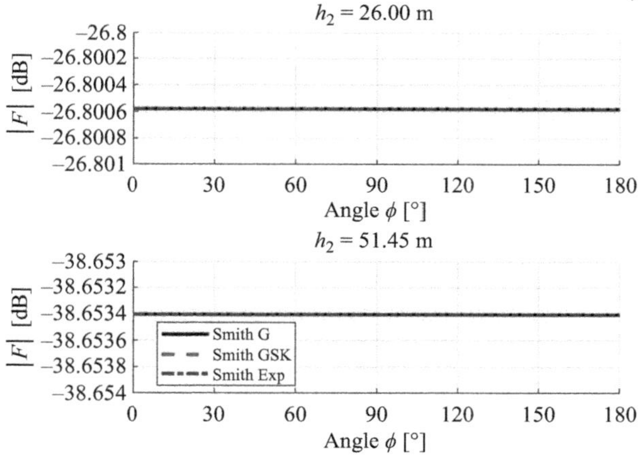

Figure 4.10 Propagation factor modulus in dB scale ($20\log_{10}|F|$) versus the azimuthal direction ϕ and for a given receiver height $z' = z_0 = z'_{S,\min}$ given in the title of the sub-figures. The frequency is $f = 3$ GHz, the wind speed is $u_{10} = 5$ m/s, the transmitter height is $z = 10$ m and the abscissa of the receiver is 5 km. Otherwise, the simulation parameters are the same as in Figure 4.9

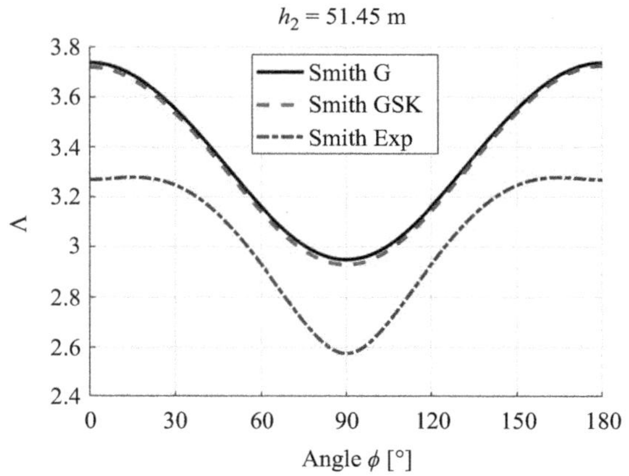

Figure 4.11 Λ versus ϕ (the simulation parameters are the same as in Figure 4.10)

the transmitter height is $z = 10$ m and the abscissa of the receiver is 5 km. Otherwise, the simulation parameters are the same as in Figure 4.9. $z' = z'_{S,\min}$ to have the best sensitivity (level very low) with respect to ϕ.

Figure 4.10 shows clearly that the propagation factor can be considered as independent of ϕ. To explain this behaviour, Figure 4.11 plots Λ versus ϕ (the simulation parameters are the same as in Figure 4.10).

4.3 Sea surface infrared radiation

Remote sensing of the Earth's environment by passive infrared sensing systems is important because it provides a tool in the study of weather forecasting, weather modification, pollution studies and storm warning. These studies need a collection of environmental data over wide range of space within a limited time. In this section, we discuss remote sensing of geophysical data in the ocean environment and the problems associated with passive infrared systems. In passive infrared equipment, the observing camera simply receives the natural radiation from the environment, such as the radiation from gas and aerosol, the radiation from the Earth and the sun, and the radiation reflected by the earth. An example is thermal camera techniques.

In an active system, the signal is sent out from the transmitter, interacts with the target or the clutter, and after the interaction, this signal is observed and measured. Examples of active systems are lidar and radar.

Thermal emissions in the infrared bands can be used to probe the sea surface parameters: wave height and direction, surface temperature, foam distribution and sea state. Passive thermal imaging of the ocean surface is currently measured from an aerial platform. If the platform is located near the sea surface, it becomes difficult to directly obtain the intrinsic radiation of the sea surface due to the fact that for grazing angles, the emissivity is of the same order as the hemispherical reflectivity. Therefore, we have to take into account the reflected sky background. In this case, it is necessary to derived both the emissivity and reflectivity in order to determine the intrinsic sea-surface temperature (SST) from the emissivity knowledge by subtracting the reflected sky background flux.

A recent Ph.D. thesis (written in English) [32] has been published, in which the derivations of the infrared emissivity and reflectivity with multiple reflections are addressed.

Contrary to the scattering problem in the microwave band, the GO approximation is valid because the curvature radius of the capillary wave is much larger than the infrared wavelength. For microwave frequencies, the emissivity is obtained by integrating the scattering coefficient over the half-space [48], where the scattering coefficient may be computed from the small perturbation method [48,49].

4.3.1 Zero-order emissivity from 2D sea surfaces

4.3.1.1 Definition

The intrinsic thermal radiation of a body is characterized by two quantities: its emissivity and the spectral radiance of a black body. A black body is a body which totally absorbs any incident radiation. For any material, Kirchhoff [77,78] showed that its radiance is equal to the radiance of a black body which would radiate at the same temperature multiplied by a coefficient named emissivity. The mathematical expression of the black body radiance as a function of the temperature and wavelength is given by Planck's distribution, involving that the body at surrounding temperature radiates in the near infrared (Wien's law).

The emissivity is difficult to model, it depends on the surface parameters (temperature, roughness) and on the incident beam characteristics (wavelength, incidence angle and polarization).

The goal of this section is to derive the 2D infrared emissivity of the sea surface.

The isotherms of Planck's distribution (for instance, see Figure 3.3 of [50]) defined as

$$L(\lambda, T) = \frac{C_1 \lambda^{-5}}{e^{(C_2/\lambda T)} - 1}, \tag{4.31}$$

where $C_1 = 1.192 \times 10^{-16}$ W m^2 and $C_2 = 1.439 \times 10^{-2}$ m K, reach a maximum $\lambda_{\max}$ with respect to the wavelength satisfying the following relation:

$$\lambda_{\max} T = \text{cste} = 2897\ \mu\text{m K}. \tag{4.32}$$

The above equation, named Wien's law, shows that the hotter the body the more its maximum $\lambda_{\max}$ shifts to a small wavelength. For wavelengths ranging from 5.6 μm to 14 μm, the temperatures are defined from −66°C to 244°C; this region is named the thermal infrared (the limit values may change according to the authors). For a temperature of $T = 37°\text{C} = 310.15\text{K}$, $\lambda_{\max} \approx 9.34$ μm, which implies why the infrared cameras work around this value to observe the human body. But as shown latter, this value can change due to the emissivity.

Physically a black body does not exist, but its theoretical model can be applied to anybody. Indeed, from the Kirchhoff law, for a real body, its brightness $l(\lambda, T)$ is obtained from that of a black body $L(\lambda, T)$ radiating at the same temperature multiplied by a coefficient ε:

$$l(\lambda, T) = \varepsilon L(\lambda, T), \tag{4.33}$$

where ε stands for the emissivity of the material. For a sea surface, it depends on the wavelength λ, the temperature T, the sea roughness, the emission angle θ (similar to the incidence angle) and on the wind direction ϕ. Since the emissivity ranges from 0 to 1, the radiation of a real body is always smaller than that of a black body at the same temperature.

An apparent temperature T_a of a real body can be defined as

$$L(\lambda, T_a) = \varepsilon L(\lambda, T) \Rightarrow T_a = \frac{C_2}{\lambda} \frac{1}{\ln\left((e^{C_2/\lambda T} - 1 + \varepsilon)/\varepsilon\right)}, \tag{4.34}$$

which depends on the emissivity. For $\lambda = 9.34\ \mu\text{m} \approx \lambda_{\max}$ for $T_0 = 37°\text{C} = 310.15\text{K}$, if $\varepsilon \in [0.7; 1]$, then $T_a \in [289.45; 310.15]\text{K} \leq T_0$. As the emissivity decreases, the body is colder.

For Radar frequencies, the ratio $x = C_2/(\lambda T) \ll 1$, and since $\exp(x) \approx 1 + x$ and $\ln(1 + x/\varepsilon) \approx x/\varepsilon$, $T_a \approx \varepsilon T$, which means that the apparent temperature is directly proportional to the emissivity.

4.3.1.2 Brief review

Early models derived the sea surface emissivity without considering sea surface reflections (named direct emissivity or zero-order emissivity contribution). By contrast, the shadowing effect was usually considered. Masuda *et al.* [51] calculated the unpolarized sea surface infrared emissivity by modelling the sea as a 2D surface with Gaussian surface slope PDF. A normalization factor was introduced to estimate the shadowing effect. Instead of using the normalization factor, Yoshimori *et al.* [52,53] took the shadowing effect into account in their emissivity model by using the Smith illumination function [5] (and Chapter 1). Freund *et al.* [54] calculated the sea surface emissivity from the hemispherical ensemble average. Bourlier [23] took a step forward by considering a non-Gaussian surface slope distribution introduced by Cox and Munk [21], which takes the skewness and kurtosis effects into account.

However, Smith *et al.* [62] reported a difference of about 0.02–0.03 between the measurements and the direct emissivity model of Masuda *et al.* [51] for a zenith angle of 73.5°, because surface reflections were ignored. The model of Watts *et al.* [31] and that of Wu and Smith [27] both defined an empirical cut-off angle to calculate the surface-emitted surface-reflected (SESR) emissivity (or named first-order emissivity contribution, as one reflection is considered). Because of the difficulty in defining the cut-off angle, the result has a large uncertainty. The model of Henderson *et al.* [28] developed a ray-tracing Monte Carlo algorithm to calculate the sea surface emissivity with up to ten surface reflections. This method may be valuable reference, but it needs a long computation time.

Masuda [29] took into account the first-order emissivity contribution (SESR) by using a weighting function, which avoided defining an exact cut-off angle. More rigorously, Bourlier [55] evaluated the first-order emissivity contribution by developing a first-order illumination function (with one reflection), which estimates the probability that a surface-emitted ray is reflected once by another point of the surface into the observation direction (Chapter 3). Masuda [29] and Bourlier [55] developed analytical models, but they do not agree well with the results of the ray-tracing Monte Carlo method [33].

Nalli *et al.* [56] shared the idea of Masuda [29] which used a weighting function to calculate the first-order emissivity contribution but replaced the shadowing term used in Masuda [29] by that of Saunders [57]. The most recent model was developed by Li *et al.* [33], in which one surface reflection was considered. They showed that the agreement with measurements is greatly improved by considering one surface reflection.

Recently, Li *et al.* [58] derived the sea surface infrared emissivity by taking both the zero- (direct) and first-order (SESR) emissivity contributions into account. The zero-order emissivity contribution is calculated following the model of [23], where the Smith illumination function [5] (or Chapter 1) is used. When deriving the first-order contribution, they extended the model of Li *et al.* [33] to a 2D sea surface.

To sum up, in the derivation of the emissivity, the more relevant parameter is the shadowing function.

In this section, the surface is 2D and anisotropic with respect to the wind direction ϕ. In addition, the multiple reflections are neglected (named direct emissivity or zero-order emissivity contribution). The extension to multiple reflections is investigated in Section 4.3.2 [32,58].

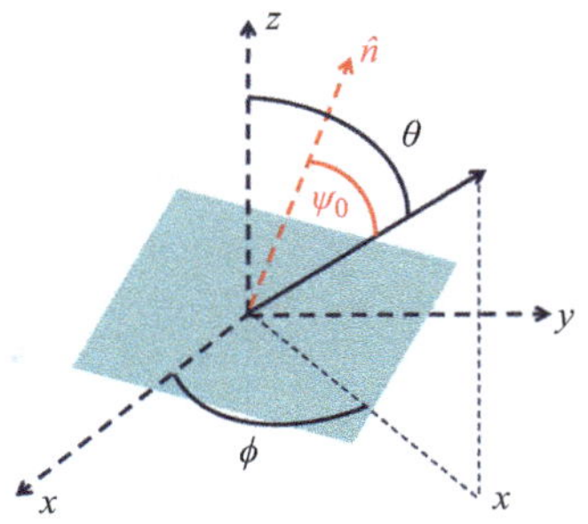

Figure 4.12 Emissivity of a facet

4.3.1.3 Derivation

For a horizontal plane, the unpolarized emissivity is given by the Kirchhoff law expressed as $1 - \mathscr{R}(\psi_0)$, where $\mathscr{R} = (|\mathscr{R}_V^2| + |\mathscr{R}_H^2|)/2$ is the unpolarized Fresnel coefficient in intensity expressed from the Fresnel coefficients in V and H polarizations (Equation (4.2)).

For a tilted facet, the projected intensity equals the incident intensity multiplied by $\cos\psi_0 = \hat{s}\cdot\hat{n}$, where ψ_0 is the angle between the observation direction $\hat{s}$ and the normal to the facet $\hat{n}$ (Figure 4.12). The symbol hat indicates that the vector is unitary. The local emissivity of the facet of area $\mathrm{d}S$ projected into the reference plane (Ox, Oy) is then $\varepsilon_{0,l} = \mathrm{d}S\cos(\psi_0)/(\mathrm{d}x\mathrm{d}y\cos\theta)[1 - \mathscr{R}(\psi_0)]$, for which $\cos\theta$ is a projection term related to the sensor. In Cartesian coordinates $(\hat{x}, \hat{y}, \hat{z})$, $\mathrm{d}S/(\mathrm{d}x\mathrm{d}y) = \sqrt{1 + \gamma_{0x}^2 + \gamma_{0y}^2}$, $\hat{s} = (\sin\theta\cos\phi, \sin\theta\sin\phi, \cos\theta)$ and $\hat{n} = (-\gamma_{0x}, -\gamma_{0y}, +1)/\sqrt{1 + \gamma_{0x}^2 + \gamma_{0y}^2}$, where $(\gamma_{0x}, \gamma_{0y})$ are the slopes of an arbitrary point on the surface in the directions (Ox, Oy), respectively. Then

$$\varepsilon_{0,l}(\theta, \phi; \gamma_{0x}, \gamma_{0y}) = [1 - \mathscr{R}(\psi_0)]\, g_0, \tag{4.35}$$

where

$$g_0 = \frac{\cos\psi_0\sqrt{1 + \gamma_{0x}^2 + \gamma_{0y}^2}}{\cos\theta} = 1 - \frac{\mathrm{s}_\theta}{\mu}(\gamma_{0x}\cos\phi + \gamma_{0y}\sin\phi), \tag{4.36}$$

and

$$\cos[\psi_0(\theta, \phi; \gamma_{0x}, \gamma_{0y})] = \hat{n}\cdot\hat{s} = \cos\theta\frac{1 - (\mathrm{s}_\theta/\mu)(\gamma_{0x}\cos\phi + \gamma_{0y}\sin\phi)}{\sqrt{1 + \gamma_{0x}^2 + \gamma_{0y}^2}}, \tag{4.37}$$

where $\mathrm{s}_\theta = \mathrm{sign}(\theta)$ et $\mu = |\cot\theta|$.

The local emissivity statistically depends on the surface slopes and it is independent of the surface height ζ. It is consistent with the GO approximation, which is an incoherent approach meaning that the phase is not accounted for.

The emissivity measured by a sensor results of the mean value of the local emissivity averaged over the random variables, which are the surface slopes $(\gamma_{0x}, \gamma_{0y})$.

To include the shadow and apply the results of Chapter 2, a rotation of an angle ϕ expressed by (2.1) is made. Moreover, since the emissivity does not depend on ζ, the monostatic statistical shadowing function is averaged over ζ, and it is given by (2.29) omitted of the first term, corresponding to the average over the surface slopes. For any (not even) surface, from the Smith uncorrelated formulation, the monostatic statistical shadowing function is then

$$\bar{S}^0_{\rm M}(\theta, \phi; \Gamma_{0X}) = \frac{\Upsilon\,(\mu - {\rm s}_\theta \gamma_{0X})}{1 + \Lambda_{{\rm s}_\theta}(\theta, \phi)}, \tag{4.38}$$

where $\Lambda_{{\rm s}_\theta}$ is defined as (Equation (1.115))

$$\Lambda_{{\rm s}_\theta}(\theta, \phi) = \frac{1}{\mu} \int_\mu^\infty (\gamma_{0X} - \mu) p_{\gamma_X}({\rm s}_\theta \gamma_{0X}) {\rm d}\gamma_{0X}, \tag{4.39}$$

where $\mu = |\cot\theta| \geq 0$. The average emissivity is

$$\begin{aligned} \varepsilon_0(\theta, \phi) = {} & \frac{1}{1 + \Lambda_{{\rm s}_\theta}(\theta, \phi)} \int_{-\infty}^{\mu} \left(1 - \frac{\gamma_{0X}}{\mu}\right) \\ & \times \left\{ \int_{-\infty}^{+\infty} [1 - \mathscr{R}\,(\psi_0(\theta, \phi; \gamma_{0X}, \gamma_{0Y}))] p_{\gamma_{XY}}({\rm s}_\theta \gamma_{0X}, \gamma_{0Y}) {\rm d}\gamma_{0Y} \right\} {\rm d}\gamma_{0X}, \end{aligned} \tag{4.40}$$

where

$$\cos\left[\psi_0(\theta, \phi; \gamma_{0X}, \gamma_{0Y})\right] = \hat{n} \cdot \hat{s} = \cos\theta \frac{1 - \gamma_{0X}/\mu}{\sqrt{1 + \gamma_{0X}^2 + \gamma_{0Y}^2}}, \tag{4.41}$$

and $p_{\gamma_{XY}}$ is the surface slope PDF obtained from that of $p_{\gamma_{xy}}(\gamma_{0x}, \gamma_{0y})$ by making a rotation of an angle ϕ.

For a centred Gaussian process, $\Lambda_{{\rm s}_\theta} = \Lambda$ is expressed by (4.10) where $v = |\cot\theta|/(\sqrt{2}\sigma_{\gamma_X})$, in which the surface slope standard deviation is $\sigma_{\gamma_X} = \sqrt{\sigma^2_{\gamma_x} \cos^2\phi + \sigma^2_{\gamma_y} \sin^2\phi}$.

4.3.1.4 Numerical results

Numerical implementation

From (4.40), the emissivity requires 2-fold numerical integrations over the slopes $\{\gamma_{0X}, \gamma_{0Y}\}$. To evaluate accurately and shortly this double integral, we use the method reported in [59] and summarized in this section. In (4.40), the significant ranges of integrations over $\{\gamma_{0X}, \gamma_{0Y}\}$ are determined by the exponential function expressed as

$$\exp\left(-\frac{\gamma_{0x}^2}{2\sigma^2_{\gamma_x}} - \frac{\gamma_{0y}^2}{2\sigma^2_{\gamma_y}}\right), \tag{4.42}$$

for a Gaussian slope PDF. The slopes $\{\gamma_{0x}, \gamma_{0y}\}$ are substituted by $\{\gamma_{0X}, \gamma_Y\}$ from the variable transformations given by (2.1). The exponential term then becomes $\exp(-a\gamma_{0Y}^2 - 2b\gamma_{0Y}\gamma_{0X} - c\gamma_{0X}^2)$, in which

$$a = \frac{\alpha + \beta\cos(2\phi)}{2(\alpha^2 - \beta^2)}, \; b = \frac{\beta\sin(2\phi)}{2(\alpha^2 - \beta^2)}, \; c = \frac{\alpha - \beta\cos(2\phi)}{2(\alpha^2 - \beta^2)}, \tag{4.43}$$

where $\alpha = (\sigma_{\gamma_x}^2 + \sigma_{\gamma_y}^2)/2$, $\beta = (\sigma_{\gamma_x}^2 - \sigma_{\gamma_y}^2)/2$. All these quantities are positive because $\sigma_{\gamma_x} \geq \sigma_{\gamma_y} > 0$. The integration limits over γ_{0Y} are then chosen as $\gamma_{0Y} \in [-s_0/\sqrt{a}; s_0/\sqrt{a}]$ since $\exp(-s_0^2) \approx 0$, and over γ_{0X} as $\gamma_{0X} \in [-s_0/\sqrt{c}; \mu]$. If $\mu > s_0/\sqrt{c}$, then the upper limit is $s_0/\sqrt{c}$. Typically, s_0 ranges from 3 to 4. A study of the number of samples, N, for each integration over $\{\gamma_{0X}, \gamma_{0Y}\}$ showed that the value $N = 80$ is sufficient.

If the slope $1 + \gamma_{0Y}^2 \ll 1$, then $\sqrt{1 + \gamma_{0X}^2 + \gamma_{0Y}^2} \approx \sqrt{1 + \gamma_{0X}^2}$. Then, in (4.40), the kernel depends only on γ_{0Y} via the slope PDF. The integration over γ_{0Y} leads to

$$\varepsilon_0(\theta, \phi) \approx \frac{1}{1 + \Lambda_{s_\theta}(\theta, \phi)} \int_{-\infty}^{\mu} \left(1 - \frac{\gamma_{0X}}{\mu}\right) [1 - \mathscr{R}(\psi_0(\theta, \phi; \gamma_{0X}))] p_{\gamma_X}(s_\theta \gamma_{0X}) d\gamma_{0X}, \tag{4.44}$$

where $\cos[\psi_0(\theta, \phi; \gamma_{0X})] = (1 - \gamma_{0X}/\mu)\cos\theta/\sqrt{1 + \gamma_{0X}^2}$. The above equation needs only one numerical integration.

Emissivity versus the emission angle

The values of the surface slope variances $\sigma_{\gamma_x}^2$ and $\sigma_{\gamma_y}^2$ are expressed from the Cox and Munk Model [21] as

$$\sigma_{\gamma_x}^2 = 3.16 \times 10^{-3} u_{12} \quad \sigma_{\gamma_y}^2 = 1.92 \times 10^{-3} u_{12} + 0.003, \tag{4.45}$$

where u_{12} is the wind speed at 12.5 m above the sea surface mean plane. This model is consistent with the Elfouhaily *et al.* spectrum [46].

Figure 4.13 plots the unpolarized emissivity versus the emission angle θ. The wind speed $u_{12} = \{5, 10\}$ m/s, the wind direction is $\phi = \{0, 90\}°$ and the wavelength is $\lambda = 4$ μm.

As the emission angle θ grows or/and the with speed u_{12} decreases, the emissivity decreases. In the legend, the label '(2)' means that (4.40) (2-fold numerical integrations) is applied, whereas for the label '(1)', (4.44) (one numerical integrations) is applied. As we can see, (4.44) is a good approximation and we allow us to reduce significantly the computing time. Figure 4.13 also shows as the wind direction grows, the emissivity decreases because the slope standard deviation is smaller.

Inasmuch the shadowing effect carries a restriction over the surface slope γ_X, the upper limit in the integral over γ_{0X} of (4.44) is $\mu = |\cot\theta|$ instead of $+\infty$. Then, $\gamma_{0X} \leq \mu$, which implies that $1 - (\gamma_X/\mu) \geq 0$, which ensures that the emissivity is positive.

In addition, when the emission angle θ tends towards 90° (grazing angle), $|\tan\theta| = 1/\mu$ tends towards infinity and the emissivity diverges, due to the term $1 - \gamma_{0X}/\mu$ in (4.40), which has no physical meaning. However, if the shadowing is

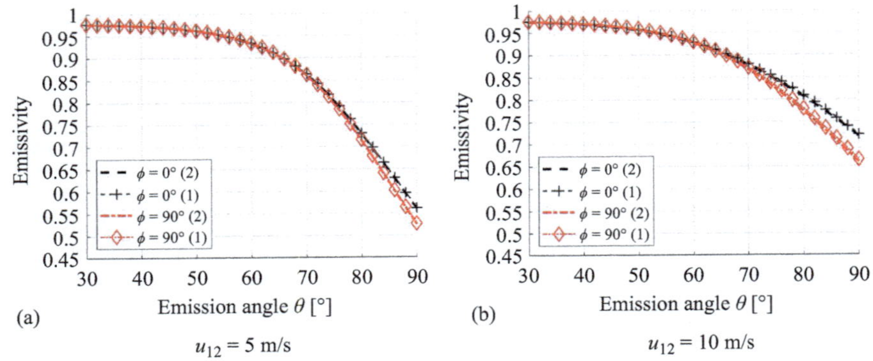

Figure 4.13 *Unpolarized emissivity versus the emission angle θ. The wind speed $u_{12} = \{5,10\}$ m/s, the wind direction is $\phi = \{0,90\}^o$ and the wavelength is $\lambda = 4$ μm. $u_{12} = 5$ m/s in (a) and $u_{12} = 10$ m/s in (b)*

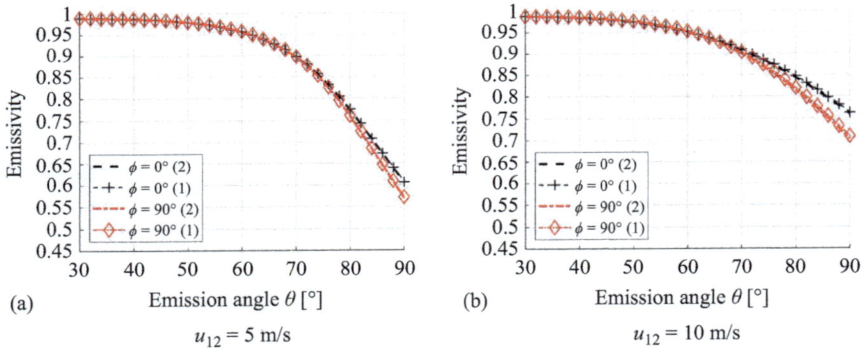

Figure 4.14 *Same variations as in Figure 4.13 but the wavelength is $\lambda = 10$ μm. $u_{12} = 5$ m/s in (a) and $u_{12} = 10$ m/s in (b)*

taken into account, the emissivity converges towards a constant, because from (1.115), we have

$$\lim_{\theta \to (\pi/2)} \frac{\tan\theta}{1 + \Lambda_{s_\theta}(\theta, \phi)} = \left[\int_0^{+\infty} \gamma_{0X} p_{\gamma_X}(\gamma_X) \mathrm{d}\gamma_{0X} \right]^{-1}. \tag{4.46}$$

For instance, for a Gaussian slope PDF, the above limit equals $\sqrt{2\pi}/\sigma_{\gamma_X}$. Hence, for grazing emission angles, it is relevant to account for the shadowing effect.

Figure 4.14 plots the same variations as in Figure 4.13, but the wavelength is $\lambda = 10$ μm.

In (4.40), for θ close to zero, $\mu \to \infty$ and $\Lambda_{s_\theta} \to 0$, which implies that

$$\varepsilon_0(\theta,\phi) \approx [1 - \mathscr{R}(0)] \int_{-\infty}^{+\infty} \int_{-\infty}^{+\infty} p_{YXY}(s_\theta \gamma_{0X}, \gamma_{0Y}) \mathrm{d}\gamma_{0Y} \mathrm{d}\gamma_{0X} \approx 1 - \left|\frac{n-1}{n+1}\right|^2. \tag{4.47}$$

For $\lambda = \{4, 10\}$ μm, the sea refraction index $n = \{1.351 + 0.005i, 1.218 + 0.051i\}$ [60], and $\varepsilon(0,\phi) = \{0.978, 0.990\}$ for θ close to zero.

Infrared sensors can measure the SST, T_a, defined from (4.34). We can show that

$$\frac{\Delta\varepsilon}{\varepsilon} \approx \frac{d\varepsilon}{\varepsilon} = \Delta T \frac{C_2}{\lambda T^2} \frac{e^x}{e^x - 1}, \tag{4.48}$$

where $x = C_2/(\lambda T)$. For $x \gg 1$ (case of the infrared band), we obtain that $\Delta\varepsilon/\varepsilon \approx \Delta T C_2/(\lambda T^2)$. With $T = 15°$C, an accuracy in ΔT of 0.1K (corresponds approximately to the resolution of infrared cameras) implies an accuracy in the relative emissivity, $\Delta\varepsilon/\varepsilon$, of 0.42% for a wavelength of $\lambda = 4$ μm and of 0.17% for $\lambda = 10$ μm. This means that the accuracy increases with the wavelength and the SST is very sensitive to a small variation of the emissivity.

As depicted in Fig. 7 of [23], from this criteria, an upper emission angle $\theta_{\max}$ can be computed, for which for $\theta \leq \theta_{\max}$, the surface can be considered as smooth, which implies that (4.47) is valid.

Comparison with a Monte Carlo method

As the validation of the shadowing function presented in Chapter 1, the emissivity can be compared with a ray-tracing Monte Carlo method to check if the shadow is well accounted for. For a 1D surface, this method is detailed in Chapter 3 from [32,55], in which the multiple reflections are included. Here, we consider a 2D surface, and the MC results are obtained from the paper of Henderson *et al.* [28], in which Gaussian statistics of the surfaces are assumed. In addition, the wind direction is $\phi = 0$ and the wavelength is $\lambda = 4$ μm. The emission angle ranges from 0° to 85° with a step of 5°.

On the left, Figure 4.15 plots the unpolarized emissivity versus the emission angle θ. The wind speed $u_{12} = 5$ m/s, the wind direction is $\phi = 0$, and the wavelength is $\lambda = 4$ μm. On the right, the difference against the results obtained from the analytical model (Equation (4.40)) is plotted. Figure 4.16 plots the same variations as in Figure 4.15, but the wind speed $u_{12} = 10$ m/s. In the legend, the labels mean

- 'Ana: 0', emissivity computed from (4.40).
- 'MC: 0', emissivity computed from the results provided in [28], in which the multiple reflections are ignored.
- 'MC: N', emissivity computed from the results provided in [28], in which the multiple reflections are accounted for.

Figures 4.15 and 4.16 show a very good agreement between $\varepsilon_{\text{Ana: 0}}$ and $\varepsilon_{\text{MC: 0}}$, which means that the modelling of the shadowing in the derivation of ε_0 is correct. As the emission angle increases, the contribution of the multiple reflections increases, but decreases from 85° to 90°. The maximum of the deviation is 0.03 for an emission angle equal to 80°.

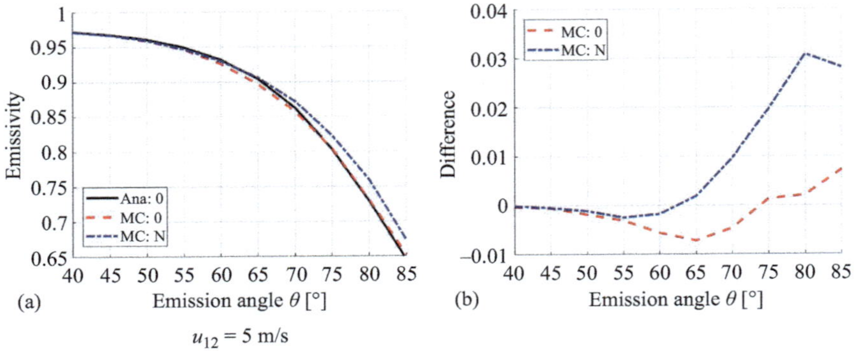

Figure 4.15 *(a) Unpolarized emissivity versus the emission angle θ. The wind speed $u_{12} = 5$ m/s, the wind direction is $\phi = 0$ and the wavelength is $\lambda = 4\ \mu m$. (b) The difference against the results obtained from the analytical model (Equation (40))*

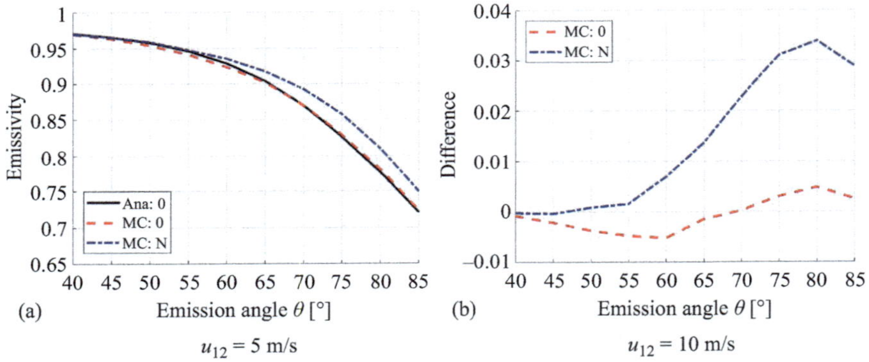

Figure 4.16 *Same variations as in Figure 4.15, but the wind speed $u_{12} = 10$ m/s*

These results are in agreement with those obtained for a 1D surface [32,55]. In addition, Li [32] showed that the emissivity computed from 1D and 2D surfaces agreed well, which means that a cross section of the surface with respect to the wind direction ϕ (1D surface) is a good approximation (it is equivalent to apply (4.44)). This assumption allows us to reduce significantly the computing time since the number of the fold integrations is divided by two.

Comparison with data

A complete description of the spectral radiance measurements and the technique used for deriving the spectral distribution of ocean emissivity is provided by Niclòs *et al.* [61] and Smith *et al.* [62].

In [61], the emissivity experimental values are determined from thermal infrared radiometric measurements carried out from an oil rig under open Mediterranean

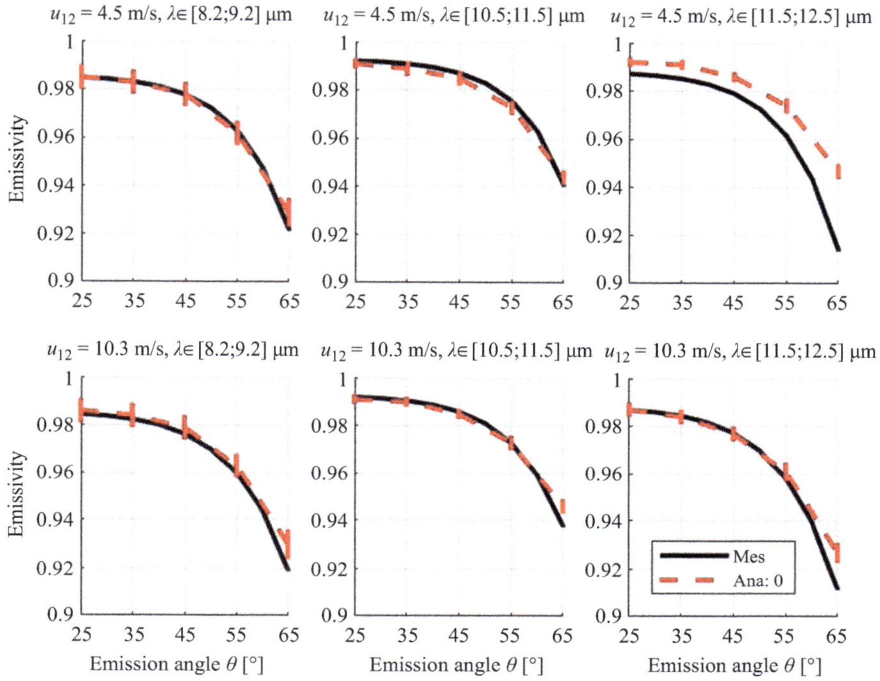

Figure 4.17 Unpolarized emissivity versus the emission angle θ over different wavelength windows given in the titles and for two wind speeds $u_{12} = \{4.5 \pm 0.9, 10.3 \pm 1.1\}$ *m/s and the wind direction* $\phi = 284^o$

conditions during the WInd and Salinity Experiment 2000 campaign (WISE 2000) founded by ESA. The methodology consists of quasi-simultaneous measurements of the radiance coming from the sea surface and the downwelling sky radiance, in addition to the corresponding sea temperature as reference. Radiometric data were taken by a CE 312 radiometer, with four channels placed in the 8–14 μm interval. The wind speeds are $u_{12} = \{4.5 \pm 0.9, 10.3 \pm 1.1\}$ m/s and the wind direction is $\phi = 284 \pm 32^\circ$. The emission angle θ ranges from 25° to 65° with a sampling step of 10°.

In Figure 4.17, the unpolarized emissivity is compared with measurements [61] versus the emission angle θ. At the top, the wind speed is $u_{12} = 4.5$ m/s, whereas at the bottom $u_{12} = 10.3$ m/s. On the left, the unpolarized emissivity is averaged over the range 8.2–9.2 μm, at the middle over 10.5–11.5 μm and on the right, over 11.5–12.5 μm. The wind direction is $\phi = 284^\circ$. The vertical lines around the data curve indicate the error bars. The numerical results are averaged over the wavelength with a sampling step of 0.2 μm, which is enough because the sea refraction index weakly varies with the wavelength. In addition, for each point of the curve, the error bar is displayed. Its minimum value is obtained for $u_{12} = \{3.6, 9.2\}$ m/s and $\phi = 252^\circ$, whereas its maximum value is obtained for $u_{12} = \{5.4, 11.4\}$ m/s and $\phi = 316^\circ$.

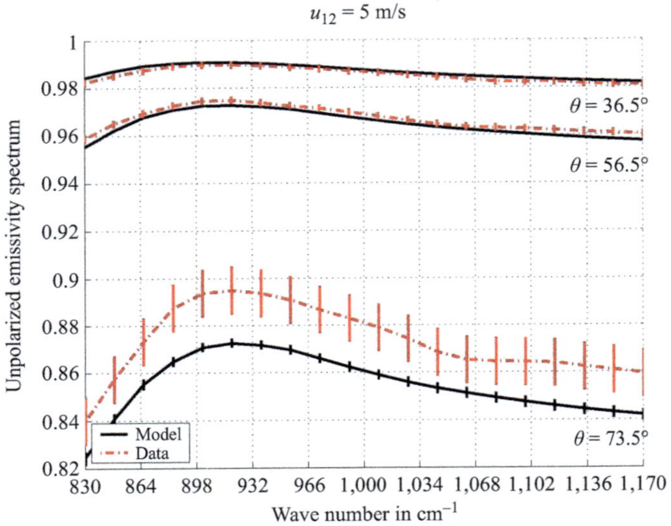

Figure 4.18 Comparison of the unpolarized emissivity spectrum with measurements [62] versus the wavenumber 1/λ in cm^{-1}. The wind speed is $u_{12} = 5$ m/s and the wind direction is $\phi = 0°$. Figure obtained, with permission, from [55]

For the windows 8.2–9.2 μm and 10.5–11.5 μm, there is a good agreement between the model and the measurements. Nevertheless, for the window 11.5–12.5 μm and for the wind speed $u_{12} = 4.5$ m/s, the curve of the model goes below the data curve. The possible causes are explained in [55].

In [62], the emissivity was measured on 16 January 1995 in the Gulf of Mexico with a Fourier transform spectrometer. On the day of measurement, the sky was clear and the wind calmed down to approximately 5 m/s for the entire day. The instrument was mounted aboard the side of an oceanographic research vessel, the R. V. Pelican, operated by the Louisiana University Marine Consortium. Spectral radiance was measured, with a spectral resolution of 0.5 cm^{-1}, at several emission angles 36.5°, 56.5° and 73.5°.

In Figure 4.18 (obtained from [55]), the unpolarized emissivity spectrum is compared with measurements [62] versus the wavenumber $1/\lambda$ in cm^{-1}. The corresponding range of the wavelength λ is 8.55–12.05 μm. The wind speed is $u_{12} = 5$ m/s and the wind direction is $\phi = 0$. Since the wind direction is not given by Smith *et al.* [62] (an isotropic sea surface is assumed), for each point of the curve, an error bar obtained for wind directions $\phi = \{0, 90\}°$ is displayed. For an emission angle $\theta = 73.5°$, a deviation is noted between the data and the model. As explained in [55], the cause is attributed to the multiple reflections, not accounted for in the model. Indeed, later, Li *et al.* [58] (and Section 4.3.2) did comparisons by including the first reflection, leading to a better agreement with the measurements [62].

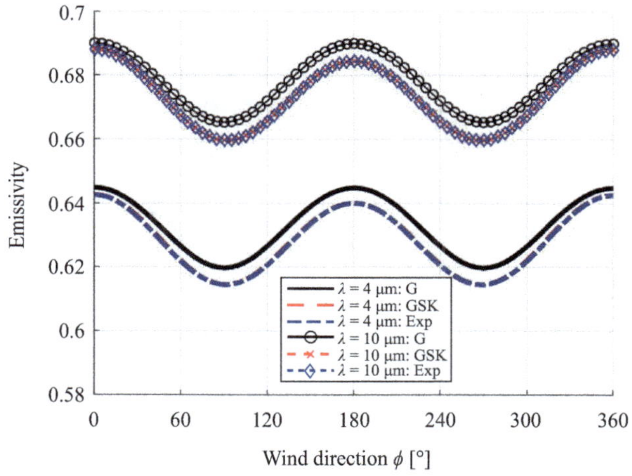

Figure 4.19 *Unpolarized emissivity versus the wind direction ϕ. The emission angle $\theta = 85^o$, the wind speed $u_{12} = 5$ m/s and the wavelengths are $\lambda = \{4, 10\}$ μm*

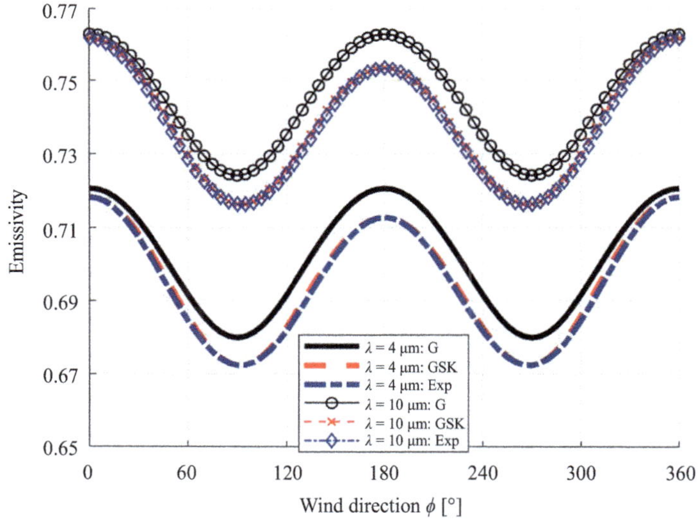

Figure 4.20 *Same variations as in Figure 4.19, but the wind speed is $u_{12} = 10$ m/s*

Emissivity versus the wind direction

Figure 4.19 plots the unpolarized emissivity versus the wind direction ϕ. The emission angle $\theta = 85^o$, the wind speed $u_{12} = 5$ m/s and the wavelengths are $\lambda = \{4, 10\}$ µm. Figure 4.20 plots the same variations as in Figure 4.19, but the wind speed is $u_{12} = 10$ m/s.

For any statistics, Figures 4.19 and 4.20 reveal that the emissivity is symmetric along the downwind direction, i.e. $\varepsilon(\theta, \pi + \phi) = (\theta, \pi - \phi)$. This means that the emissivity keeps the symmetry properties of the Cox and Munk slope distribution.

For $\phi \in [0; \pi]^{\circ}$, when the wind speed increases, the emissivity becomes more dependent of the wind direction, i.e. of the sea surface anisotropy. For Gaussian statistics (label 'G'), the emissivity varies like a cosine function with a non-zero mean value (i.e. $\varepsilon(\theta, \phi) \approx \varepsilon_0(\theta) + \varepsilon_2(\theta)\cos(2\phi)$), where its minimum occurs in the crosswind direction ($\phi = 90^{\circ}$ and the slope RMS $\sigma_{\gamma_X} = \sigma_{\gamma_y}$ is minimum). Conversely, its maximum occurs in the up- ($\phi = 0^{\circ}$) and downwind directions ($\phi = 180^{\circ}$), for which $\sigma_{\gamma_X} = \sigma_{\gamma_x}$ is maximum. In addition, one can see that $\varepsilon(\theta, 0) = \varepsilon(\theta, \pi)$ and $\varepsilon(\theta, \pi/2) = \varepsilon(\theta, 3\pi/2)$ since for a Gaussian process, the slope occurrence is symmetric.

When the higher order statistics are taken into account (label 'GSK'), the emissivity is more sensitive to the wind direction for high-wind speeds ($u_{12} = 10$ m/s).

The emissivity can be modelled with respect to the wind direction ϕ as

$$\varepsilon(\theta, \phi) \approx \varepsilon_0(\theta) + \varepsilon_1(\theta)\cos(\phi) + \varepsilon_2(\theta)\cos(2\phi), \tag{4.49}$$

where $\{\varepsilon_{0,1,2}(\theta)\}$ can be found from the values of $\varepsilon(\theta, 0)$, $\varepsilon(\theta, \pi/2)$ and $\varepsilon(\theta, \pi)$ as

$$\begin{cases} \varepsilon_0(\theta) = [\varepsilon(\theta, 0) + \varepsilon(\theta, \pi) + 2\varepsilon(\theta, \pi/2)]/4 \\ \varepsilon_1(\theta) = [\varepsilon(\theta, 0) - \varepsilon(\theta, \pi)]/2 \\ \varepsilon_2(\theta) = [\varepsilon(\theta, 0) + \varepsilon(\theta, \pi) - 2\varepsilon(\theta, \pi/2)]/4 \end{cases}, \tag{4.50}$$

$\varepsilon_0(\theta)$ corresponds to the emissivity of an isotropic surface, $\varepsilon_1(\theta)$ corresponds to the asymmetry between the up- and the downwind directions (for Gaussian statistics, it vanishes since $\varepsilon(\theta, 0) = \varepsilon(\theta, \pi)$) and $\varepsilon_2(\theta)$ corresponds to the asymmetry between the up- and the crosswind directions

For non-Gaussian statistics, the expansion given by (4.49) is plotted in Figures 4.19 and 4.20 (label 'Exp'). One can see a good agreement between the emissivity and its expansion.

Like Gaussian statistics, the emissivity for non-Gaussian statistics reaches its maximum in the downwind direction ($\phi = 180^{\circ}$), whereas for the up-wind direction ($\phi = 0$), the emissivity is smaller than the one obtained with Gaussian statistics. We can show from (4.49) that the minimum of $\varepsilon(\theta, \phi)$ according to $\phi \in [0; \pi]$ is given by $\phi_{\min} \approx \pi/2 + \varepsilon_1(\theta)/[4\varepsilon_2(\theta)]$. For Gaussian statistics, $\phi_{\min} = 90^{\circ}$ for any wind speed, any wavelength and any emission angle.

The values of $\{\varepsilon_0, \varepsilon_1, \varepsilon_2, \phi_{\min}\}$ are reported in Table 4.1.

Figure 4.21 plots the SST, defined by (4.34), versus the wind direction ϕ. The emission angle $\theta = 85^{\circ}$, the wind speed $u_{12} = 5$ m/s, the wavelengths are $\lambda = \{4, 10\}$ µm and the SST is $T = 15^{\circ}$C.

As predicted theoretically (Equation (4.48)), a small variation of the emissivity implies a large variation of the SST.

Table 4.1 *Coefficients* $\{\varepsilon_{0,1,2}(\theta)\}$ *of the emissivity expansion defined as* $\varepsilon(\theta,\phi) = \varepsilon_0(\theta) + \varepsilon_1(\theta)\cos(\phi) + \varepsilon_2(\theta)\cos(2\phi)$ *and the angle* $\phi_{\min} \approx \pi/2 + \varepsilon_1(\theta)/[4\varepsilon_2(\theta)]$ *in degrees giving the minimum of* $\varepsilon(\theta,\phi)$ *for* $\phi \in [0;\pi]$ *and for a given* θ

θ [°], u_{12} [m/s], λ [μm]	$\varepsilon_0(\theta)$	$\varepsilon_1(\theta)$	$\varepsilon_2(\theta)$	$\phi_{\min}$
85, 5, 4	0.6279	0.0013	0.0134	91.36
85, 5, 10	0.6730	0.0018	0.0132	91.95
85, 10, 4	0.6939	0.0027	0.0215	91.83
85, 10, 10	0.7369	0.0041	0.0207	92.87

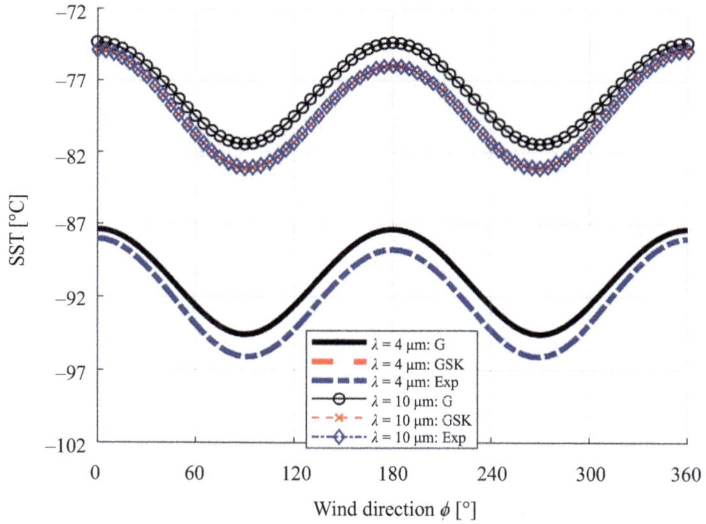

Figure 4.21 *SST, defined by (34), versus the wind direction* ϕ*. The emission angle* $\theta = 85^o$*, the wind speed* $u_{12} = 5$ *m/s, the wavelengths are* $\lambda = \{4, 10\}$ μm *and the sea surface temperature is* $T = 15^o C$

4.3.2 One-order emissivity from 1D sea surfaces

4.3.2.1 Introduction

As shown in Figure 4.18, to improve the comparison with the measurements, the multiple reflections must be accounted for. For a 2D surface, the addition of a reflection in the emissivity derivation requires the calculation of a 4-fold numerical integrations. In the previous section, we showed that the 2D integral of the zero-order emissivity can be converted into 1D integral by introducing a simplifying assumption. This is similar to make a cross section of the 2D sea surface along the ϕ direction, and the surface becomes 1D.

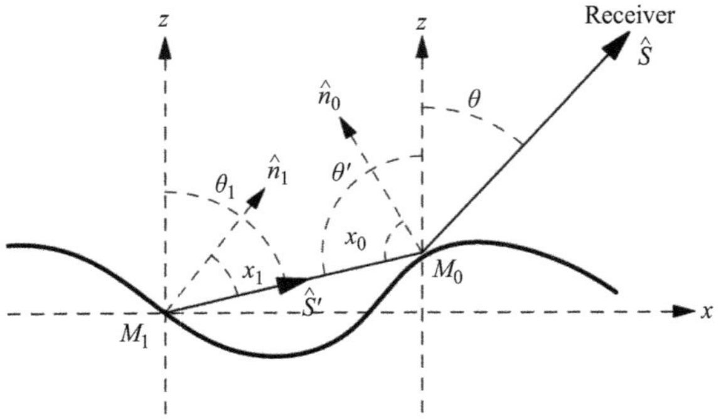

Figure 4.22 Illustration of one surface reflection: the ray emitted from the source point M_1 propagates along the $\hat{s}'$ direction and intersects the surface at M_0 where it is reflected towards the receiver along the direction $\hat{s}$

In this section, the surface is assumed to be 1D to calculate the one-order emissivity with one surface reflection.

4.3.2.2 Mathematical expression

Emissivity with one reflection, or called first-order emissivity, corresponds to the emission energy which is reflected once on the surface before it is captured by the receiver. In order to calculate the emissivity with one reflection, the probability that such situation occurs is evaluated by the shadowing function with one reflection. This issue is addressed in [29,55,63]. In fact, these models differ only on the choice of the derivation of the shadowing function. In this section, the shadowing function by Li *et al.* [33] will be applied, and its derivation is thoroughly presented in Chapter 3.

A surface reflection from a sea surface is shown in Figure 4.22. The facet M_1 on the surface emits a ray which propagates along the direction $\hat{s}'$. It intersects the surface at M_0 where it is reflected towards the receiver along the direction $\hat{s}$. The first-order local emissivity at point M_0 is given by [63]

$$\begin{cases} \varepsilon_{1,H}^{\text{local}} = \left[1 - |r_H(\chi_1)|^2\right] |r_H(\chi_0)|^2 \\ \varepsilon_{1,V}^{\text{local}} = \left[1 - |r_V(\chi_1)|^2\right] |r_V(\chi_0)|^2 \end{cases}, \tag{4.51}$$

where χ_0 and χ_1 are the local incidence angle at the points M_0 and M_1, respectively. The term $r(\chi)$ stands for the Fresnel reflection coefficient which depends on the local incidence angle and the sea refraction index. In (4.51), the first term $1 - |r(\chi_1)|^2$ corresponds to the intensity emitted by the source point M_1. The second term $|r(\chi_0)|^2$ is the reflection coefficient in intensity at the point M_0.

For a 1D rough surface, the emission and the reflection rays both belong to the xz plane (no depolarization), as shown in Figure 4.22. The polarization state of the emission ray is not changed during the propagation. In other words, an emitted ray of H polarization is still H polarized when it intersects the facet at the point M_0. As a result, there is not a cross-polarized component.

The first-order emissivity is obtained by averaging the local emissivity over the heights and slopes of the points M_0 and M_1, leading to

$$\begin{cases} \varepsilon_{1,H} = \int_{-\infty}^{+\infty}\int_{-\infty}^{+\infty}\int_{-\infty}^{+\infty}\int_{-\infty}^{+\infty} \varepsilon_{1,H}^{\text{local}} g_0 S_M^1 p_4(\zeta_1, \gamma_1, \zeta_0, \gamma_0)\mathrm{d}\zeta_1\mathrm{d}\gamma_1\mathrm{d}\zeta_0\mathrm{d}\gamma_0 \\ \varepsilon_{1,V} = \int_{-\infty}^{+\infty}\int_{-\infty}^{+\infty}\int_{-\infty}^{+\infty}\int_{-\infty}^{+\infty} \varepsilon_{1,V}^{\text{local}} g_0 S_M^1 p_4(\zeta_1, \gamma_1, \zeta_0, \gamma_0)\mathrm{d}\zeta_1\mathrm{d}\gamma_1\mathrm{d}\zeta_0\mathrm{d}\gamma_0 \end{cases}, \quad (4.52)$$

where S_M^1 is the monostatic shadowing function with one reflection and $p_4(\zeta_1, \gamma_1, \zeta_0, \gamma_0)$ is the joint PDF of the heights and the slopes of points M_1 and M_0. The term g_0 results from the area projection of the facet M_0 onto the orthogonal direction of the observation direction. It is expressed as

$$g_0 = 1 - \gamma_0 \tan\theta. \quad (4.53)$$

The monostatic shadowing function with one reflection is thoroughly derived in Section 3.1.5. In this section, a Gaussian rough surface is considered, which means that the surface heights and slopes follow a Gaussian distribution.

From the model of Li *et al.* [33] and neglecting the correlation between the points M_0 and M_1, we have

$$S_M^{1,\text{unco}}(\theta, \gamma_0, \zeta_0) = \Upsilon(\mu - \gamma_0)F(\zeta_0)^{\Lambda(\mu)} \times \begin{cases} 1 & \text{if } |\theta_1| > 90^\circ \\ 0 & \text{if } 0^\circ < \theta_1 \leq \theta \\ 1 - F(\zeta_0)^{\Lambda(\mu_1)} & \text{if } \theta < \theta_1 \leq 90^\circ \\ 1 - F(\zeta_0)^{\Lambda_-(\mu_1)} & \text{if } -90^\circ \leq \theta_1 < 0^\circ \end{cases}, \quad (4.54)$$

where $\mu = \cot\theta$ is the slope of the incident ray, $\mu_1 = \cot\theta_1$ is the slope of the reflected ray and

$$\begin{cases} v = \dfrac{\cot\theta}{\sigma_\gamma\sqrt{2}} \qquad v_1 = \dfrac{\mu_1}{\sigma_\gamma\sqrt{2}} \\ \Lambda(\mu) = \dfrac{\exp(-v^2) - v\sqrt{\pi}\,\text{erfc}(v)}{2v\sqrt{\pi}} \\ \Lambda_-(\mu_1) = \Lambda(|\mu_1|) \\ \text{erfc}(x) = \dfrac{2}{\sqrt{\pi}}\displaystyle\int_x^{+\infty} \exp(-t^2)\mathrm{d}t \end{cases}. \quad (4.55)$$

It is noticeable that if the correlation between M_0 and M_1 is neglected, the surface emissivity does not depend on the surface heights ζ_0 and ζ_1. In addition, the surface slope PDF of γ_0 is assumed to be Gaussian, leading to

$$p_\gamma(\gamma_0) = \frac{1}{\sqrt{2\pi}\sigma_\gamma} \exp\left(-\frac{\gamma_0^2}{2\sigma_\gamma^2}\right), \quad (4.56)$$

with σ_γ the surface root-mean-square slope.

In this section, σ_γ is determined by the wind speed u_{12} recorded at 12.5 m above the sea level, given as [21]

$$\sigma_\gamma^2 \approx 3.16 \times 10^{-3} u_{12}. \tag{4.57}$$

The only undetermined term is the distribution of the slope γ_1. As discussed in Section 3.1.6.1, the slope γ_1 should at least meet the requirement that the local angle of incidence χ_1 is smaller than 90°. Since no further information is available, the slope PDF of γ_1 is assumed to be

$$p_{\gamma_1}(\gamma_1) = \begin{cases} \dfrac{p_\gamma(\gamma_1)}{F_\gamma(\mu_1)} \Upsilon(\mu_1 - \gamma_1) & \text{if } \theta_1 < 0^\circ \\ \dfrac{p_\gamma(\gamma_1)}{1 - F_\gamma(\mu_1)} \Upsilon(\gamma_1 - \mu_1) & \text{if } \theta_1 > 0^\circ \end{cases}, \tag{4.58}$$

where p_γ is the surface slope PDF and F_γ its cumulative density function.

4.3.2.3 Numerical results

In this section, numerical results of the sea surface first-order emissivity are presented. Infrared wavelengths are considered so that the Geometric Optics Approximation is valid and the sea refraction index is given by [60].

As the shadowing function, a Monte Carlo ray-tracing algorithm is used as reference to evaluate the accuracy of the model. The ray-tracing process is the same as the one explained in Section 3.1.2. In a ray-tracing method, the local incidence angle at every intersection is known. For a given rough surface, the first-order emissivity is then expressed as

$$\begin{cases} \varepsilon_{\mathrm{MC},V} = \dfrac{1}{N} \displaystyle\sum_{i=1}^{i=N_i} \left[1 - |r_V(\chi_{1,i})|^2\right] |r_V(\chi_{0,i})|^2 g_0 \\ \varepsilon_{\mathrm{MC},H} = \dfrac{1}{N} \displaystyle\sum_{i=1}^{i=N_i} \left[1 - |r_H(\chi_{1,i})|^2\right] |r_H(\chi_{0,i})|^2 g_0 \end{cases}, \tag{4.59}$$

where N_i equals the number of groups of M_0 and M_1 obtained by the ray-tracing process and N is the number of samples of the generated surface. In addition, to improve the convergence, the above equation can be applied on several realizations.

Figure 4.23 plots the sea surface emissivity with one reflection (or first-order emissivity) for the H and V polarizations. The wind speed $u_{12} = 5$ m/s in (a) and (b), and $u_{12} = 10$ in (c) and (d). The wavelength $\lambda = 10$ μm.

As we can see, the first-order emissivity is significant at large incidence angle θ. As expected, the monostatic shadowing function with one reflection shares a similar trend. The maximum is found at about $\theta \approx 85°$ with a level of about 0.025. In general, the model of Li *et al.* [33] agrees well with the Monte Carlo ray-tracing algorithm for both polarizations.

The emissivity with two reflections obtained from the Monte Carlo ray-tracing algorithm is also shown to give an idea of its level. A maximum of about 0.0025 is found around $\theta \approx 80°$, which is about one-tenth of that of the emissivity with

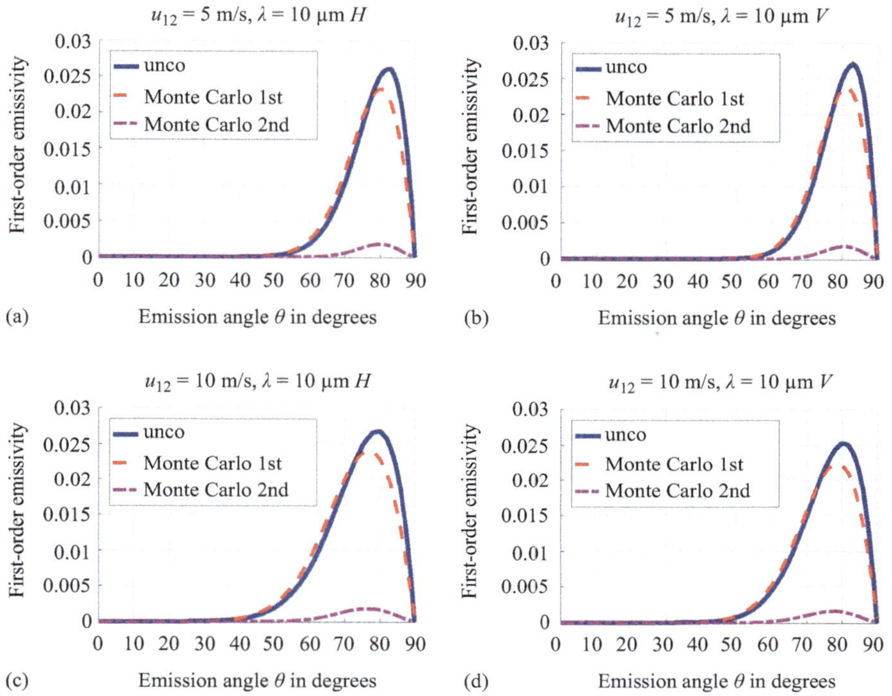

Figure 4.23 *Sea surface emissivity with one reflection (or first-order emissivity) for the* H *and* V *polarizations. The wind speed* $u_{12} = 5$ *m/s in (a) and (b), and* $u_{12} = 10$ *m/s in (c) and (d). The wavelength* $\lambda = 10$ *μm*

one reflection. As a result, when calculating the sea surface infrared emissivity, one reflection is enough.

Figure 4.24 compares the sea unpolarized (mean value of the H and V polarized ones) emissivity considering one reflection with measurement obtained by Niclòs *et al.* [61]. The measurement is obtained over the open Mediterranean Sea surface from an oil rig. Radiometric data were taken for four channels of wavelength: 8–14, 8.2–9.2, 10.5–11.5 and 11.5–12.5 μm, but comparisons are made only for the latter three channels. The wind speeds were approximately $u_{12} = \{4.5 \pm 0.9; 10.3 \pm 1.1\}$ m/s during the experiment. See also Figure 4.17 for the comparison with the zero-order emissivity of a 2D sea surface.

The sea refraction index is given by the model of Hale and Querry [60]. For the numerical results, wind speeds of $u_{12} = 4.5$ m/s and $u_{12} = 10.3$ m/s are assumed.

It is observed that the zero-order emissivity agrees well with measurements for $\theta \leq 55°$, but an underestimation is found for $\theta = 65°$. By taking into account the first-order emissivity, the agreement is improved, especially for the cases when $u_{12} = 10.3$ m/s, when surface reflections are more significant.

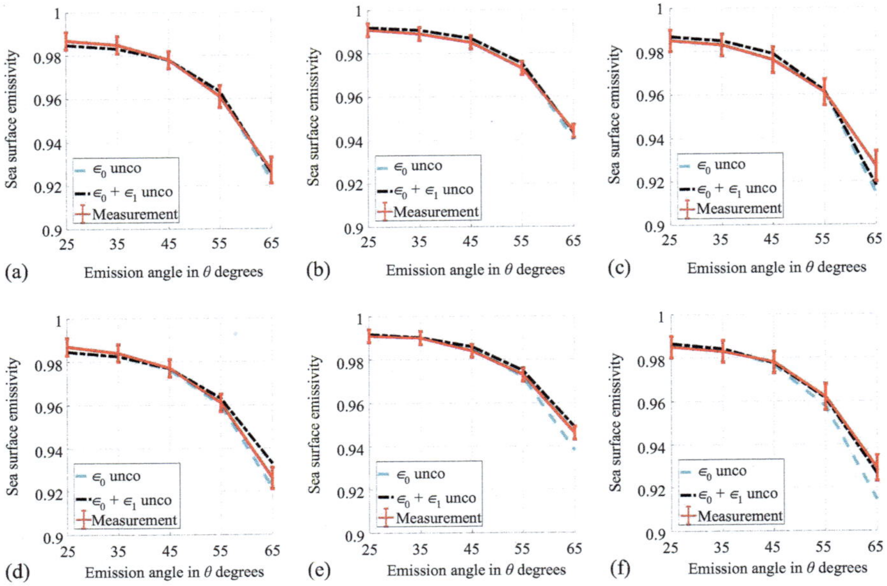

Figure 4.24 Sea surface unpolarized infrared emissivity compared with measurements by Niclòs et al. [61] at wind speeds of 4.5 m/s at the first row, and 10.3 m/s at the second one. Three wavelength channels are considered: 8.2–9.2 μm at the first column, 10.5–11.5 μm at the second column, 11.5–12.5 μm at the third column

Figure 4.25 shows a comparison between the calculated sea emissivity considering one reflection and the measurement obtained by Smith *et al.* [62] at three observation angles $\theta = \{36.5°, 56.5°, 73.5°\}$. The measurement is obtained in the Gulf of Mexico at 16 January 1995. During the experiment, the wind speed ranges from 2 to 8 m/s. For the simulation, the wind speed is set to the average wind speed of the experiment $u_{12} = 5$ m/s. See also Figure 4.18 for the comparison with the zero-order emissivity of a 2D sea surface.

As we can see, the agreement between the model and the measurement obtained by Smith *et al.* [62] is significantly improved for $\theta = 73.5°$ and the minor difference can be attributed to the estimation of the wind speed. For observation angles $\theta = 36.5°$ and $\theta = 56.5°$, the contribution of the first-order emissivity is minor since the surface reflections can be ignored for this range of angles.

4.3.3 One-order reflectivity from 2D surfaces

4.3.3.1 Brief review

Sea surface infrared reflectivity ρ receives wide attention in many fields of oceanic remote sensing, e.g. in the estimation of SST [62], the determination of the sea surface BRDF [64–66] and the vessel detection [67]. It corresponds to the ability of

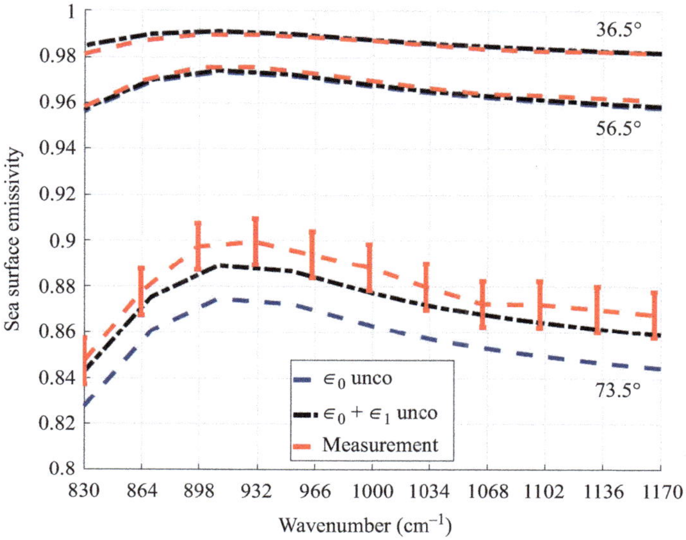

Figure 4.25 Unpolarized emissivity compared with measurements by Smith et al. at emission angles $\theta = 36.5°$, $56.5°$ and $73.5°$. Simulations are performed for a wind speed of 5 m/s

the sea surface to reflect the incident energy; thus, it depends on the wavelength of the incident (ray) wave, the incidence and the observation directions, and the surface roughness. While reflectivity is closely related to the emissivity ε, their derivations for sea surfaces differ. Specifically, the emissivity is derived under a monostatic configuration (one receiver and no emitter), but the derivation of the reflectivity employs a bistatic configuration (one receiver and one emitter).

According to the law of energy conservation, the sum of the reflected and absorbed energies equals the incident energy (for an opaque body), meaning that the sum of the emissivity ε and hemispherical reflectivity ρ_{hemi} equals 1. Nalli *et al.* [56] and Watts *et al.* [31] derived the sea surface infrared reflectivity from the equation $\rho_{\text{hemi}} = 1 - \varepsilon$. This method avoids the calculation of the probability of having single or multiple reflections, and it is then simpler. However, the bidirectional characteristics of the reflectivity cannot be studied with this method. As a result, the surface reflectivity is usually calculated from facet models under a bistatic configuration without deriving the emissivity.

Published analytical facet models of sea surface infrared reflectivity, e.g. Bourlier *et al.* [68], Caillault *et al.* [64], Fauqueux *et al.* [65], Ross *et al.* [66] and Yoshimori *et al.* [52], considered single surface reflection. The shadowing effect due to the surface roughness was evaluated by using a bistatic illumination function (Chapter 1). However, when examining the energy conservation criterion with analytical facet models of the sea surface infrared emissivity and reflectivity, a loss of energy was reported for large zenith angles by Yoshimori *et al.* [52]. A maximum of the loss of energy of the order of 0.04 was found around $\theta \approx 80^o$, meaning that about 4% of the incident energy is 'lost'. The cause is that the energy undergoing multiple surface reflections is not taken into account.

Multiple surface reflections are seldom studied because the probability of their occurrence, which is determined by a bistatic illumination, is hard to evaluate. The Monte Carlo ray-tracing method, such as the model of Henderson *et al.* [28] for the emissivity and that of Schott *et al.* [69] for the reflectivity, is a direct way to study multiple surface reflections and is a good reference for analytical models. As written in Chapter 3, Bourlier *et al.* [35] derived an analytical bistatic illumination function with n reflections, which was still not validated from a Monte Carlo method. Lynch and Wagner [34] derived an analytical bistatic illumination function with two reflections for the calculation of the reflected field of an incident wave on a perfectly conducting rough surface. They proved that the law of energy conservation was better satisfied by considering the second reflection. Li *et al.* [30] derived the sea surface infrared reflectivity with two reflections by introducing a new bistatic illumination function addressed in Chapter 3. They also showed that the energy conservation is better satisfied after taking into account multiple surface reflections. Their model is extended to a 2D surface in [36].

In this section, the unpolarized reflectivity with a single reflection is investigated for a 2D sea surface. The next section will extend the results to multiple reflections by considering a 1D surface.

4.3.3.2 Derivation

As the derivation of the emissivity, the geometric optics approximation is assumed to be valid, and then the surface is modelled as a collection of facets. For a given facet, the local unpolarized reflectivity $\rho_{1,l}$ is expressed as

$$\rho_{1,l} = \mathscr{R}(\psi)g, \tag{4.60}$$

where $\mathscr{R} = (|\mathscr{R}_V^2| + |\mathscr{R}_H^2|)/2$ is the unpolarized Fresnel coefficient in intensity expressed from the Fresnel coefficients in V and H polarizations (Equation (4.2)). In addition, g is a projection term similar to that introduced in the derivation of the emissivity and expressed as

$$g(\gamma_x, \gamma_y; \theta_2, \phi_2) = \frac{\hat{n} \cdot \hat{s}_2}{\left(\hat{n} \cdot \hat{z}\right)\left(\hat{z} \cdot \hat{s}_2\right)} = 1 - \left(\gamma_x \cos\phi_2 + \gamma_y \sin\phi_2\right)\tan\theta_2, \tag{4.61}$$

where γ_x and γ_y are the surface slopes with respect to the (Ox) and (Oy) directions of the facet. The unitary vector $\hat{s}_2 = \hat{s}(\theta_2, \phi_2) = (\cos\phi_2 \sin\theta_2, \sin\phi_2 \sin\theta_2, \cos\theta_2)$ stands for the direction of the receiver and ψ is the angle between the normal to the facet (Figure 4.26) defined as $\hat{n} = (1, -\gamma_x, -\gamma_y)/\sqrt{1+\gamma_x^2+\gamma_y^2}$. It is expressed by

$$\cos\left[\psi_0(\gamma_x, \gamma_y; \theta_2, \phi_2)\right] = \hat{n} \cdot \hat{s}_2 = \frac{\cos\theta_2 - \sin\theta_2\left(\gamma_x \cos\phi_2 + \gamma_y \sin\phi_2\right)}{\sqrt{1+\gamma_x^2+\gamma_y^2}}. \tag{4.62}$$

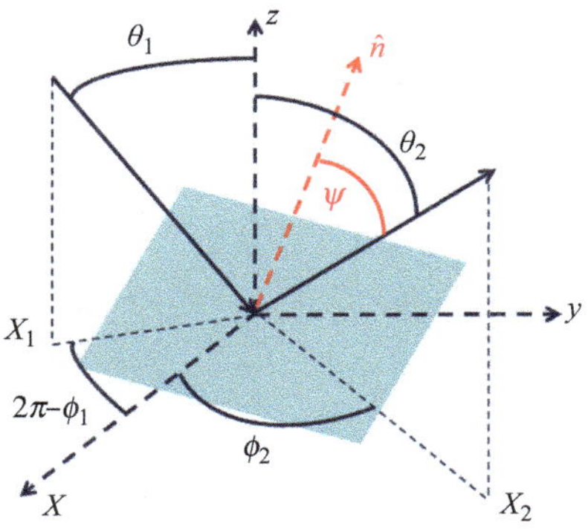

Figure 4.26 Reflectivity of a facet: the angles $\theta_1 \in [0;\pi/2]$, $\phi_1 \in [0;2\pi]$, $\theta_2 \in [0;\pi/2]$ and $\phi_2 \in [0;2\pi]$

The direction of the transmitter $\hat{s}_1 = \hat{s}(\theta_1, \phi_1) = (\cos\phi_1 \sin\theta_1, \sin\phi_1 \sin\theta_1, \cos\theta_1)$ is not arbitrary because it must obey a Snell–Descartes law giving the specular direction. This implies that

$$\hat{s}_1 = 2\left(\hat{n}\cdot\hat{s}_2\right)\hat{n} - \hat{s}_2 = 2\cos(\psi)\,\hat{n} - \hat{s}_2. \tag{4.63}$$

From the above equation, the slopes of the facet satisfying the specular directions are

$$\gamma_{0x} = -\frac{\cos\phi_2 \sin\theta_2 + \cos\phi_1 \sin\theta_1}{\cos\theta_1 + \cos\theta_2}, \quad \gamma_{0y} = -\frac{\sin\phi_2 \sin\theta_2 + \sin\phi_1 \sin\theta_1}{\cos\theta_1 + \cos\theta_2}. \tag{4.64}$$

The unpolarized reflectivity ρ_1 measured by a sensor results of the mean value of $\rho_{1,l}$, for which the random variables are γ_x and γ_y. Introducing the shadowing, this leads from (4.60) to

$$\rho_1(\theta_1, \phi_1, \theta_2, \phi_2) = \int_{-\infty}^{+\infty}\int_{-\infty}^{+\infty} \mathscr{R}(\psi) g \bar{S}_{\mathrm{B}}^1 p_{\gamma_{xy}}\left(\gamma_x, \gamma_y\right) \mathrm{d}\gamma_x \mathrm{d}\gamma_y, \tag{4.65}$$

where $\bar{S}_{\mathrm{B}}^1$ is the statistical bistatic shadowing function averaged over the surface heights ζ since the local reflectivity is independent of ζ. Besides, $p_{\gamma_{xy}}$ stands for the joint surface slope PDF. For a given direction $\hat{s}_1$, the slopes satisfying (4.64) are unique giving the Dirac delta function $\delta(\hat{s}_2 - \hat{s}_1(\gamma_{0x}, \gamma_{0y}))$. Then, the integrations over γ_x and γ_y yield

$$\begin{aligned}\rho_1(\theta_1, \phi_1, \theta_2, \phi_2) &= \left[\mathscr{R}(\psi) g \bar{S}_{\mathrm{B}}^1 p_{\gamma_{xy}}\left(\gamma_x, \gamma_y\right)\right]_{\gamma_x = \gamma_{0x}, \gamma_y = \gamma_{0y}} J \\ &= \mathscr{R}(\psi_0) g_0 p_{\gamma_{xy}}\left(\gamma_{0x}, \gamma_{0y}\right) \bar{S}_{\mathrm{B}}^1\Big|_{\gamma_x = \gamma_{0x}, \gamma_y = \gamma_{0y}} J,\end{aligned} \tag{4.66}$$

where

$$\begin{cases} p = 1 + \cos\theta_1 \cos\theta_2 + \sin\theta_1 \sin\theta_2 \cos(\phi_2 - \phi_1) \\ \cos\psi_0 = \sqrt{\dfrac{p}{2}} \\ \dfrac{J}{\sin\theta_1} = \left| \dfrac{\partial\gamma_{0x}}{\partial\theta_1}\dfrac{\partial\gamma_{0y}}{\partial\phi_1} - \dfrac{\partial\gamma_{0x}}{\partial\phi_1}\dfrac{\partial\gamma_{0y}}{\partial\theta_1} \right| = \dfrac{p}{(\cos\theta_1 + \cos\theta_2)^3} \\ g_0 = \dfrac{p}{\cos\theta_2(\cos\theta_1 + \cos\theta_2)} \end{cases}. \tag{4.67}$$

The term J is the Jacobian of the variables transformations from (γ_x, γ_y) to (θ_1, ϕ_1).

In addition, we showed in Chapter 2 that the statistical bistatic shadowing function averaged over the surface heights and evaluated at the slopes $(\gamma_{0x}, \gamma_{0y})$ (Equation (2.67)) is

$$\bar{S}^1_{\mathrm{B}}\Big|_{\gamma_x=\gamma_{0x},\gamma_y=\gamma_{0y}} = \frac{1}{1 + \tilde{c}_0(v_p, v_q, \phi)\Lambda(v_q) + \Lambda(v_p)}, \tag{4.68}$$

where $\tilde{c}_0(v_p, v_q, \phi) \approx c_0(v_p, v_q, \phi, h_0 = 0)$ defined by (2.71), $\phi = |\phi_2 - \phi_1|$ and $(p, q) = (1, 2)$ if $0 \leq \mu_1 = |\cot\theta_1| \leq \mu_2 = |\cot\theta_2|$, $(p, q) = (2, 1)$ otherwise. Moreover, for a Gaussian process, Λ is expressed from (4.7), in which $v_p = |\cot\theta_p|/(\sqrt{2}\sigma_{\gamma_{X_p}})$ and $\sigma^2_{\gamma_{X_p}} = \sigma^2_{\gamma_x}\cos^2\phi_p + \sigma^2_{\gamma_y}\sin^2\phi_p$.

In the specular direction, (4.68) shows that there is no restriction (no Heaviside function Υ) over the surface slopes.

4.3.3.3 Numerical results

Figure 4.27 plots the normalized unpolarized reflectivity $\rho_1/\rho_{1,\max}$ in dB scale, $20\log_{10}(\rho_1/\rho_{1,\max})$, versus the angles ϕ_1 and θ_1 for given θ_2 and ϕ_2, whose values are indicated in the subtitles. In addition, the next two numbers are the maximum of ρ_1, $\rho_{1,\max}$, and the value of $\rho_{1,\mathrm{hem}}$ of the unpolarized hemispherical reflectivity. The cross gives the position of the specular direction defined as $\theta_1 = \theta_2$ and $\phi_1 = \phi_2 - \pi$. The wind speed $u_{12} = 5$ m/s (model of Cox and Munk, (4.45)).

The specular direction corresponds to the case when the slopes γ_{0x} and γ_{0y} vanish. The unpolarized hemispherical reflectivity is expressed as

$$\rho_{1,\mathrm{hem}}(\theta_2, \phi_2) = \int_0^{\pi/2} \int_0^{2\pi} \rho_1(\theta_1, \phi_1, \theta_2, \phi_2) \mathrm{d}\phi_1 \mathrm{d}\theta_1. \tag{4.69}$$

As the angle θ_2 increases, the width of the lobe defined near the specular direction decreases, whereas the maximum of ρ_1 and $\rho_{1,\mathrm{hem}}$ increase. In other words, the intensity is reflected in the narrower angular sector with a larger strength. For $\phi_1 = 90^\circ$, the width of the lobe decreases slightly in comparison to the results obtained for $\phi_1 = 0^\circ$, because the surface slope standard deviation is smaller and then, the intensity is less scattered in all the directions.

In (4.66), the specular direction is obtained when the argument of the slope PDF $p_{\gamma_{xy}}$ vanishes. Due to the multiplicative factors, and especially the term

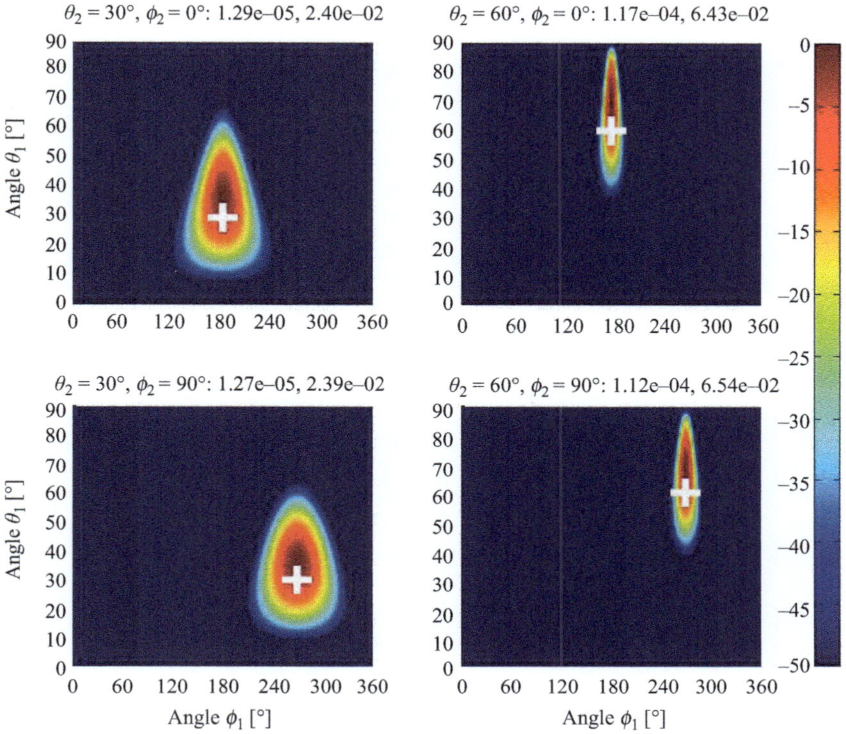

Figure 4.27 *Normalized unpolarized reflectivity $\rho_1/\rho_{1,max}$ in dB scale, $20log_{10}(\rho_1/\rho_{1,max})$, versus the angles ϕ_1 and θ_1 for given θ_2 and ϕ_2, whose values are indicated in the subtitles. In addition, the next two numbers are the maximum of ρ_1 and the value of $\rho_{1,hem}$ of the unpolarized hemispherical reflectivity. The cross gives the position of the specular direction defined as $\theta_1 = \theta_2$ and $\phi_1 = \phi_2 - \pi$. The wind speed $u_{12} = 5$ m/s (model of Cox and Munk, (45))*

$1/(\cos\theta_1 + \cos\theta_2)^4$, the maximum location of ρ_1 deviates from the specular directions and this deviation increases as θ_2 increases.

Figure 4.28 plots the same variations as in Figure 4.27 but the Cox and Munk [21] distribution is used, in which the skewness and the kurtosis are accounted for.

In comparison to Figure 4.27, the maxima of ρ_1 are larger, whereas the values of $\rho_{1,\mathrm{hem}}$ slightly vary. This means that the total scattered intensity is nearly constant, but the angular distribution is directly related to the slope PDF, as shown by (4.66). It is important to note that the Cox and Munk distribution was performed from optics measurements of the sun glitter, strongly related to the reflectivity.

4.3.3.4 Energy conservation

According to the law of energy conservation, under thermal equilibrium, the energy absorbed by the sea surface equals the energy it radiated. It is assumed that the sea

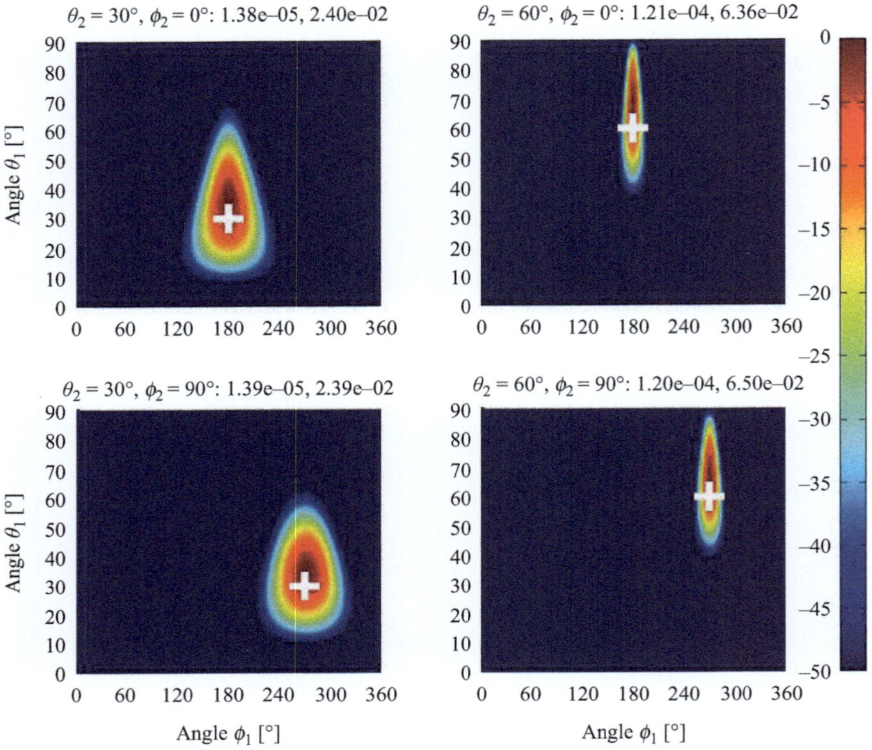

Figure 4.28 Same variations as in Figure 4.28 but the Cox and Munk [21] distribution is used, in which the skewness and the kurtosis are accounted for

surface is opaque, which means that all the energy of the refractive rays is absorbed by the sea. Then, the law of energy conservation states that the sum of the surface emissivity and hemispherical reflectivity equals 1, given by

$$\varepsilon_0(\theta_2, \phi_2) + \rho_{1,\text{hem}}(\theta_2, \phi_2) = 1 = c. \tag{4.70}$$

This criterion function c is a means to check the validity of a model.

Figure 4.29 plots the unpolarized criterion function $c = \varepsilon_0(\theta_2, \phi_2) + \rho_{1,\text{hem}}(\theta_2, \phi_2)$ versus the angle θ_2 for different wind speeds $u_{12} = \{5, 10, 15\}$ m/s (model of Cox and Munk, (4.45)). The angle $\phi_2 = 0$.

As we can see, for small angles θ_2, the criterion is satisfied. From $\theta_2 = \theta_{2,0} \approx 20°$, the function c decreases to reach the minimum value $c_{\min}$ of the order of 0.94–0.95 and next increases to tend to unity. As the wind speed increases, $\theta_{2,0}$ and $c_{\min}$ decrease and the angle for which $c_{\min}$ is reached is shifted towards smaller values.

Then, (4.70) is not fulfilled when only the direct emissivity ε_0 and the first-order hemispherical reflectivity $\rho_{1,\text{hem}}$ are considered, for which $c \leq 1$.

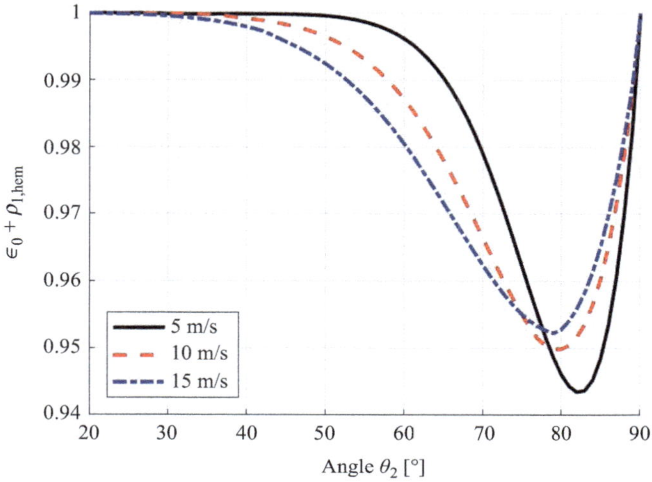

Figure 4.29 Unpolarized criterion function $c = \varepsilon_0(\theta_2, \phi_2) + \rho_{1,hem}(\theta_2, \phi_2)$ *versus the angle* θ_2 *for different wind speeds* $u_{12} = \{5, 10, 15\}$ *m/s (model of Cox and Munk, (45)). The angle* $\phi_2 = 0$

As presented in the next section, Li *et al.* [32,36] (Thesis in English) showed by accounting for both an additional reflection in the calculation of the emissivity and reflectivity, the criterion is better satisfied. Typically, the minimum $c_{\min}$ becomes 0.98.

4.3.4 Second-order reflectivity from 1D surfaces

As explained in Section 4.3.2.1, the sea surface is assumed to be 1D.

4.3.4.1 Derivation

Reflectivity with two surface reflections, also known as the second-order reflectivity, is related to the reflection twice by the surface, as illustrated in Figure 4.30. This issue is addressed in [30,36] and the derivation of the bistatic shadowing function with two reflections is needed. In this section, the model of Li *et al.* [30], thoroughly discussed in Section 3.2.4, is employed.

An incident ray coming from the upper half-space first intersects the surface at the point M_1 with a local incidence angle χ_1, then the reflected ray intersects the surface again at the point M_0 with a local incidence χ_0. The local reflectivity with two reflections is then

$$\begin{cases} \rho_{2,H}^{\text{local}} = |r_H(\chi_1)|^2 |r_H(\chi_0)|^2 \\ \rho_{2,V}^{\text{local}} = |r_V(\chi_1)|^2 |r_V(\chi_0)|^2 \end{cases}. \tag{4.71}$$

In (4.71), the first and second terms correspond to the reflection coefficients in intensity at the first and the second intersection points, respectively. For a 1D rough surface, as discussed in Section 4.3.2.2, no cross-polarization occurs during the reflections.

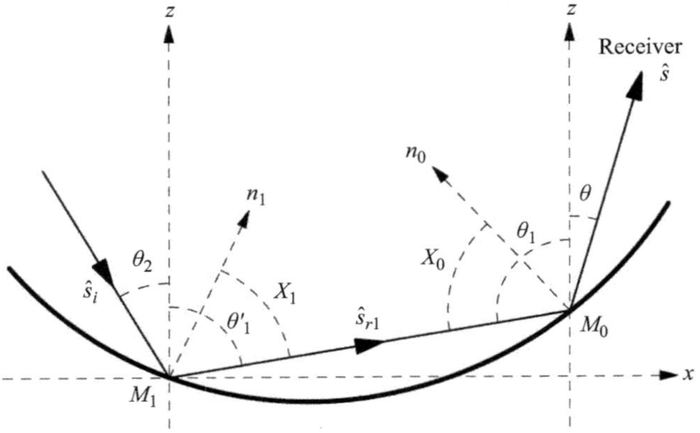

Figure 4.30 Illustration of two surface reflection: the emitted ray coming from the upper half-space is reflected twice at the points M_1 and M_0 before it is reflected towards the receiver

The hemispherical reflectivity with two reflections can be obtained by averaging the local reflectivity over the heights and slopes of the points M_0 and M_1. This leads to

$$\begin{cases} \rho_{2,H} = \int_{-\infty}^{+\infty}\int_{-\infty}^{+\infty}\int_{-\infty}^{+\infty}\int_{-\infty}^{+\infty} \rho_{2,H}^{\text{local}} g_0 S_B^2 p_4(\zeta_1, \gamma_1, \zeta_0, \gamma_0) \mathrm{d}\zeta_1 \mathrm{d}\gamma_1 \mathrm{d}\zeta_0 \mathrm{d}\gamma_0 \\ \rho_{2,V} = \int_{-\infty}^{+\infty}\int_{-\infty}^{+\infty}\int_{-\infty}^{+\infty}\int_{-\infty}^{+\infty} \rho_{2,V}^{\text{local}} g_0 S_B^2 p_4(\zeta_1, \gamma_1, \zeta_0, \gamma_0) \mathrm{d}\zeta_1 \mathrm{d}\gamma_1 \mathrm{d}\zeta_0 \mathrm{d}\gamma_0 \end{cases}, \quad (4.72)$$

where S_B^2 is the statistical bistatic shadowing function with two reflections, $p_4(\zeta_1, \gamma_1, \zeta_0, \gamma_0)$ is the joint PDF of the heights and the slopes of the points M_1 and M_0, and g_0 is the area projection term given by (4.53).

The statistical bistatic shadowing function with two reflections is addressed in Section 4.3.4.1. In this section, the model of Li *et al.* [30] is employed. Similar to the surface emissivity, a Gaussian sea surface is assumed and the correlation between the surface heights and slopes is ignored.

Then, S_B^2 is the probability of occurrence of the four events defined at the beginning of section and

$$S_B^2 = p(abcd) = p(ab)p(c|ab)p(d|abc). \quad (4.73)$$

The first term $p(ab)$ equals the statistical monostatic shadowing function with one reflection $S_M^{1,\text{unco}}$ [30] given by (4.54). The second term is related to the probability that the slope of M_1 equals the desired value so that the reflected ray propagates towards the given direction. It is expressed as

$$p(c|ab) = \delta\left(\gamma_1 - \gamma_1^{\text{spe}}\right), \quad (4.74)$$

where γ_1^{spe} is the slope that reflects the ray $\hat{s}_i$ into the specular direction $\hat{s}_{r1}$ (Figure 4.30).

The third term $p(d|abc)$ is expressed from the statistical monostatic shadowing function without reflection [30]. It is expressed as

$$p(d|abc) = S_M^0 = \begin{cases} F(\zeta_1)^{\Lambda(\mu_2)} & \text{for } \theta_2 > 0 \\ F(\zeta_1)^{\Lambda_-(\mu_2)} & \text{for } \theta_2 < 0. \end{cases} \tag{4.75}$$

Besides, similar to the shadowing function, the bidirectional reflectivity with two reflections ρ_2^{spe} can also be calculated by modifying the integration range of γ_1 to a small range so that the actual reflected angle θ_r is within a small range around a given value θ_2. In this section, we consider the range $\theta_r \in [\theta_2 - 0.1; \theta_2 + 0.1]^\circ$.

In (4.72), the integration over the heights ζ_1 and ζ_0 can be done analytically since the local reflectivity does not depend on the surface heights. The slope PDF of M_0 is assumed to be Gaussian and the slope PDF of M_1 is given by (4.58).

4.3.4.2 Numerical results

A Monte Carlo ray-tracing algorithm is applied as reference to evaluate the accuracy of the model. The ray-tracing process is the same as the one explained in Section 3.2.2. In a ray-tracing algorithm, the local incidence angle at every intersection is known. Thus, the second-order reflectivity is expressed as

$$\begin{cases} \rho_{\text{MC},V} &= \dfrac{1}{N} \displaystyle\sum_{i=1}^{i=N_i} |r_V(\chi_{1,i})|^2 |r_V(\chi_{0,i})|^2 g_0 \\ \rho_{\text{MC},H} &= \dfrac{1}{N} \displaystyle\sum_{i=1}^{i=N_i} |r_H(\chi_{1,i})|^2 |r_H(\chi_{0,i})|^2 g_0 \end{cases}, \tag{4.76}$$

where N is the number of groups of M_0 and M_1 obtained by the ray-tracing process. In addition, to improve the convergence, several surface realizations can be generated.

For the computation of the bidirectional reflectivity, the pairs of M_0 and M_1 are chosen so that the reflected angle θ_r is around a given value θ_2. For the hemispherical reflectivity, the reflected angle can be any value.

Figure 4.31 plots the sea surface bidirectional reflectivity with two reflections versus the angle θ_2. The wind speed $u_{12} = 10$ m/s, the wavelength $\lambda = 10$ μm, $\theta_r \in [\theta_2 - 0.1; \theta_2 + 0.1]^\circ$, and the observation angles $\theta = 30^\circ$, 60° and 80°. The results are also compared with those obtained from the Monte Carlo ray-tracing algorithm.

Some overestimation is found when the uncorrelated model is compared with the Monte Carlo ray-tracing algorithm, especially when the observation angle θ_2 is small. The best agreement is found when $\theta_2 \approx 80^\circ$, even though an overestimation still occurs. This is not surprisingly, since the same situation is found in Section 3.2.4.4 when the histogram of the second reflected angle θ_2 (inverse path) is studied from the bistatic shadowing function S_B^2. As a result, the bidirectional reflectivity with two reflections is strongly related to the bistatic shadowing function S_B^2.

Figure 4.32 plots the sea surface hemispherical reflectivity with two reflections. The wind speed is $u_{12} = 10$ m/s and the wavelength $\lambda = 10$ μm. The results are also compared with those obtained from the Monte Carlo ray-tracing algorithm.

The hemispherical reflectivity with two reflections shows a very similar shape as that of the averaged bistatic shadowing function S_B^2. In general, the model agrees well

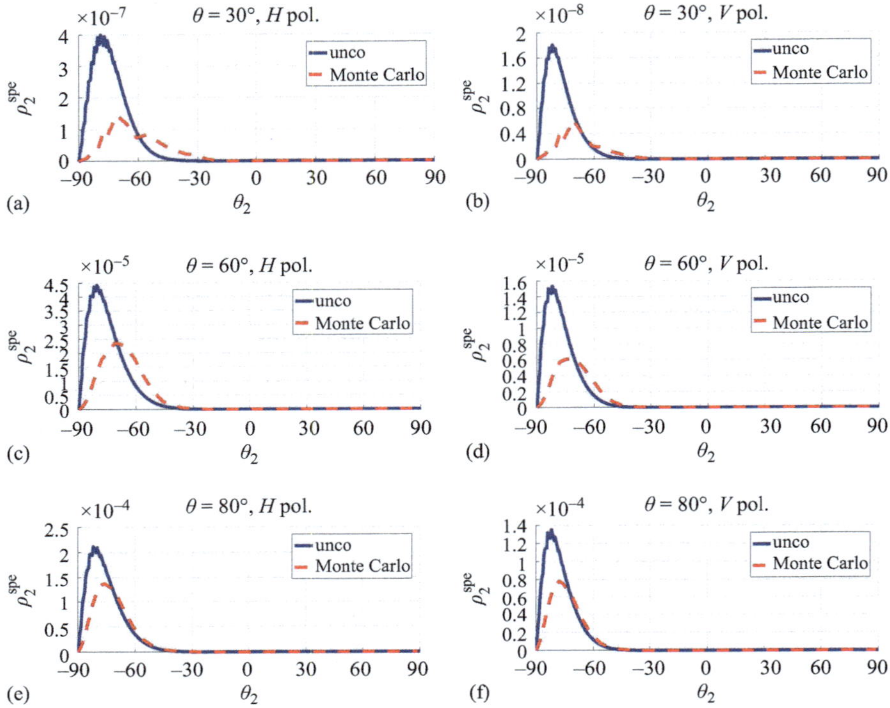

Figure 4.31 *Sea surface bidirectional reflectivity with two reflections $\lambda = 10\ \mu m$ versus the angle θ_2. The wind speed $u_{12} = 10$ m/s, the wavelength $\lambda = 10\ \mu m$, $\theta_r \in [\theta_2 - 0.1; \theta_2 + 0.1]°$, and the observation angles $\theta = 30°$, $60°$ and $80°$. Results for H polarization are shown on the left column, while results for V polarization are on the right*

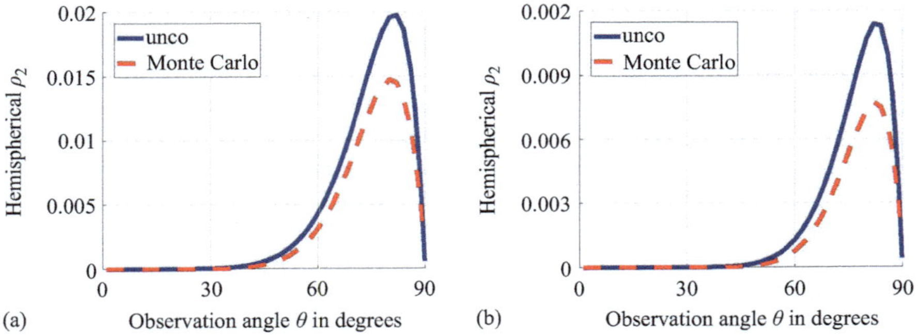

Figure 4.32 *Sea surface hemispherical reflectivity with two reflections. The wavelength $\lambda = 10\ \mu m$ and the wind speed $u_{12} = 10$ m/s. Results for H polarization are shown on the left, while results for V polarization are on the right*

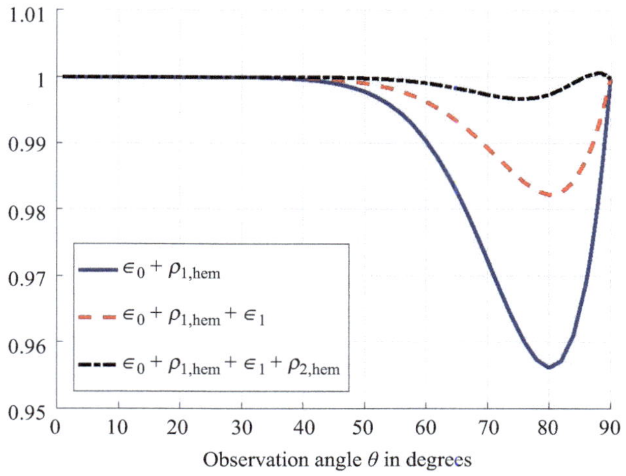

Figure 4.33 Energy conservation criterion versus θ. The wind speed u_{12} = 10 m/s and the wavelength λ = 10 μm

with the Monte Carlo results with a slightly overestimation. The main contribution occurs for large observation angle θ and maxima are found around $\theta = 80°$.

For the calculations of the sea surface emissivity and reflectivity, Figures 4.23 and 4.32 clearly show that the surface reflections cannot be neglected. Remember that a 'loss' of energy is found in Section 4.3.2.4 when only the direct emissivity ε_0 and the first-order hemispherical reflectivity $\rho_{1,\text{hem}}$ are considered. The first-order emissivity ε_1 and the second-order hemispherical reflectivity $\rho_{2,\text{hem}}$ must be then accounted for.

Figure 4.33 plots the energy conservation criterion versus θ when surface reflections are taken into account. The wind speed $u_{12} = 10$ m/s and the wavelength $\lambda = 10$ µm. The light is assumed to be unpolarized (mean value of the H and V polarized contributions). The emissivity and reflectivity are computed from the uncorrelated model of Li *et al.* [30,33].

When only the direct emissivity ε_0 and the first-order hemispherical reflectivity $\rho_{1,\text{hem}}$ are considered, their sum is one only for small observation angles θ. For moderate and large θ, the sum is smaller than one, with a minimum of about 0.96 around $\theta \approx 80°$, meaning that 4% of power is 'lost'. By taking into account the emissivity with one reflection, the energy conservation criterion is better met. The peak of the 'loss' of power is reduced by about 2%. By accounting for the second-order hemispherical reflectivity $\rho_{2,\text{hem}}$, the peak is still reduced by 1.5% and, the energy conservation is almost achieved for all θ. Note that for $\theta > 85°$, the sum is slightly over one, which is most likely due to numerical problems on the second-order reflectivity computation.

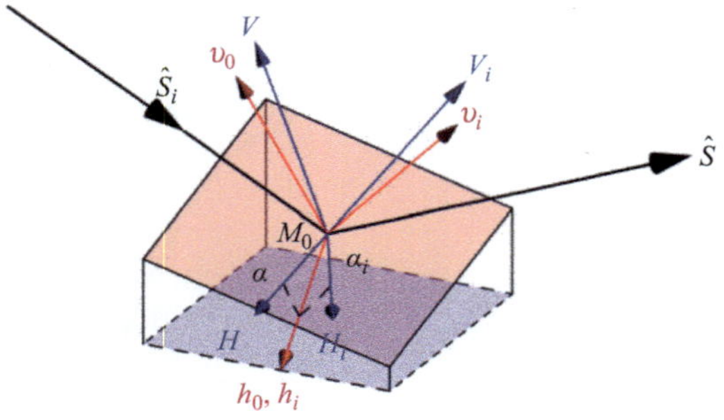

Figure 4.34 *Single surface reflection by an arbitrary surface point M_0. There is a rotation angle α_i between the local polarization direction h_i (or v_i) and the global one H_i (or V_i). Similarly, there is a rotation angle α between h_0 (or v_0) and H (or V)*

4.3.5 Polarization effect

In this chapter, numerical results of the optical properties of a wind-roughened sea surface are presented by considering the unpolarized emissivity and reflectivity. In this case, the reflection coefficient is defined as the arithmetic sum of the square modulus of the Fresnel coefficients in V and H polarization.

First, to quantify the polarization effect, the conventional degree of polarization (DOP) expressed as

$$\mathrm{DOP} = \frac{q_H - q_{VV}}{q_H + q_V}, \tag{4.77}$$

can be performed, where q is either the emissivity or the reflectivity. It is related to the difference between the H and V components and can be computed from 1D or 2D surfaces.

For rough surfaces, the orientations of the surface facets are arbitrary. As a result, the directions of the local horizontal and vertical polarizations, defined by the local normal $\hat{n}$ to the facet and the propagation direction of the incident ray $\hat{s}_i$ (denoted as h_i and v_i) or that of the reflected ray $\hat{s}$ (denoted as h_0 and v_0), are different from one surface point to another.

To describe the polarization state of the sea surface reflectivity (see Figure 4.34), global horizontal and vertical polarizations are introduced by the average sea surface (horizontal plane), or its normal, the zenith direction. In the same way, the propagation direction of the incident ray $\hat{s}_i$ or the reflected ray $\hat{s}$ are denoted as H_i and V_i, or H and V, respectively.

Because of the arbitrariness of the local polarization directions, there is an angle α_1 between h_i and H_i, or equally between v_i and V_i, if the facet is different from the horizontal plane (case of a rough surface). Similarly, there is an angle α_2 between h_0 and H, or equally between v_0 and V. In this case, one part of the incident energy in global H_i (or V_i) polarization has to be projected into the local v_i (or h_i) polarization by the angle α_1. Similarly, one part of the reflected energy locally h_i (or v_i) polarized has to be projected into the global V (or H) polarization by the angle α_2 . In other words, cross polarizations occur. Cross polarizations never occur for 1D surfaces, because α_1 and α_2 always vanish.

Li *et al.* [36,58] showed that the depolarization effect is weak on ε_0 and ρ_1, whereas it becomes significant when the higher reflections are accounted for. Then, it is a practical means to quantify the importance of the multiple reflections.

If sensors sensitive to polarization are available, it would be possible to measure optical properties in four polarizations by emitting both the H_i and V_i polarizations and by receiving those in H and V. To study deeply the polarization effect, the Mueller matrix, obtained from the Stokes vectors, can be measured from optical retarders and polarisers [70]. Since the emissivity and reflectivity have different behaviours with respect to the polarization, the measurements of the 16 elements of the Mueller matrix will be a power tool to discriminate the intrinsic radiation (related to the emissivity) of the sea surface and the sun glitter (related to the reflectivity).

4.3.6 Multi-resolution model

Analytical expressions of optical properties of a wind-roughened sea surface have been derived in the previous sections. Statistically, they depend on the slope PDF, for which the mean values vanish. This means that the surface area is enough large to meet this property. This model is named 'Low resolution'. More precisely, the two lengths (with respect to the (Ox) and (Oy) directions) of the surface must be larger than the surface slope correlation lengths (it is not those of the surface heights because the emissivity and the reflectivity depend on the slope PDF).

For a 1D surface, Fauqueux *et al.* [65] and Caillault *et al.* [64] showed that the minimum length ranged from 1 to 4 m for wind speeds ranging from 5 to 20 m/s. For resolution smaller that this length, the previous models cannot be longer applied.

To include resolution smaller than 1 m, a first simple solution is to generate a realization of the surface and compute the optical properties from a Monte Carlo process, as already presented in the previous sections. Assuming that the low resolution model can be applied for surface area of 100 m^2 and the surface sampling steps along the (Ox) and (Oy) directions are $\Delta x = \Delta y = 1$ mm to account for the wave capillary regime, for a 2D surface, the surface has $(100/0.001)^2 = 10^{10}$ samples. This is a solution that requires huge memory and computing time.

An alternative solution is then proposed in [64,65], which is based on a two-scale model:

1. The surface is decomposed into two scales of roughness, $\hat{S}(k,\psi) = \hat{S}_C(k,\psi) + \hat{S}_G(k,\psi)$. The small scale corresponds to the capillarity waves, whose the sea height spectrum is defined as

$$\hat{S}_C(k,\psi) = \begin{cases} \hat{S}(k,\psi) & \text{if } k \geq k_c \\ 0 & \text{if } k < k_c \end{cases}, \tag{4.78}$$

whereas the large scale corresponds to the gravity waves, whose sea height spectrum is defined as

$$\hat{S}_G(k,\psi) = \begin{cases} 0 & \text{if } k \geq k_c \\ \hat{S}(k,\psi) & \text{if } k < k_c \end{cases}. \tag{4.79}$$

The cut-off wave number k_c is chosen such as the sampling steps along $(0x)$ and $(0y)$ directions, $\{\Delta_{xG,yG}\}$, of the gravity waves equal the expected maximal resolution, typically 1 m. Since $\Delta_{xG,yG} = \pi/k_{xc,yc} \Rightarrow k_{xc,yc} = \pi/\Delta_{xG,yG} = \pi$ rad/m with $\Delta_{xG,yG} = 1$ m.

2. The capillarity slope variances, $\{\sigma^2_{\gamma_{xC},\gamma_{yC}}\}$, are computed from the spectrum (4.78).
3. For a sea surface of area $L_x \times L_y$ m^2 (the lengths L_x and L_y are multiple of 1 meter, giving the sampling step), a realization of the sea surface gravity slopes $\gamma_{xG,yG}$ is generated from the spectrum (4.79).
4. The low resolution model is then applied, for which the slope variances equal those of the capillarity regime and the mean slopes equal those of the gravity regime $\gamma_{xG,yG}$ (deterministic variables). Then, the low resolution model must be extended by considering a Gaussian process with non-zeros mean values [64].

To establish this model, the authors [64,65] are referred to the two-scale models [71,72] developed in the context of the microwaves scattering from sea surfaces. Its main drawback is the introduction of a cut-off wave number separating the small scales from the large scales. In the infrared band, this condition does not exist because the sea optical properties do not depend on the surface height autocorrelation function. The only constraint to check is that the slope PDF fit to a Gaussian process on the smallest resolution (1 m). This means that the number of samples is enough to apply a statistical approach. They showed [64,65] that this condition is satisfied from 1 m and the high-resolution model converges to the low one from a surface length of 256 m (1D surface), which depends on the wind speed.

Assuming that the atmosphere is isotropic and non-absorbing, the luminance received by a sensor in the directions (θ_2,ϕ_2) is expressed as

$$\begin{aligned} L_r(\theta_2,\phi_2) = {} & \rho(\theta_1,\phi_1,\theta_2,\phi_2)L_{\text{sun}}(\theta_1,\phi_1) + [1-\varepsilon(\theta_2,\phi_2)]\,L_{\text{atm}} \\ & + \varepsilon(\theta_2,\phi_2)L_{\text{bb}}(T_{\text{sea}}), \end{aligned} \tag{4.80}$$

where L_{sun} is the sun luminance assumed to be collimated of angles (θ_1,ϕ_1), L_{atm} that of the atmosphere and L_{bb} that of the black body evaluated at the sea temperature T_{sea}.

Images of sea-surface radiance of effective optical properties obtained with this approach are published in [64]. The model is implemented as a prototype before its integration in the Matisse-v2.0 code. A multi-resolution architectural solution, based on clipmaps method [73], is also implemented. It efficiently deals with levels of details consistent with the scene observation context. It relies on nested regular grids, composed of facets whose sizes are proportional to the sensor distance. Thus, objects far away from the sensor are not described with the same level of details as the ones closed by.

Figures 4.35–4.37 show images of the sea surface radiance obtained in [64] for a wavelength $\lambda = 4\,\mu$m and for a wind direction orthogonal to the observation direction. The SST T_{sea} is 288K.

In Figure 4.35, the wind speed is set to $u_{10} = 10$ m/s. Sun being at zenith and sensor looking at nadir, the rendering of solar glitter pattern is well done. Due to the observational configuration, two levels corresponding to high resolution case are displayed: 1 m in the central part of the image and 2 m otherwise. Figure 4.38(a) shows a representation of these levels with the associated domain sizes. The outer square corresponds to the edge of the image displayed in Figure 4.35.

In Figure 4.36, the wind speed is $u_{10} = 15$ m/s. The sensor (altitude 0.5 km) is opposite to the sun, and the zenith angles are $\theta_2 = 36^\circ$ and $\theta_1 = 40^\circ$. In that case, three levels of spatial resolution can be observed: 2 m in the lower part of the image, 4 m in the central part, and 8 m in the upper part. Figure 4.38 is obtained with the same configuration as in Figure 4.36, except that the sensor altitude is higher (1.5 km). Thus, the levels of spatial resolutions displayed are larger: 8, 16 and 32 m from the lower part of the image to the upper part (see Figure 4.38(b) and (c) for the associated diagrams and sizes). On these images, we can see that the spatial variability is well retrieved. They also show that the multi-resolution is correctly taken into account in the sensor field of view and that the transition between two levels of resolution is smooth and does not add artefacts.

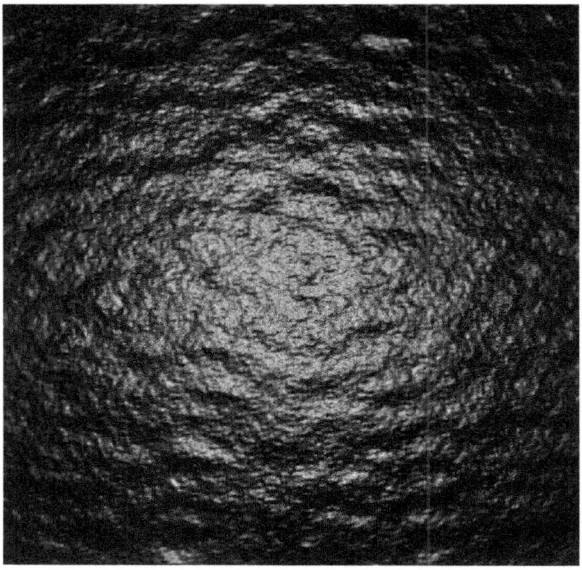

Figure 4.35 Image of sea surface radiance for a wavelength $\lambda = 4\,\mu m$ composed of 500 × 500 pixel. The sensor is located at an altitude of 0.389 km and looks at nadir. The sun is at zenith. The wind speed is $u_{10} = 10$ m/s. Figure obtained, with permission, from [64]

Figure 4.36 Same variations as in Figure 4.35 but with the sensor located at an altitude of 0.5 km. Sensor observation angles are $\theta_2 = 36^o$, $\phi_2 = 0^o$ and sun angles are $\theta_1 = -40^o$ and $\phi_1 = 180^o$. The wind speed is $u_{10} = 15$ m/s. Figure obtained, with permission, from [64]

Figure 4.37 Same variations as in Figure 4.36 but with the sensor located at an altitude of 1.5 km. Figure obtained, with permission, from [64]

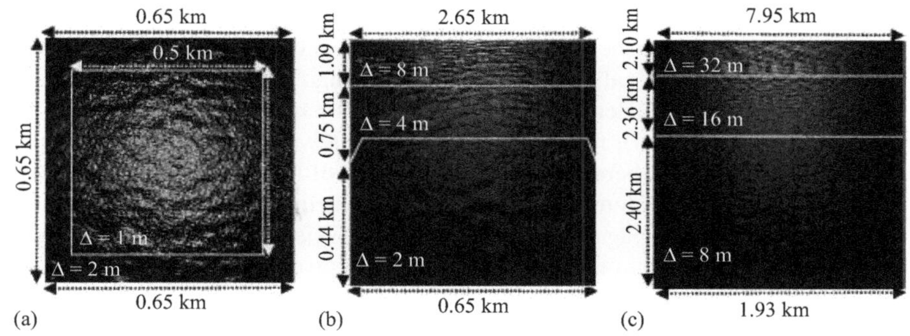

Figure 4.38 Levels of spatial resolution associated with images of sea surface radiance. Sizes of each domain and related spatial resolutions are also indicated. Diagrams (a), (b) and (c) give domains and sizes related to Figures 4.35, 4.36 and 4.37, respectively. Figure obtained, with permission, from [64]

4.4 Computer graphics

4.4.1 Introduction

In the graphics community, the masking and shadowing functions play an important role in the derivation of the BRDFs. The BRDF is strongly related to the reflectivity calculated in the infrared band from the geometric optics approximation. In addition, the masking and shadowing functions are equivalent to the bistatic shadowing function, terminology uses in the Radar community.

In this context, a recent review [74] provided a new presentation of the masking-shadowing functions (or geometric attenuation factors) in microfacet-based BRDFs and answer some common questions about their applications. The motivation is to define a correct (geometrically indicated), physically based masking function for application in microfacet models, as well as the properties that function should exhibit.

The Smith shadowing function is one of the most famous ones from the computer graphics literature. At the end of the article, Smith [5] pointed out that his function has the property of ensuring the conversation of the visible area, a fundamental property that is expected from a masking function. Later, this feature has been re-demonstrated independently by Ashikmin *et al.* [75].

4.4.2 Smith shadowing function versus that of Ashikmin et al.

In this section, the approach of Ashikmin *et al.* [75] is presented by adopting our notation illustrated in Figure 4.12.

The self-shadowing corresponds to the condition $\hat{s} \cdot \hat{n} > 0$ requiring that the angle ψ_0 between the ray $\hat{s}(\theta, \phi)$ and the normal to the surface $\hat{n}$ must be less than 90°. Then

$$\hat{s} \cdot \hat{n} > 0 = \cos \psi_0 > 0 \Leftrightarrow |\psi_0| < 90^{\circ}. \tag{4.81}$$

The angle ψ_0 is named 'local angle of incidence' in the model we used. This condition corresponds to the basic requirement that a microfacet is seen by the sensor. As a result, this is also called self-shadowing or angular shadowing in the classic models of shadowing function. In our model, this part is represented by an unit step function, $\Upsilon(\mu - \gamma_X)$, for a ray propagating rightward and $\Upsilon(\gamma_X - \mu)$, for that propagating leftward, where $\mu = \cot\theta$ is the slope of the ray $\hat{s}(\theta, \phi)$ and γ_X is the slope of the microfacet along the direction X or the azimuthal direction ϕ.

In addition, the propagation shadowing term $P(\hat{s})$ is defined. It does not depend on the microfacet orientation, given that self-shadowing does not occur ($\hat{s} \cdot \hat{n} > 0$). As will show later, its expression depends only on the direction $\hat{s}$, and it is very similar to that obtained from the Smith model expressed as $P = 1/(1 + \Lambda)$ or $P = 1/(1 + \Lambda_-)$.

Although P does not depend on the microfacet orientation $\hat{n}$, it does depend on the distribution of $\hat{n}$. This is also true from the Smith model.

The model of Ashikmin *et al.* [75] states for $\hat{s} \cdot \hat{n} > 0$ that

$$P = \frac{(\hat{s} \cdot \hat{z})\langle \hat{z} \cdot \hat{n} \rangle}{\langle \hat{s} \cdot \hat{n} \rangle} = \frac{1}{\frac{\langle \hat{s} \cdot \hat{n} \rangle}{(\hat{s} \cdot \hat{z})\langle \hat{z} \cdot \hat{n} \rangle}} = \left(\frac{1}{\cos\theta} \frac{\langle \cos\psi_0 \rangle}{\langle \cos\theta_n \rangle} \right)^{-1}. \tag{4.82}$$

In what follows, (4.82) is expressed in our original notations so as to compared easily with the model we used. The angles $\theta_n = \theta - \psi_0$ and ψ_0 can be both expressed by the observation direction θ and the slope (γ_x, γ_y) of the microfacet, given by

$$\begin{cases} \cos\theta_n = \dfrac{1}{\sqrt{1 + \gamma_x^2 + \gamma_y^2}} \\ \cos\psi_0 = \dfrac{\cos\theta - (\gamma_x \cos\phi + \gamma_y \sin\phi)\sin\theta}{\sqrt{1 + \gamma_x^2 + \gamma_y^2}} \end{cases}. \tag{4.83}$$

If the surface slopes γ_x and γ_y satisfy the condition $1 + \gamma_x^2 + \gamma_y^2 \approx 1$, $\cos\theta_n$ is few dependent of the orientation of the microfacet. Then, $\langle \cos\theta_n \rangle \approx 1$, which implies that $\langle \cos\psi_0 \rangle \, / \, \langle \cos\theta_n \rangle \approx \langle \cos\psi_0 / \cos\theta_n \rangle$. Equation (4.82) can be expressed approximately as

$$\begin{aligned} P &\approx \left(\frac{1}{\cos\theta} \left\langle \frac{\cos\psi_0}{\cos\theta_n} \right\rangle \right)^{-1}, \quad |\psi_0| < 90^\circ \\ &= \left[\int_{-\infty}^{+\infty} \int_{-\infty}^{+\infty} \frac{\cos\psi_0}{\cos\theta \cos\theta_n} p_{\gamma_{xy}}(\gamma_x, \gamma_y) \mathrm{d}\gamma_x \mathrm{d}\gamma_y \right]^{-1}, \quad |\psi_0| < 90^\circ. \end{aligned} \tag{4.84}$$

From (4.83) and (2.1), P is simplified as

$$\begin{aligned} \frac{1}{P} &= \int_{-\infty}^{+\infty} \int_{-\infty}^{+\infty} \Upsilon(\mu - \gamma_X) \frac{\cos\theta - \gamma_X \sin\theta}{\cos\theta} p_{\gamma_{XY}}(\gamma_X, \gamma_Y) \mathrm{d}\gamma_X \mathrm{d}\gamma_Y \\ &= \int_{-\infty}^{\mu} \left(1 - \frac{\gamma_X}{\mu} \right) p_{\gamma_X}(\gamma_X) \mathrm{d}\gamma_X, \end{aligned} \tag{4.85}$$

where p_{γ_X} is the marginal slope PDF derived by calculating the integration over $\gamma_Y \in \mathbb{R}$ of $p_{\gamma_{xy}}$, in which the slopes (γ_x, γ_y) are expressed from those of (γ_X, γ_Y) by using (2.1).

For a surface with an even slope PDF $p(\gamma_X)$, we have

$$\int_{-\infty}^{+\infty} \gamma_X p(\gamma_X) \mathrm{d}\gamma_X = 0, \tag{4.86}$$

which implies that

$$\begin{aligned}\int_{-\infty}^{\mu} \left(1 - \frac{\gamma_X}{\mu}\right) p_{\gamma_X}(\gamma_X) \mathrm{d}\gamma_X &= \int_{-\infty}^{+\infty} \left(1 - \frac{\gamma_X}{\mu}\right) p_{\gamma_X}(\gamma_X) \mathrm{d}\gamma_X \\ &\quad + \int_{+\infty}^{\mu} \left(1 - \frac{\gamma_X}{\mu}\right) p_{\gamma_X}(\gamma_X) \mathrm{d}\gamma_X \\ &= 1 + \int_{\mu}^{\infty} \left(\frac{\gamma_X}{\mu} - 1\right) p_{\gamma_X}(\gamma_X) \mathrm{d}\gamma_X. \end{aligned} \tag{4.87}$$

Under this condition, P is given by

$$P(\theta) = \left[1 + \frac{1}{\mu} \int_{\mu}^{+\infty} (\gamma_X - \mu) p_{\gamma_X}(\gamma_X) \mathrm{d}\gamma_X\right]^{-1} = \frac{1}{1 + \Lambda(\mu)}, \tag{4.88}$$

which is exactly the model of Smith.

As a result, the model of Ashikmin *et al.* is equivalent to ours if the condition $1 + \gamma_x^2 + \gamma_y^2 \approx 1$ is satisfied. Later Ashikmin *et al.* [75] also observed that the visible projected area is a quantity that is conserved from the geometric surface to the microsurface. They used this knowledge to derive the general equation for a correct masking term, which ensures correct normalization and energy conservation. By doing this, they were actually reinventing the Smith masking function without being aware of it.

4.4.3 One-order reflectivity versus the BRDF

Typically, the BRDF with single scattering from a rough surface modelled as a collection of microfacets is expressed as (see [76] for instance)

$$\mathrm{BRDF}(\hat{s}_i, \hat{s}_s) = \frac{F(\hat{s}_i, \hat{s}_s) D(\hat{s}_i, \hat{s}_s) G(\hat{s}_i, \hat{s}_s)}{4 \cos\theta_i \cos\theta_s}, \tag{4.89}$$

where F is related to the Fresnel coefficients in intensity, D stands for the distribution of the normal to the microfacets that satisfy the specular direction, and G the bistatic shadowing function. Moreover, $\hat{s}_i = \hat{s}(\theta_i, \phi_i)$ and $\hat{s}_s = \hat{s}(\theta_s, \phi_s)$ denote the directions of the transmitter and receiver, respectively. The unitary vector normal $\hat{n}$ to the facet is related to their slopes (γ_x, γ_y) as

$$\hat{n} = \frac{1}{\sqrt{1 + \gamma_x^2 + \gamma_y^2}} \begin{bmatrix} -\gamma_x \\ -\gamma_y \\ 1 \end{bmatrix} = \begin{bmatrix} n_x \\ n_y \\ n_z \end{bmatrix}. \tag{4.90}$$

Then, the normal PDF $p_{n_{xy}}$ is related to the slope PDF $p_{\gamma_{xy}}$ as

$$p_{n_{xy}}(n_x, n_y) = |J_1| p_{\gamma_{xy}}(\gamma_x, \gamma_y) = \left(1 + \gamma_x^2 + \gamma_y^2\right)^2 p_{\gamma_{xy}}(\gamma_x, \gamma_y), \tag{4.91}$$

where the Jacobian of the variable transformations is defined as

$$J_1 = \frac{\partial \gamma_x}{\partial n_x}\frac{\partial \gamma_y}{\partial n_y} - \frac{\partial \gamma_y}{\partial n_x}\frac{\partial \gamma_x}{\partial n_y} = \frac{1}{\left(1 - n_x^2 - n_y^2\right)^2}$$
$$= \left(\frac{\partial n_x}{\partial \gamma_x}\frac{\partial n_y}{\partial \gamma_y} - \frac{\partial n_y}{\partial \gamma_x}\frac{\partial n_x}{\partial \gamma_y}\right)^{-1} = \left(1 + \gamma_x^2 + \gamma_y^2\right)^2. \tag{4.92}$$

In (4.89), D is defined in the specular direction whose slopes are expressed from (4.64). This leads to

$$J_1(\hat{s}_i, \hat{s}_s) = \frac{4p^2}{(\cos\theta_1 + \cos\theta_2)^2}, \tag{4.93}$$

where p is expressed from (4.67).

The one-order reflectivity given by (4.66) becomes

$$\rho_1(\hat{s}_i, \hat{s}_s) = \mathscr{R}(\psi_0) p_{\gamma_{xy}}\left(\gamma_{0x}, \gamma_{0y}\right) \bar{S}_{\mathrm{B}}^1\Big|_{\gamma_x=\gamma_{0x}, \gamma_y=\gamma_{0y}} \frac{J_1(\hat{s}_i, \hat{s}_s)}{4}\frac{\sin\theta_1}{\cos\theta_2}$$
$$= \mathrm{BRDF}(\hat{s}_i, \hat{s}_s)\cos\theta_1 \sin\theta_1 \text{ from (4.91) and (4.89)}, \tag{4.94}$$

where $F = \mathscr{R}(\psi_0)$ and $G = \bar{S}_{\mathrm{B}}^1\Big|_{\gamma_x=\gamma_{0x}, \gamma_y=\gamma_{0y}}$.

In conclusion, the BRDF is strongly related to the zero-order (infrared) reflectivity of a sea surface.

Equation (4.89) is named the 'Smith-GGX' model, and it is used in all games that have been coming out for a few years, for example:

- The Order 1886 (page 4): https://blog.selfshadow.com/publications/s2013-shading-course/rad/s2013_pbs_rad_slides.pdf.
- The Frostbite engine used for Battlefields, Need for Speed, etc.: https://blog.selfshadow.com/publications/s2014-shading-course/frostbite/s2014_pbs_frostbite_slides.pdf.
- In ImageWorks movies (Spiderman, Smurfs, etc.) (page 127): https://blog.selfshadow.com/publications/s2017-shading-course/imageworks/s2017_pbs_imageworks_slides_v2.pdf.

All the above links come from courses of ACM SIGGRAPH.

4.4.4 Energy conservation

In graphics community, a means to test the validity of the BRDF is to introduce the criterion of energy conservation [74]. This principle is named the 'The White Furnace Test' and it is equivalent to test that $\varepsilon + \rho_{\mathrm{hem}} = 1$, where ε and ρ_{hem} are the emissivity and the hemispherical reflectivity with multiple reflections, respectively. Since the emissivity is not needed in computer graphics, the surface is assumed to be perfectly conducting (mirror) and the criterion becomes $\rho_{\mathrm{hem}} = 1$.

Nowadays, the inclusion of the multiple reflections remains an open issue in computer graphics. Then, the authors hope that the work presented in this book could help this community to progress in this topic. In addition, the computation time of the models must be very short, and then they must be simple with a good precision. It remains a challenge.

References

[1] Bass FG, Fuks IM. Wave scattering from statistically rough surfaces. Natural Philosophy. Oxford: Pergamon Press; 1979.

[2] Kuznetsov PI, Stratonovich VL, Tikhonov VI. The duration of random function overshoots. Sov. Phys. Tech. Phys. 1954;24(103).

[3] Bourlier C, Berginc G, Saillard J. Monostatic and bistatic statistical shadowing functions from one-dimensional stationary randomly rough surface according to the observation length: Part I. Single scattering. Waves Random Complex Media. 2002;12:145–174.

[4] Wagner RJ. Shadowing of randomly rough surfaces. J. Acoust. Soc. Am.1967;41(1):138–147.

[5] Smith BG. Geometrical shadowing of a random rough surface. IEEE Trans. Antennas Propag. 1967;15(5):668–671.

[6] Beckman P. Shadowing of random rough surfaces. IEEE Trans. Antennas Propag. 1965;13:384–388.

[7] Bourlier C, Berginc G. Shadowing function with single reflection from anisotropic Gaussian rough surface. Application to Gaussian, Lorentzian and sea correlations. Waves Random Complex Media. 2003;13:27–58.

[8] Heitz E, Bourlier C, Pinel N. Correlation effect between transmitter and receiver azimuthal directions on the illumination function from a random rough surface. Waves Random Complex Media. 2013;3:318–335.

[9] Bass FG, Fuks IM. Calculation of shadowing for wave scattering from a statistically rough surface. Sov. Radiophys. 1964;7:101–112.

[10] Ricciardi LM, Sato S. A note on first passage time problems for Gaussian processes and varying boundaries. IEEE Trans. Inf. Theory. 1983;29: 454–457.

[11] Ricciardi LM, Sato S. On the evaluation of first passage time densities for Gaussian processes. Signal Process. 1986;11:339–357.

[12] Abramowitz M, Stegun IA. Handbook of mathematical functions. New York: Dover Publications, Inc.; 1970.

[13] Brokelman RA, Hagfors T. Note of the effect of shadowing on the backscattering of waves from a random rough surface. IEEE Trans. Antennas Propag. 1966;14:621–627.

[14] Kunt M. Techniques modernes de traitement numériques des signaux. France: Presses Polytechniques et Universitaires Romandes; 1991.

[15] Tsang L, Kong JA, Ding KH, *et al.* Scattering of electromagnetic waves, numerical simulations. New York: John Wiley & Sons, INC.; 2000.

[16] Bourlier C, Pinel N, Kubické G. Method of moments for 2D scattering problems. Basic concepts and applications. Focus Series. London: Wiley-ISTE; 2013.
[17] Pinel N, Bourlier C. Electromagnetic wave scattering from random rough surfaces – Asymptotic models. Focus Series. London: Wiley-ISTE; 2013.
[18] Tsang L, Kong JA. Scattering of electromagnetic waves, advances topics. New York: John Wiley & Sons, INC.; 2000.
[19] Longuet-Higgins MS. The effect of non-linearities on statistical distributions in the theory of sea waves. J. Fluid Mech. 1963;17:459–480.
[20] Longuet-Higgins MS. On the skewness of the sea surface slopes. J. Phys. Oceanogr. 1982;12:1283–1291.
[21] Cox C, Munk W. Statistics of the sea surface derived from sun glitter. J. Mar. Res. 1954;13(2):198–226.
[22] Bourlier C. Azimuthal harmonic coefficients of the microwave backscattering from a non-Gaussian ocean surface with the first-order SSA model. IEEE Trans. Geosci. Remote Sens. 2004;42(11):2600–2611.
[23] Bourlier C. Unpolarized infrared emissivity with shadow from anisotropic rough sea surfaces with non-Gaussian statistics. Appl. Opt. 2006;44(20):4335–4349.
[24] Bourlier C. Upwind-downwind asymmetry of the sea backscattering normalized Radar cross section versus the skewness function. IEEE Trans. Geosci. Remote Sens. 2018;51(1):17–24.
[25] Bourlier C, Berginc G, Saillard J. One- and two-dimensional shadowing functions for any height and slope stationary uncorrelated surface in the monostatic and bistatic configurations. IEEE Trans. Antennas Propag. 2002;50(3): 312–324.
[26] Gradshteyn IS, Ryzhik IM. Table of integrals, series, and products. New York: Academic Press; 2000.
[27] Wu X, Smith WL. Emissivity of rough sea surface for 8–13 μm: Modeling and verification. Appl. Opt. 1997;36:2609–2619.
[28] Henderson BG, Theiler J, Villeneuve P. The polarized emissivity of a wind-roughened sea surface: A Monte Carlo model. Remote Sens. Environ. 2003;88:453–467.
[29] Masuda K. Infrared sea surface emissivity including multiple reflection effect for isotropic Gaussian slope distribution model. Remote Sens. Environ. 2006;103:488–496.
[30] Li H, Pinel N, Bourlier C. Polarized infrared reflectivity of one-dimensional Gaussian sea surfaces with surface reflections. Appl. Opt. 2013;52(25):6100–6111.
[31] Watts PD, Allen MR, Nightingale TJ. Wind speed effects on sea surface emission and reflection for the along track scanning radiometer. J. Atmos. Oceanic Technol. 1996;13:126–141.
[32] Li H. Emissivité et réflectivité infrerouges de la surface de mer avec ombre et réflexions multiples. France: University of Nantes; 2012.

[33] Li H, Pinel N, Bourlier C. A monostatic illumination function with surface reflections from one-dimensional rough surfaces. Waves Random Complex Media. 2011;21(1):105–134.

[34] Lynch PJ, Wagner RJ. Rough-surface scattering: Shadowing, multiple scatter, and energy conservation. J. Math. Phys. 1970;11:3032–3042.

[35] Bourlier C, Berginc G, Saillard J. Monostatic and bistatic statistical shadowing functions from a one-dimensional stationary randomly rough surface: II. Multiple scattering. Waves Random Complex Media. 2002;12:175–200.

[36] Li H, Pinel N, Bourlier C. Polarized infrared reflectivity of 2D sea surfaces with two surface reflections. Remote Sens. Environ. 2014;147:145–155.

[37] Bowman JL, Senior TBA, Uslenghi PLE. Electromagnetic and acoustic scattering by simple shapes. New York: Hemisphere Publishing Corporation; 1987.

[38] Bourlier C. Radar propagation and scattering in a complex maritime environment: Modeling and simulation from MatLab. London: ISTE Press; 2018.

[39] Bouche D, Molinet F. Méthodes asymptotiques en électromagnétisme. Berlin: Springer-Verlag; 1994.

[40] Ament WS. Toward a theory of reflection by a rough surface. Proc. IRE. 1953;41(1):142–146.

[41] Miller AR, Brown RM, Vegh E. New derivation for the rough-surface reflection coefficient and for the distribution of sea-wave elevations. Inst. Electr. Eng. Proc. 1984;131:114–116.

[42] Fabbro V, Bourlier C, Combes PF. Forward propagation modeling above Gaussian rough surfaces by the parabolic wave equation: Introduction of the shadowing effect. Prog. Electromagn. Res. 2006;58:243–269.

[43] Freund DE, Woods NE, Ku HC, *et al.* Forward radar propagation over a rough sea surface: A numerical assessment of the Miller-Brown Approximation using a horizontally polarized 3-GHz line source. IEEE Trans. Antennas Propag. 2006;54(4):1192–1304.

[44] Freund DE, Woods NE, Ku HC, *et al.* The effects of shadowing on modelling forward radar propagation over a rough sea surface. Waves Random Complex Media. 2008;18(3):387–408.

[45] Hristov TS, Anderson KD, Friehe CA. Scattering properties of the ocean surface: The Miller–Brown–Vegh model revisited. IEEE Trans. Antennas Propag. 2008;56(4):1103–1109.

[46] Elfouhaily T, Chapron B, Katsaros K, *et al.* A unified directional spectrum for long and short wind-driven waves. J. Geophys. Res. 1997;102(C7): 781–796.

[47] Collin RE. Hertzian dipole radiating over a lossy earth or sea: Some early and late 20th-century controversies. IEEE Antennas Propag. Mag. 2004;46(2): 64–79.

[48] Ulaby FT, Moore RK, Fung AK. Microwave remote sensing in theory to applications. vol. I-II. New York: Artech House; 1986.

[49] Johnson JT, Zang M. Theoretical study of the small slope approximation for ocean polarimetric thermal emission. IEEE Trans. Geosci. Remote Sens. 1999;37:2305–2316.

[50] Bourlier C, Saillard J, Berginc G. Intrinsic infrared radiation of the sea surface. Prog. Electromagn. Res. 2000;27:185–335.

[51] Masuda K, Takashima T, Takayama Y. Emissivity of pure and sea waters for the model sea surface in the infrared window regions. Remote Sens. Environ. 1988;24:313–329.

[52] Yoshimori K, Itoh K, Ichioka Y. Thermal radiative and reflective characteristics of a wind-roughened water surface. J. Opt. Soc. Am. A. 1994;11:1886–1893.

[53] Yoshimori K, Itoh K, Ichioka Y. Optical characteristics of a wind-roughened water surface: A two-dimensional theory. Appl. Opt. 1995;34:6236–6247.

[54] Freund DE, Joseph RI, Donohue DJ, *et al.* Numerical computations of rough sea surface emissivity using the interaction probability density. J. Opt. Soc. Am. A. 1997;14:1836–1849.

[55] Bourlier C. Unpolarized emissivity with shadow and multiple reflections from random rough surfaces with the geometric optics approximation: Application to Gaussian sea surfaces in the infrared band. Appl. Opt. 2006;45:6241–6254.

[56] Nalli NR, Minnett PJ, Van Delst P. Emissivity and reflection model for calculating unpolarized isotropic water surface-leaving radiance in the infrared. I: Theoretical development and calculations. Appl. Opt. 2008;47:3701–3721.

[57] Saunders PM. Radiance of sea and sky in the infrared window 800–1200 cm^{-1}. J. Opt. Soc. Am.1968;58:645–652.

[58] Li H, Pinel N, Bourlier C. Polarized infrared emissivity of 2D sea surfaces with one surface reflection. Remote Sens. Environ. 2012;124:299–309.

[59] Bourlier C, Saillard J, Berginc G. Effect of the observation length on the two-dimensional shadowing function of the sea surface: Application on infrared 3–13 μm emissivity. Appl. Opt. 2000;39:3433–3442.

[60] Hale GM, Querry MR. Optical constants of water in the 200-nm to 200-mm wavelength region. Appl. Opt. 1973;12:555–563.

[61] Niclòs R, Valor E, Caselles V, *et al.* Sea surface emissivity angular measurements. Comparison with theoretical models. SPIE. 2003;5233:348–356.

[62] Smith WL, Knuteson RO, Revercomb HE, *et al.* Observations of the infrared radiative properties of the ocean-implications for the measurement of sea surface temperature via satellite remote sensing. Bull. Am. Meteorol. Soc. 1996;77:41–51.

[63] Li H, Pinel N, Bourlier C. Polarized infrared emissivity of one-dimensional Gaussian sea surfaces with surface reflections. Appl. Opt. 2011;50(23): 4611–4621.

[64] Caillault K, Fauqueux S, Bourlier C, *et al.* Multiresolution optical characteristics of rough sea surface in the infrared. Appl. Opt. 2007;46:5471–5481.

[65] Fauqueux S, Caillault K, Simoneau P, *et al.* Multiresolution infrared optical properties for Gaussian sea surfaces: Theoretical validation in the one-dimensional case. Appl. Opt. 2009;48:5337–5347.

[66] Ross V, Dion D, Potvin G. Detailed analytical approach to the Gaussian surface bidirectional reflectance distribution function specular component applied to the sea surface. J. Opt. Soc. Am. A. 2005;22:2442–2453.

[67] Vaitekunas DA, Alexan K, Lawrence OE, *et al.* SHIPIR/NTCS: A naval ship infrared signature countermeasure and threat engagement simulator. Proc. SPIE. 1996;2744:411–424.

[68] Bourlier C, Saillard J, Berginc G. Theoretical study on two-dimensional Gaussian rough sea surface emission and reflection in the infrared frequencies with shadowing effect. IEEE Trans. Geosci. Remote Sens. 2001;39:379–392.

[69] Schott P, De Beaucoudrey N, Bourlier C. Reflectivity of one-dimensional rough surfaces using the ray tracing technique with multiple reflections. In: Geoscience, Remote Sensing Symposium. Proceedings. 2003 IEEE International. vol. 7; 2003. p. 4214–4216.

[70] Collett E. Polarized light: Fundamental and applications. New York: Marcel Dekker; 1993.

[71] Kur'yanov BF. The scattering of sound at a rough surface with two types of irregularity. Sov. Phys. Acoust. 1963;8(3):252–257.

[72] Wright JW. A new model for sea clutter. IEEE Trans. Antennas Propag. 1968;16(2):217–223.

[73] Losasso F, Hoppe H. Geometry clipmaps: Terrain rendering using nested regular grids. ACM Trans. Graphics (SIGGRAPH). 2004;23:769–776.

[74] Heitz E. Understanding the masking-shadowing function in microfacet-based BRDFs. J. Comput. Graphics Tech. 2014;3(2):48–107.

[75] Ashikmin M, Premoze S, Shirley P. A microfacet-based BRDF generator. In: SIGGRAPH Proc; 2000. p. 65–74.

[76] Walter B, Marschner SR, Li H, *et al.* Microfacet models for refraction through rough surfaces. In: Proc. Eurographics Symposium on Rendering; 2007. p. 195–206.

[77] Gaussorgue G. La thermographie infrarouge: principes, technologies, applications. Paris: Technique et documentation Lavoisier; 1989.

[78] Papini F, Gallet P. Thermographie infrarouge, image et mesure. Masson; 1994.

Index

Ament 162–163, 169–171, 173–174
angular shadowing 212
apparent temperature 177
Ashikmin M 211–213
average 26–28, 34, 42–44, 52, 54–56, 125, 147–148
average shadowing function 1, 12–13, 16, 25–27, 32, 34–35, 38–39, 48–49, 59–60, 71, 82–84, 86–88, 132, 143–146, 150–152, 168
azimuthal 2, 63–64, 70, 79–81, 154, 157, 173–175, 212

bistatic 45–56, 61, 79–96, 126–154, 157–159, 202–203
Black body 176–177, 208
BRDF (bidirectional reflectance distribution function) 211–214

CDF (cumulative density function) 71, 79
conditional probability 5–8, 116–118, 139–140, 147–148, 153
correlation 1–2, 17–28, 34, 40–41, 45, 54–58, 63
correlation coefficient 80–81, 83–86
correlation functions 45, 64–66
covariance matrix 17–18, 22–23, 75, 78, 83
Cox and Munk 68, 173–174, 178, 181, 188, 198–199, 200–201
cut-off angle 105–110, 112, 178

downwind 188

emissivity 97, 104–105, 108, 110, 112, 154–155, 176, 180–183, 185–188, 190–192, 194–196, 200, 205
empirical 105–110
energy conservation 195–196, 199–201, 205, 214
event 5–6, 36–37, 79–80, 98, 114, 116–117, 124, 128, 140, 146–148, 153
exponential 13, 30–32, 67–70, 72–74, 173–174, 180–181

Fourier 29–30, 32, 186

Gaussian 1–2, 6–7, 13, 16–17, 19–20, 22, 24, 28, 30–32, 34, 39–45, 49–50, 53–60, 67–75, 78, 83, 85, 93, 99, 103, 107, 111, 115, 119, 128–129, 135, 138–139, 156–157, 166–167, 169, 172, 174, 178, 180–183, 188, 191, 198, 202–203, 208
geometrical optics (GO) 161–162
Gram–Charlier 57–58, 68–70, 72–74, 167, 174
grazing 11, 16, 97, 161–165, 169, 176, 181–182

hemispherical 143–146, 150–152
histogram 30, 99, 102–104, 108, 110–112, 118–122, 131–132, 134–137, 141–145, 148–150

illuminated height PDF 163–169
infinite surface length 13, 26, 39, 47–48
infrared 97, 104–106, 108, 110, 112, 123, 126, 147, 154–155, 158, 161, 176–178, 183–184, 192–196, 208, 211, 214
intersection 98–99, 101–104, 114, 124, 126, 131–136, 141–143, 145–150, 153–154, 159, 192, 201, 203
isotropic 83, 165, 173, 186, 188, 208

Kirchhoff 161, 176–177, 179
kurtosis 57–59, 68–69, 173, 178, 199–200

Li H 113–123, 146–148
limit angle 16
Lorentzian 19–20, 22, 30–34, 78
luminance 208
Lynch 138–146, 148–152, 159, 196

marginal 68, 108–109, 111–112, 118–122
Masuda 110–113, 118, 120–121, 125, 178
Matisse 208
measurements 178, 184–186, 189, 193–195, 199, 207
moments 34, 161–162, 168, 170
monostatic 2–7, 58–61, 63–68, 97–126, 154–157
Monte Carlo (MC) 28, 34, 99, 103–104, 108–113, 115, 128, 141, 143, 192, 196, 203
multiple reflections, 97–159

non-Gaussian 56–61

one-order emissivity 189–194
one-order reflectivity 194–201
one reflection 98, 101–102, 104–107, 109–114, 118, 121

PDF (probability density function) 1, 6–8, 11–13, 17–18, 22, 24, 28–29, 34, 38–41, 48–49, 52, 57–58, 63–64, 66–74, 79, 81, 85, 107, 110, 124–125, 147, 163–165
Planck 176–177
polarization 190, 206–207
power spectral density (PSD) 29
propagation factor 170–175

radar 161–162, 170, 176–177
radiation 3, 176–177
ray-tracing 32–34, 41–42, 52, 99–102
reflectivity 97, 110, 126–127, 147, 158, 161, 176, 194–207, 211, 213–214
refraction index 183, 185, 190, 192–193
resolution 207–209, 211
Rice 2, 7–8, 13, 15–16
rotation 63–64, 67, 154, 180, 206

scattering 1, 118, 161–176, 208, 213
sea 162, 165, 176–210
second-order reflectivity 201–205
shadowing 1, 2–4, 11, 97–159, 169, 178, 181–183, 212
skewness 58–59, 68–69, 73, 167–169, 172–173, 178, 199–200
Smith BG 2, 4–5, 8, 15, 21, 26, 40, 43, 50–51, 53, 55–56, 106–107, 109–110, 118, 121, 138–139, 146–147, 166–167, 194–195, 211–213
special function 13, 88–96, 161
spectrum 29, 76, 65, 165, 173, 181, 186, 207–208
specular 100–101, 129–131, 140, 147, 153, 158, 162–164, 197–199, 202, 213–214
SST (sea-surface temperature) 176, 183, 188–189, 194
statistical shadowing function 1, 4–5, 10–11, 16, 21–22, 40–41, 49–41, 70–71, 78, 82, 87, 98, 113–114, 121, 123, 127, 180

temperature 176–177, 185, 208
transmission 1, 45–56
two-dimensional 63–96, 154–159
two reflections 102, 122–128, 131, 133, 137–138, 148, 150–152, 157–159, 192, 196, 201–204
two-scale model 207–208

uncorrelated 2, 8–16, 51–52, 58, 66, 115, 146–148, 150–152, 154, 156
up-wind 188

Wagner RJ 2–6, 8–9, 16, 18–22, 25–27, 35, 38, 40, 42–45, 47–48, 51, 53–55, 70–71, 78, 138–146, 148, 150–152, 166, 196
Watts PD 105–106, 108, 110
Wien 176–177
Wu and Smith 106, 108–110, 118, 178

zero-order emissivity 176–188